HANDBOOK OF TECHNOLOGY TRANSFER

Handbook of Technology Transfer

Edited by

David B. Audretsch

Indiana University, USA and the Department of Innovation Management and Entrepreneurship, University of Klagenfurt, Austria

Erik E. Lehmann

Professor of Management and Organization, University of Augsburg, Germany

Albert N. Link

Virginia Batte Phillips Distinguished Professor of Economics, University of North Carolina at Greensboro, USA

EE Edward Elgar PUBLISHING

Cheltenham, UK • Northampton, MA, USA

Published by
Edward Elgar Publishing Limited
The Lypiatts
15 Lansdown Road
Cheltenham
Glos GL50 2JA
UK

Edward Elgar Publishing, Inc.
William Pratt House
9 Dewey Court
Northampton
Massachusetts 01060
USA

A catalogue record for this book
is available from the British Library

Library of Congress Control Number: 2022941178

This book is available electronically in the **Elgar**online
Business subject collection
http://dx.doi.org/10.4337/9781800374409

Printed on elemental chlorine free (ECF)
recycled paper containing 30% Post-Consumer Waste

ISBN 978 1 80037 439 3 (cased)
ISBN 978 1 80037 440 9 (eBook)

Printed and bound in the USA

Contents

List of contributors

Cristiano Antonelli is Full Professor of Economics at the Dipartimento di Economia e Statistica Cognetti de Martiis of the Università di Torino and Fellow of the Collegio Carlo Alberto. He has been a junior economist at the OECD and a Rockefeller Fellow at the MIT. He is the Managing Editor of *Economics of Innovation and New Technology*.

David B. Audretsch is Distinguished Professor and Ameritech Chair of Economic Development at Indiana University, where he also serves as Director of the Institute for Development Strategies. He is co-founder and Editor-in-Chief of *Small Business Economics: An Entrepreneurship Journal*. He was recognized as a Clarivate Citation Laureate in 2021 and awarded the Global Award for Entrepreneurship Research by the Swedish Foundation for Entrepreneurship. He has received honorary doctorate degrees from the University of Augsburg, Jönköping University and the University of Siegen. Audretsch was also awarded the Schumpeter Prize from the University of Wuppertal.

Laís Bandeira Barros is currently a second year PhD student in Business Economics at Ghent University. Her doctorate is part of the SMARTINCS project under the Marie Sklodowska-Curie grant agreement. Her research focuses on technology transfer and commercialization of self-healing technologies in the construction sector. She obtained her Master's degree in Construction Engineering at Universidade de Brasília.

Maksim Belitski is Professor in Entrepreneurship and Innovation at the Henley Business School, University of Reading. Prior to joining Henley, Belitski was a Research Fellow at the Institute for Development Strategies, Indiana University Bloomington, and a Contract Professor of Econometrics at the University of Bolzano. Previously he has held appointments at Loughborough University, University College London, University of Leicester, University of Economics Bratislava, Belarusian State University. His research interests lie in the area of entrepreneurship, innovation and regional economics, with a particular focus on entrepreneurship as a spillover of knowledge and creativity.

Shiri M. Breznitz is Professor and Director of Research at the Munk School of Global Affairs and Public Policy at the University of Toronto. An economic geographer, specializing in innovation, technology commercialization, and regional economic development. She is a member of the Advisory Board of Monash University's Better Governance and Policy research focus area and the *Journal of Technology Transfer*. Her research is at the critical intersection of theory and policy to fit the new realities of globalization: Breznitz's work has informed policymaking at the local, national, and international levels. She has advised on the role of universities in the larger story of innovation, on the economic impact of biotechnology and intellectual property, and on the role of clusters in driving innovation. Her current projects include a study on the economic impact of entrepreneurship education, the impact of gender and work experience on entrepreneurship, the geography of crowdfunding, and a study on the Greater Toronto ecosystems of innovation. Breznitz's book, *The Fountain of Knowledge* with Stanford University Press (July 2014), analyzes universities' relationships with government and industry, focusing on the biotechnology industry as a case study. She has also co-edited the book, *University Technology Transfer: The Globalization of Academic Innovation*, with Routledge Press (September 2015).

Mara Brumana is Assistant Professor at the Department of Management, Information and Production Engineering, University of Bergamo (SSD ING-IND/35) and Affiliated Researcher at the Institute for Change Management and Management Development (CMMD) at WU Vienna University of Economics and Business (WU Wien). She is also a research member of the Research Center for Young and Family Enterprise (CYFE) at the University of Bergamo. Brumana has recently published in main management and entrepreneurship journals such as *Strategic Entrepreneurship Journal*, *Journal of Small Business Management*, *Journal of Family Business Strategy*, and *European Management Review*. Her research interests revolve around the understanding of how and why the embeddedness of firms, and of family firms in particular, in a social and institutional context shapes their decision-making process and behavior. Her teaching is about international business, business administration, and change management.

Rosa Caiazza (PhD) is Professor of Management at Parthenope University of Naples, Visiting Professor at Wharton School at in the University of Pennsylvania. She has been included in the Top 100,000 Scientists worldwide across all knowledge areas, according to a study of Stanford University published in *PloS Biology*. She has been included in the ICSB's Educator 300, the International Council for Small Business list of 300 of the world's most well-renowned professors of small business and entrepreneurship. Caiazza serves as advisory board member to a number of top-tier academic journals, including *Academy of Management Perspectives*, *Journal of Management Studies*, *Journal of Technology Transfer*, *Small Business Economics*, *Trends in Food Science & Technology*, *Management Decision*, *Journal of Small Business Management*, *British Food Journal: An International Multi-disciplinary Journal for the Dissemination of Food-related Research*, *Journal of Management Development*, *Journal of Organizational Change Management*, *Corporate Governance: The International Journal of Business in Society*, *The Journal of Organizational Effectiveness: People and Performance*, *Technological Forecasting and Social Change*, *International Entrepreneurship and Management Journal*, *Thunderbird International Business Review*, *International Journal of Entrepreneurial Venturing*. Her research and teaching activity is focused on strategy, corporate governance, entrepreneurship, innovation, and operation management. She has published four books and over 100 journal articles and book chapters. Caiazza has been chairman of many international conferences and has won several Literati Network Awards for Excellence – Outstanding Paper Award.

Alice Civera is Post-doc Research Fellow at the Department of Management, Information and Production Engineering, University of Bergamo. Her teaching activities concern finance and public management. She is member of the CISAlpino Institute for Comparative Studies in Europe (CCSE), University of Bergamo and University of Augsburg, responsible for the Research Group on Higher Education. She is Research Fellow at the University of Augsburg. Her research interests include academic entrepreneurship, higher education, academic career paths and science policy. Her research has been published in leading academic journals such as *Research Policy* and *European Economic Review*, among others.

James A. Cunningham is Professor of Entrepreneurship and Innovation at Newcastle University Business School, Newcastle University, and is an affiliated member of the Centre for Innovation Research at Lund University. He has held academic positions at University College Dublin and National University of Ireland Galway (NUI Galway) and Northumbria University. Cunningham's research intersects the fields of strategic management, innovation, and entrepreneurship. His research focuses on strategy issues concerning scientists as principal investigators, university technology transfer commercialization, academic, public sector and technology entrepreneurship, entrepreneurial universities, and business failure. He has papers published in leading international journals such as *Research Policy, Small Business Economics, R&D Management, Long Range Planning, Journal of Small Business Management, Journal of Technology Transfer, Technological Forecasting and Social Change, International Marketing Management, Journal of Business Research*, and the *Journal of Rural Studies*, among others. Awards for his research include seven best paper conference awards and two case study international competition awards. Cunningham has published several books on the themes of strategy, entrepreneurship, technology transfer, and technology entrepreneurship with leading publishers such as Oxford University Press, Palgrave, Macmillian, Routledge, Springer, and World Scientific Publishing.

Manlio Del Giudice is Full Professor of Management at Link Campus University in Rome, where he serves as Deputy Rector (Erasmus Affairs), Coordinator of the PhD Programme "Tech for Good," Director of the Master in Smart Public Administration and Director of the CERMES Research Centre. He holds the position of Editor-in-Chief of the *Journal of Knowledge Management* and holds key editorial positions in several top journals in the management sector such as *Journal of Intellectual Capital, Journal of Business Research, International Entrepreneurship and Management Journal, Technological Forecasting and Social Change, IEEE Technology on Engineering Management, Journal of International Management.* His research has been published in some of the most important scientific journals in the world in the management sector, including *MIS Quarterly, Journal of Product*

Innovation Management, Journal of Organizational Behavior, Journal of World Business, Long Range Planning, IEEE Transactions on Engineering Management, Journal of Technology Transfer, Journal of Business Research, R&D Management, Technological Forecasting and Social Change, Product Planning & Control, International Marketing Review.

Henry Etzkowitz is CEO of the International Triple Helix Institute in Palo Alto and President of the Triple Helix Association. He is author of iconic articles on the entrepreneurial university, triple helix and women in science, research policy, and technology forecasting and social change. Etzkowitz is author of *MIT and the Rise of Entrepreneurial Science* (2002), co-author of *Triple Helix: University-Industry-Government Innovation and Entrepreneurship* (2008), and co-author of *Athena Unbound: The Advancement of Women in Science and Technology* (2000).

João Ricardo Faria is Full Professor of Economics at Florida Atlantic University. He worked as consultant for the World Bank and European Central Bank. His research focuses on theoretical aspects of entrepreneurship, where he has introduced dynamic models on the relation between entrepreneurship and unemployment, location, and innovation.

Rajeev K. Goel is Professor of Economics at Illinois State University and a member of the Kiel Institute for the World Economy, Germany. He serves on the editorial boards of several economics journals. Goel has made more than 100 professional presentations worldwide, is the author of two scholarly books and has published more than 200 research articles in refereed journals. He has held positions at the Indian Institute of Management, ZEW (Germany), Bank of Finland, Tata Energy Research Institute (India), University of Rome, LUMS (Pakistan), Oak Ridge National Laboratory (USA), Okayama University (Japan), and Zhejiang Gongshang University (China), among others. His research interests lie in public economics, the economics of technological change and applied microeconomics. Goel is a recipient of the Outstanding University Researcher award at Illinois State University and the Bergson Prize for the best paper published in Comparative Economic Studies. More details on his research can be found at https://econpapers.repec.org/RAS/pgo308.htm.

Devrim Göktepe-Hultén is Associate Professor at the Department of Business Administration at Lund University and Project Team Leader for Research Management Subprogram for University of Rwanda-Sweden at the Swedish International Development Agency (SIDA), and she is affiliated with KnowScience Research Group at Lund University. She was a Marie Curie-Skłodowska research associate and the principal investigator of the Commercialization of Academic Research Results project (Vinnova). Before joining Lund University, she was a post-doctoral research fellow at Max Planck Institute of Economics in Germany. She has received her PhD degree in Innovation Engineering from Lund Institute of

Technology. Göktepe-Hultén's research lies at the intersection of university-industry relations, academic entrepreneurship, and the strategic use of intellectual property rights. She has a special interest in how policies, programs, and relationships between science and industry can be designed to accelerate the productive role of universities in innovation systems both in developed and developing countries. Her research has been published in leading academic journals and distinguished books.

Maribel Guerrero is based at Facultad de Economía y Negocios, Universidad del Desarrollo, Chile and Global Center for Technology Transfer, School of Public Affairs, Arizona State University. She holds a PhD and MPhil in Business Economics at the Department of Business of the Autonomous University of Barcelona. Her main research interests are: 1 the determinants and impacts of entrepreneurial activities developed by individuals, public organizations, and private organizations; 2 the configuration/evolution of entrepreneurship, innovation, and digital ecosystems; and 3 the role of diversity and minority on public policies. She is an active research fellow from the outstanding international projects that measure entrepreneurship: Global Entrepreneurship Monitor (GEM – Spain, Chile, and Belarus), Panel Studies of Entrepreneurial Dynamics (PSED – Spain), Global University Entrepreneurial Spirit Students' Survey (GUESSS – Belarus), and Project for Family Enterprising (STEP – Chile and Mexico). She also serves on the board of the *Journal of Technology Transfer*, *Journal of Small Business Management*, *Technology Forecasting and Social Change*, *Entrepreneurship Theory and Practice*, *Entrepreneurship & Regional Development*, and *Small Business Economics*. Moreover, she serves the AoM Entrepreneurship Division Executive Committee (re-elected treasurer 2018–23), a board member of the Global Entrepreneurship Monitor Association (GERA 2021–23), Technology Transfer Society, and others. For further details, review the ORCID: https://orcid.org/0000–0001–7387–1999.

Samaa Kazerouni is an MGA1 candidate at the Munk School of Global Affairs & Public Policy. She obtained her Bachelor of Arts and Science in International Development and Psychology from McGill University. Her research interests include refugee settlement, labor migration, and human trafficking. She has a particular interest in the intersection of psychology and human migration. Kazerouni has spent time in Toronto's newcomer settlement sector, working to integrate refugees into Canadian society, and is part of the Global Migration Lab.

Mirjam Knockaert is Associate Professor in Entrepreneurship at Ghent University and an Adjunct Professor at the TUM School of Management (Germany). She obtained her Master in Business Economics from KULeuven (Belgium) and her PhD in Business Economics from Ghent University, and worked in financial audit before joining academia. Knockaert worked at the University of Oslo as an adjunct Associate Professor for ten years. Her research focuses on two specific research streams. First, she addresses research questions in the area of academic entrepreneurship and technology transfer. Second, she is interested in the human capital side of entrepreneurship, investigating the role of team members, boards of directors, and core employees for entrepreneurial ventures. Her research has been published in internationally leading entrepreneurship journals such as *Academy of Management Journal*, *Technovation*, *Journal of Management Studies*, *Journal of Business Venturing*, and *Entrepreneurship Theory and Practice*.

Thomas Lauvås is Associate Professor in Innovation at Nord University Business School. He received his PhD from Nord University Business School, focusing on university-industry collaboration in research centers. His main research interests include open innovation, university-industry collaboration, sustainability and circular economy. His research has been published in leading international journals such as *Journal of Cleaner Production* and *Innovation: Organization & Management*.

Laura Lecluyse is a Researcher/Consultant in Innovation, Entrepreneurship and Competitiveness at IDEA Consult. Lecluyse obtained her Master's degree and PhD in Business Economics from Ghent University. Later, she also worked as a post-doctoral research fellow, funded by FWO Flanders, at Ghent University. Her PhD was awarded the 2020 Heizer Award for "Outstanding Doctoral Research" from the Academy of Management. She has published her work in leading journals such as *Technovation* and the *Journal of Technology Transfer*.

Erik E. Lehmann is Full Professor of Management and Organization at Augsburg University and Director of the Program Global Business Management (GBM). He held positions as an adjunct professor at Indiana University and as a research professor at the University of Bergamo. He received his habilitation (venia legendi) from Konstanz University in 2005 and joined the Max Planck Institute (Jena) as an assistant director (2004–05). Lehmann's research is focused on linking entrepreneurship, higher education, and corporate governance in the global context. He serves as an Associate Editor of *Small Business Economics*, among other journals. His research has been published in leading academic journals such as *Review of Finance*, *Research Policy*, *Entrepreneurship Theory and Practice*, *Journal of Economic Behavior and Organization*, *Technovation*, *Small Business Economics*, *Journal of Technology Transfer*, *Multinational Business Review*, *Economics of Innovation and New Technology*, *Review of Industrial Organization*, *Studies in Higher Education*, *European Journal of Higher Education*, among others (https://scholar.google.com/citations?user=t0bNvYkAAAAJ&hl=en). His co-authored books include *The Seven Secrets of Germany: Economic Resilience in an Era of Global Turbulence* (Oxford University Press 2015), *The Routledge Companion to the Makers of Modern Entrepreneurship* (Routledge, 2017) and *Entrepreneurship and Economic Growth* (Oxford University Press, 2006).

Dennis P. Leyden is Associate Professor Emeritus of Economics at the University of North Carolina at Greensboro. His research has focused on entrepreneurship in both the public and private sector, and on the role and behavior of universities in furthering such entrepreneurial activity and its impact on innovation and economic development. Leyden's research has been published in leading academic journals such as

Research Policy, Small Business Economics, Technovation, Economics of Innovation and New Technology, and *Journal of the Knowledge Economy*, among others.

Albert N. Link is the Virginia Batte Phillips Distinguished Professor at the University of North Carolina at Greensboro. His research areas include the economics and policy implications of public sector technology transfer activities, technology and innovation policy, and entrepreneurship policy. Link is Editor-in-Chief of the *Journal of Technology Transfer*, Co-Editor of *Foundations and Trends in Entrepreneurship*, and Editor of *Annals of Science and Technology Policy*.

Marius Tuft Mathisen is Associate Professor at the Norwegian University of Science and Technology (NTNU), associated with the NTNU School of Entrepreneurship and the Engage research center. He is also a practicing serial entrepreneur and currently the CEO and co-founder of the technology scale-up firm Appfarm based in Oslo. Mathisen's research is focused on the commercialization of science through the establishment of new spin-off firms from universities and research institutes. He was the recipient of the Heizer Dissertation Award in 2018 on this topic. His research has been published in leading academic journals such as *Small Business Economics*, and *Journal of Technology Transfer*.

Matthias Menter is Assistant Professor for Business Dynamics, Innovation, and Economic Change at the Friedrich Schiller University Jena. His research focuses on entrepreneurial and innovative ecosystems, academic entrepreneurship, technology transfer as well as science and innovation policy. He is also a senior research fellow at the Institute for Development Strategies (IDS) at Indiana University at the School of Public and Environmental Affairs (SPEA). Menter's research has been published in leading academic journals such as *Small Business Economics, Technovation, Journal of Technology Transfer, Economics of Innovation and New Technology*, and *R&D Management*, among others.

Michele Meoli is Associate Professor at the Department of Management, Information and Production Engineering, University of Bergamo, where he teaches Finance and Business Organization. He is the Deputy Director of the program in Management Engineering, and Director of the CISAlpino Institute for Comparative Studies in Europe (CCSE), University of Bergamo and University of Augsburg, where he coordinates the Research Group on Higher Education. Meoli was Marie Curie Research Fellow at the Centre for Econometric Analysis, Cass Business School (City, University of London). He is an advisor of the Italian Ministry of University and Research on Higher Education and Right to Education. He is a member of the editorial board of the *European Journal of Higher Education* and of the editorial review board of *Small Business Economics*. His research interests include corporate governance, corporate finance, academic entrepreneurship, higher education and science policy.

Tommaso Minola is Associate Professor at the Department of Management, Information and Production Engineering (DIGIP), and co-founder and Director of the Center for Young and Family Enterprise (CYFE) at the University of Bergamo. In his research, he is interested in studying how different dimensions of the enterprising individual (e.g., motivation, cognition, behavior) and of the entrepreneurial firm (e.g., goals and resource allocation) are affected by embeddedness in social contexts. In particular, he looks at the family and the university as contexts particularly relevant for venture creation, development, and performance. Minola's works have been published in leading academic journals in innovation, entrepreneurship and small business such as *Journal of Management Studies*, *Research Policy*, *Entrepreneurship Theory and Practice*, *Strategic Entrepreneurship Journal*, *Small Business Economics*, *Journal of Small Business Management*, *Entrepreneurship and Regional Development*, *Journal of Technology Transfer*, and *R&D Management*.

Melita Nicotra is Associate Professor and Qualified Full Professor of Management at the University of Catania, Department of Economics and Management, where she teaches Management. She is also a member of the teaching staff of the PhD in Economics, Management and Decision Making. She is a promoter and founder member of the Research Center ILhM. She is thematic expert in economic sustainability of large projects and R&D programs of the Regional Innovation Strategy for Smart Specialization in Sicily. Nicotra is author of numerous international publications in primary management journals. Over the years, she has carried out intense research activity in qualified Italian and foreign institutes on the topics of entrepreneurship, innovation and technology transfer.

Conor O'Kane is Associate Professor in Strategy and Innovation and Director of the Otago Business School Masters in Entrepreneurship. Current areas of research include innovation processes, research commercialization and academic entrepreneurship. O'Kane's research has been published in leading international journals such as *Research Policy*, *Technovation*, *Long Range Planning*, *Industrial Marketing Management*, *Technological Forecasting and Social Change*, *Studies in Higher Education*, *R&D Management*, and the *Journal of Technology Transfer*.

Einar Rasmussen is Professor of Technology Management at Nord University Business School. He has been visiting scholar at the University of Nottingham, University of Strathclyde, University of Twente, and University of Bologna. His main research interests are entrepreneurial processes, academic entrepreneurship, and university-industry technology transfer. Rasmussen has managed several research and development projects funded by research councils, ministries, and government agencies. He publishes regularly in international journals such as *Journal of Management Studies*, *Academy of Management Perspectives*, *Research Policy*, *Strategic Entrepreneurship Journal*, and *Small Business Economics*.

Marco Romano (PhD) is Full Professor of Management in Italy and serves as Professor in Entrepreneurship and Business Planning, Digital Innovation and Transformation Management and Management at the Department of Economics and Business, University of Catania. Romano is President of an entrepreneurial ecosystem "District Sicily 5.0" and held several leadership positions including President of Science and Technology Park of Sicily, Board Member of APSTI-Italian network of scientific and technological parks, General Manager Department of Economic Development Regione Siciliana, and General Manager, Service Health Emergency 118 in Sicily.

Alexander Starnecker is CEO of Weisser Spulenkoerper GmbH & Co. KG, a hidden champion for solutions in technical plastic parts with integrated contact options for electromagnetic applications. He is also visiting lecturer on Leadership and Restructuring at the University of Augsburg, consultant for family businesses, advisor for new venture companies as well as business coach for entrepreneurs and leaders. Current areas of research include technology transfer, leadership concepts and restructuring processes. Starnecker has published his research, among others, in the *Journal of Technology Transfer* and he is Editor of the handbook *Technology Transfer in a Global Economy* (2012).

Katerina Vasilevska is a PhD student in Technology and Innovation Management at the University of Bergamo, and a member of the CYFE center. Her research interests are related to the field of strategic management, corporate entrepreneurship and family business, mainly focused on the relationship between work-family interference and innovation.

Silvio Vismara is Vice-Chancellor for Research and Professor of Corporate Finance at the University of Bergamo. He is Research Fellow at Indiana University and an adjunct professor at the University of Augsburg. He has held prior visiting appointments at Ghent University, Manchester Business School, Cass Business School, University of Florida, and University of La Laguna. Vismara is Editor of *Small Business Economics*, co-founder and Executive Editor of the *Review of Corporate Finance*, Associate Editor of the *Financial Review*, of the *Journal of TechnologyTransfer*, of the *Management Review Quarterly*, and editorial board member of *Entrepreneurship Theory and Practice*, *Corporate Governance: An International Review*, *Venture Capital*, and *Journal of Industrial and Business Economics*. His research interests are in entrepreneurial finance and focus mainly on IPOs, equity crowdfunding, and ICOs. The research activity has benefited from his experience as a scientific consultant for the Italian Stock Exchange, as a member of the board of directors of the University of Bergamo, as well as a founder of its first academic spin-off company (Universoft). His research has been covered in media outlets around the world, including *The Economist* and the *Financial Times*.

Katharine Wirsching is Assistant Professor at the chair of management and organization at the University of Augsburg. Her research focuses on entrepreneurial ecosystems, corporate governance and innovation in family firms, female and immigrant entrepreneurship, and public policy. Wirsching is also a senior research fellow at the Institute for Development Strategies (IDS) at Indiana University at the School of Public and Environmental Affairs (SPEA). She has published her work in leading academic journals such as *Small Business Economics*, *Technovation*, and *Eurasian Business Review*, among others.

Claudia Yáñez-Valdés is a third-year doctoral student in Business Economics at Universidad del Desarrollo. She is currently working on the determinants and the impacts of digital entrepreneurship. She completed her undergraduate studies in Commercial Engineering at the Universidad Católica de la Santísima Concepción, where she also obtained her Master's Degree in Business. For further details visit ORCID: https://orcid.org/0000–0001–5437–6714.

Qiantao Zhang is Assistant Professor in Urban Economics at the Department of Urban Planning and Design, Xi'an Jiaotong-Liverpool University (XJTLU). Zhang received his PhD in Economic Geography from Cardiff University. Zhang's main research interests include entrepreneurship, innovation, and regional development. Prior to joining the XJTLU, he worked as Post-Doctoral Fellow at the University of Toronto.

Chunyan Zhou, PhD, is Director/Senior researcher of International Triple Helix Institute located in Silicon Valley. Her research interest focuses on university-industry-government triple helix for innovation, especially on government's roles, entrepreneurial university and regional innovation policy. Zhou has co-authored the book *The Triple Helix: University-Industry-Government Innovation and Entrepreneurship* (Routledge, 2018), with Henry Etzkowitz; authored *From Science to Technology: the Scientific Basis in the Technological Era* (NEU Press, 2002); co-edited *Social Entrepreneurship: Leveraging Economic, Political and Cultural Dimensions* (Springer, 2014), with Anders Lundströ́m, and translated the book *Pasteur's Quadrant:Basic Science and Technological Innovation* (by Donald E. Stokes, Brookings Institution Press, 1997; Beijing: Science Press, 1999). Her key contributions include proposing the "triple helix field," "triple helix twins", and "government-pulled triple helix" concepts.

Introduction to the *Handbook of Technology Transfer*

David B. Audretsch, Erik E. Lehmann and Albert N. Link

ORGANIZATION OF THE VOLUME

Technology transfer may be broadly defined as the transfer of research knowledge from academic research to the commercial sector, where the dissemination or transfer can occur in many different forms. The recent pandemic has drastically reflected the importance of an effective technology transfer process to cope with the major challenges in a global economy. The most effective vaccines, like Comirnaty (BioNtech/Pfizer) or Spikevax (Moderna Biotech) are the result of such an effective technology transfer process from scientific research institutions to the market.

Even introduced after World War II to spur the competitive advantage of the US economy, there has been a marked increase in various aspects of technology transfer since 1980, stimulated in large part by the Bayh-Dole Act and equivalent legislation in other countries, which provided additional incentives for research exploitation. Since then, technology transfer has become a major topic and area of interest in all academic fields, the natural sciences, and the social sciences and much of the evidence is still evident today. Just to highlight this fact, consider the special issue in *Research Policy* guest edited by A.N. Link and D. Roessner (2000). Most if not all of the articles published in this volume by eminent scholars like Jaffé (2000), Geroski (2000), Scherer and Harhoff (2000), Bozeman (2000) and Hagedoorn, Link and Vonortas (2000) have become state-of-the-art articles today. Several handbooks like *The Chicago Handbook of University Technology Transfer and Academic Entrepreneurship* by Link, Siegel and Wright (2015) and a number of volumes summarizing the state-of-the-art presented at the annual conferences of the 'Technology Transfer Society' (Audretsch et al., 2012, 2016) have been published since then, not to mention the *Journal of Technology Transfer*, edited by Al N. Link and Don Siegel, and others, since 1977.

The 14 chapters in this volume add to previous collections of papers and volumes by considering technology transfer as a complex and dynamic process involving a plethora of individuals and agents from different institutions, all of whom apply different mechanisms characterizing activities in various countries. Traditionally, technology transfer is described by at least four dimensions: the underlying mechanism of transferring knowledge, the role of individuals that trigger the transfer either as the transmitter or receiver, the role of institutions where the transfer takes place, and the role of government and politics; these four dimensions define the scope of the volume.

The chapters included in this volume focus on new aspects and developments in theory and practice within four dimensions. The first part of the volume is thus dedicated to the core of the knowledge transfer process, the knowledge base of the physical entity, the transfer object. This knowledge base is inherent, and not ancillary, as Bozeman (2000, p. 629) puts it. Cunningham et al. (2017) therefore argue that the term technology transfer has evolved to encompass a new term, namely "knowledge transfer" (Cunningham et al. 2017). The volume starts with the "knowledge transfer," and the limited transferability of knowledge as a public good. The second part of the volume is dedicated to the individual since "technology transfer is a contact sport" as Carlsson and Fridh (2002, p. 199) point out. The third part highlights the involved institutions and organizations, in particular universities. The fourth part then focuses on the government and country-specific effects leading to heterogeneous effects of the technology transfer process.

PART I: KNOWLEDGE TRANSFER

Technology transfer is often focused on the transferable object and its physical dimension, being a product or process. However, as Bozeman (2000, p. 629) expressed: "[it] is not merely the product that is transferred but also knowledge of its use and application." More than half a century ago, Arrow (1969) highlighted the problems intrinsic to the "transmission" of knowledge and the important role of the competence required to access the knowledge that is transferred by third parties. Cristiano Antonelli introduces the volume with an interpretative chapter titled "The limited transferability of knowledge" based on the large body of work on technology transfer, in an attempt to elaborate an economic theory of technology transfer to add to the rich literature on the specific mechanisms that enable advanced economic systems. He develops the notion of the limited transferability of knowledge, building on analysis of the competences required to use economic goods in general and knowledge as a particular economic good. Then he analyzes the effects of the limited transferability of knowledge on its appropriability, on the economic influence of knowledge spillovers, on the tradability of knowledge and on its recombinant production, exploiting the notion of Jacobs increasing returns to variety. He concludes that the dynamic efficiency of the technology transfer system rests on the solution to the trade-off between the limits on coordination and exploitation and the advantages of size and variety in the production of knowledge through the implementation of organizational modes of knowledge sharing and interaction protocols and mechanisms, able to combine appropriation incentives with transferability mechanisms.

Also, Dennis Leyden and Matthias Menter focus on knowledge as a key resource within the technology transfer process, but they argue that the underlying mechanisms thereof are not well understood. They criticize that the linear model of innovation posits a unidirectional sequence of basic and applied research, while current research shows that a fuller model of innovation is needed that considers the cross-fertilization of basic and applied research. The purpose of their contribution "The impact of knowledge transfer on innovation: exploring the cross-fertilization of basic and applied research" is to conceptualize a theoretical model of regional innovation output, taking into account the knowledge flows from basic to applied research and vice versa. They also emphasize the role of the government as an enabler and facilitator of knowledge creation and diffusion by public funding. Funding basic and applied research funding then act synergistically in the production of regional innovation output. Their chapter contributes to our understanding of the optimal mix of basic research funding and applied research funding and provides the basis for empirical investigations. It thereby addresses the question of how to make best use of basic and applied research resources and derives implications for science and innovation policy.

David Audretsch and Maksim Belitski focus on the role of public finance, knowledge transfer, and firm productivity and their ability to innovate. Their chapter "The role of public finance in knowledge transfer and innovation" builds and estimates a theoretical model using 17,859 innovative firms in the UK. They demonstrate the extent that various sources of external knowledge contribute to a firm's innovation and the role of access to finance for innovation. They conclude that while access to finance may limit innovation, in particular in the most productive firms, it is public finance that bestows knowledge transfer to firm innovation.

PART II: INDIVIDUALS

Two chapters are dedicated to highlight the role of individuals in the technology transfer process. James Cunningham, Manlio Del Giudice, Melita Nicotra, Conor O'Kane and Marco Romano highlight the role of principal investigators within the technology transfer process. Scientists in the principal investigator (PI) role are at the heart of knowledge creation, management and dismission. The original knowledge they create through knowledge discovery forms the basis for technology transfer and commercialization activities. There has been a growing empirical focus on different aspects of the PI role and how scientists in the role pursue knowledge creation and exploitation. However, to date, there has been no research and empirical attention on how principal investigators approach knowledge management. The purpose of their chapter "Principal investigators and knowledge management: a micro foundational conceptual framework" is to address this deficit and to present a micro foundation conceptual framework that focuses on exploration and exploitation knowledge domain and know-how, opportunity and environmental scanning and knowledge management practices.

Academic inventors present a critical case since science and entrepreneurship are often seen as radically different, not the least in terms of knowledge production (Audretsch and Göktepe-Hultén, 2015). Considering the sequential nature of nascent entrepreneurship and business ownership, Joao Ricardo Faria, Rajeev K. Goel and

Devrim Göktepe-Hultén examine the propensities of academic entrepreneurs to be business owners. In their chapter titled "Factors facilitating the inventing academics' transition from nascent entrepreneurs to business owners" they provide a theoretical model that sets up the empirical analysis based on survey data from a large German public research institute. While traditionally, scientists and entrepreneurs have been seen to occupy opposite ends of a spectrum in terms of their role in innovation, they show that in academic entrepreneurship, the two combined on a number of activities. In order to understand the ways in which academic inventors move from pure patenting to nascent entrepreneurship to business ownership and connect seemingly divergent activities, they model their behavior by looking at various factors among scientists. Bringing the analysis from the level of social behavior and roles to the level of knowledge production improves our understanding in terms like: How is knowledge in the interfaces of epistemic communities produced? How can such knowledge be organized and sustained? And how can relations between individuals on "opposing sides" be constructively managed? Their empirical results reveal that scientists' positive attitudes towards commercialization of results consistently contribute to tendencies towards academic entrepreneurship. The results of this study would contribute to a more general theory of how scientists can combine their commercial and scientific activities in spite of an alleged divergence.

A different perspective is introduced by Katerina Vasilevska, Mara Brumana and Tommaso Minola in their chapter "The role of work-family initiatives in fostering technology transfer: research opportunities on family and non-family SMEs."

They shift their focus towards family firms in the transfer process and analyze work-family initiatives as organizational changes. Exploring this relationship is of great importance, given that in firms' struggle to innovation, all factors, but especially those that are related to their employees, count and should be further explored. Until now, there are some indicative findings, even though they could be found in a more generic debate related to the effect of various generic HRM practices or employee treatment on innovation. The findings indicate that providing proper conditions to the personnel indeed leads to an improvement in the technology transfer process. They argue that future research questions should be focused on the work-family initiatives (WFIs) to investigate their role in innovation and technology transfer, how particular types of WFIs affect innovation, or what the individual-level mechanisms are that "mediate" the relationship between WFIs on the one hand and innovation and technology transfer on the other.

PART III: INSTITUTIONS

Firms have historically organized their R&D internally but since a half-decade ago are now increasingly following an open innovation model by complementing their in-house knowledge through interorganizational collaborations, in particular with universities and science institutions. This is best reflected by the growing number of university-industry collaborations by which universities may contribute relevant expertise and knowledge to a firm's technological resource base and create new possibilities for innovation through research. Technology and knowledge transfer often depend on direct collaboration between those who develop new knowledge and those who apply this knowledge, and the culture and type of institutions engaged in the transfer process.

In "University-industry collaboration: drivers and barriers," Thomas Lauvås and Einar Rasmussen provide an overview of the phenomenon of university-industry collaboration and the inherent drivers and barriers in this relationship. While often studies claim that university-industry collaboration is highly beneficial for the innovativeness and performance of firms, the authors argue that there are many obstacles that hamper effective collaboration between academia and industry. Based on the scientific research in this area, the chapter provides an overview of key drivers and barriers to university-industry collaboration. With particular emphasis on research partnerships, the chapter outlines specific drivers and barriers related to the connections between university and industry partners, differences in their organizational culture, the role of organizational characteristics and the types of relationships. Finally, they conclude by providing some suggestions for practice on how to improve university-industry collaborations.

In the chapter "Contextualizing technology transfer: a review of university-industry transfer in the construction industry," Laís Barros, Mirjam Knockaert and Laura Lecluyse provide a review of the current literature on university-industry technology transfer in the construction sector. They take the construction industry as representing an interesting setting for a review with a contextual dimension for several reasons, like, at least in Europe, the construction industry has often been considered a mature, traditional industry, often conservative and attached to familiar technology, but heavily challenged by increased global competition, urging the firms in this industry to engage in innovation. By consequence, they provide an overview of what we know about technology transfer in the construction industry based upon the literature, taking into consideration the specific nature of this industry. Furthermore, next to synthesizing current knowledge, and contrasting it to the tech transfer literature in general, this chapter also presents a research agenda for future research into this intriguing field.

In the mid-1960s German Public Universities were no longer able to supply the human capital needed by the industry or to meet the interests of students who demanded higher educational qualifications for non-scientific jobs. At the demand of industry, the German government introduced a new institution, the University of Applied Science (*Fachhochschule*), to the higher education sector, applying a well-known concept used in industry, the division of labor, to provide the economy with basic researchers as well as graduates specialized in applied science. Other European countries, like the Netherlands, Finland and Switzerland followed. In "The role of universities of applied sciences in technology transfer: the case of Germany", Alexander Starnecker and Katharine Wirsching analyze the technology process of this type of academic institution. They show that Universities of Applied Science (UAS) have been specialized in applied research, and that the technology transfer occurs almost always with small and medium-sized enterprises (SMEs). They conclude that this type of university could serve as a role model in particular for countries and regions where small and medium-sized companies dominate the industrial landscape as they do in Germany.

The importance of universities within the technology transfer is not only recognized by terms like efficiency or effectiveness, but in particular by the impact generated to society. The recent pandemic has drastically revealed the interplay between universities, industries and the government, best known as the "triple helix-model," to cope with the global crisis. Henry Etzkowitz and Rosa Caiazza highlight "The role of university in time of crisis: learn from the past to shape the future." Following the theory of knowledge spillover entrepreneurship (KSTE), opportunities for entrepreneurship are the duality of the knowledge filter and the greater the knowledge filter, the greater are the divergences in the valuation of new ideas across economic agents and the decision-making hierarchies of incumbent firms (Audretsch et al., 2019). The economic downturn brings forward a reinforced importance of the cooperation between universities, industry and government as key ingredient in economic development processes and to face this global challenge. The triple helix, thus, provides the regional capability to build upon existing resources of the university to create niches of technological innovation and secure a place within the division of labor in the global economy (Etzkowitz, 2003). Given the central role of university in the triple helix model and its relevance for knowledge spillover entrepreneurship, Etzkowitz and Caiazza focus their analysis on how this role evolved and changed in the US during the time and why it is so relevant in times of crisis.

PART IV: COUNTRIES

One of the most robust empirical facts in economics is the unequal distribution of economic and social development across location and places. Since the famous contributions by Griliches (1979), Krugman (1991), Audretsch and Feldman (1996), among others, we better understand why lasting and sizable disparities exist within and between many countries: because they differ in the creation and absorption of knowledge! Most important is the interplay between the production of knowledge on the one hand and the transmission to the economy and society on the other (Audretsch et al., 2016; Link et al., 2007). While the production of knowledge in universities, scientific institutions, laboratories, among others, does not differ in large part across countries, it is the transfer of knowledge which should then make the difference. This part is dedicated to chapters highlighting country differences in the transfer of knowledge.

One important mechanism of knowledge transfer is science-based entrepreneurial firms. They play a key role in the modern knowledge economy in the transformation of investments in basic science into economic growth, social development and competitive advantage (Acs et al., 2013). Among science-based entrepreneurial firms, academic spin-offs, companies created by academic personnel, play an important role to exploit technological knowledge that originated within universities and other science-based institutions. The chapter "Academic entrepreneurship in Italy" by Alice Civera, Michele Meoli and Silvio Vismara highlights the establishment of academic spin-offs as a multi-faceted phenomenon, involving individual-, institutional- and contextual-level factors. They review the literature on the motivation to establish academic spin-offs and link it to the country-specific regulatory and policy framework. They also document the evolution of academic spin-offs in Italy, where 1,626 academic spin-offs have been established between 1981 and 2018.

Also, Marius Tuft Mathisen and Einar Rasmussen focus on academic entrepreneurship as an important mechanism for the transfer of scientific knowledge and technology into application in society. Their chapter "Academic entrepreneurship: the performance and impacts of academic spin-offs in Norway" takes a unique approach by documenting the development and outcomes of a national population of academic spin-offs over time. By looking at spin-offs from universities and public research institutes in Norway established from 1999 to 2011, they map their profiles related to origin, technology, financing, outcomes and several performance measures. Their findings provide a comprehensive understanding of how these firms develop, emphasizing the importance of different outcomes such as acquisitions, the skewed nature of such firm portfolios, and the strong effect of time for realizing outcomes and impacts.

The next chapter highlights the importance of intellectual property (IP) rights. Since the enactment in the US in 1980 of the Bayh-Dole Act, which transferred the ownership of federally funded research to universities, a worldwide debate has been ongoing about the benefit of universities' ownership of IP and governments have implemented different policies related to the ownership of intellectual property rights (IPR) to foster technology transfer by universities and public research organizations. In "Universities' ownership of intellectual property: focus on Canada," Shiri M. Breznitz, Samaa Kazerouni and Qiantao Zhang first review different policies regarding universities' ownership of IP and their impact, and then focus on the case of Canada that does not have an IP ownership policy at either the federal or provincial levels.

The volume closes with the chapter "Technology transfer and frugal social innovations: looking inside an emerging economy" by Claudia Yañez-Valdes and Maribel Guerrero. They argue that private and public resource scarcity is a critical problem in emerging economies and that although technology transfer policy frameworks exist, innovation and entrepreneurship become complicated processes by the highest R&D opportunity-costs. They explain the emergence of frugal innovation that involves doing the best with the available resources to solve society's problems and needs. While little is known about the relationship between knowledge transfer and frugal innovations, their chapter investigates the scarcity of resources and the technology transfer framework in frugal innovation ecosystems. Their results show the frugal social innovation's obstacles and the crucial role of technology transfer policy frameworks by analyzing the Chilean case. This chapter contributes to the re-conceptualization of the frugal innovation approach and provokes discussion about the promotion of frugal social innovations through technology transfer on the road to a sustainable future.

Technology and knowledge transfer have attracted considerable attention from both the research community as well as thought leaders in public policy, business and the management of research institutions. The contents contained in the subsequent chapters of this book promise a rich and fertile field with diverse perspectives. We are looking forward to seeing how the technology and knowledge transfer will be actualized in the coming years, in particular to cope with the challenges faced by a global and interconnected world.

REFERENCES

Acs, Z.J., Audretsch, D.B., and Lehmann, E.E. (2013). The knowledge spillover theory of entrepreneurship. *Small Business Economics*, 41(4), 757–74.

Arrow, K.J. (1969). Classificatory notes on the production and transmission of technological knowledge. *American Economic Review*, 59(2), Papers and Proceedings of the Eighty-first Annual Meeting of the American Economic Association, 29–35.

Audretsch, D.B. and Feldman, M. (1996), R&D spillovers and the geography of innovation and production. *American Economic Review*, 86, 630–40.

Audretsch, D.B. and Gŏktepe-Hultén, D. (2015). University patenting in Europe. In Link, A.N., Siegel, D.S., and Wright, M. (eds), *The Chicago Handbook of University Technology Transfer and Academic Entrepreneurship*, Chicago: University of Chicago Press, 188–217.

Audretsch, D.B., Lehmann, E.E., Link, A.N., and Starnecker, A. (eds) (2012). *Technology Transfer in a Global Economy*, Heidelberg/New York: Springer.

Audretsch, D.B., Lehmann, E.E., Meoli, M., and Vismara, S. (eds) (2016). *University Evolution, Entrepreneurial Activity and Regional Competitiveness*, Heidelberg/New York: Springer.

Audretsch, D.B., Cunningham, J.A., Kuratko, D.F., Lehmann, E.E., and Menter, M. (2019). Entrepreneurial ecosystems: economic, technological, and societal impacts. *The Journal of Technology Transfer*, 44(2), 313–25.

Bozeman, B. (2000) Technology transfer and public policy: a review of research and theory. *Research Policy*, 29, 627–55.

Carlsson, B. and Fridh, A.-C. (2002). Technology transfer in United States universities: a survey and statistical analysis. *Journal of Evolutionary Economics*, 12(1–2), 199–232.

Cunningham, J.A., Menter, M., and Young, C. (2017). A review of qualitative case methods trends and themes used in technology transfer research. *Journal of Technology Transfer*, 42, 923–56.

Etzkowitz, H. (2003). Research groups as "quasi-firms": the invention of the entrepreneurial university. *Research Policy*, 32(1), 109–21.

Geroski, P.A. (2000), Models of technology diffusion. *Research Policy*, 29(4–5), 603–25.

Griliches, Z. (1979) Issues in assessing the contribution of research and development to productivity growth. *Bell Journal of Economics*, 10, 92–116.

Hagedoorn, J., Link, A.N., and Vonortas, N.S. (2000). Research partnerships. *Research Policy*, 29(4–5), 567–86.

Jaffe, A.B. (2000). The US patent system in transition: policy innovation and the innovation process, *Research Policy*, 29(4–5), 531–57.

Krugman, P. (1991), *Geography and Trade*, Cambridge, MA: MIT Press.

Link, A.N. and Roessner, D. (2000). The economics of technology policy: introduction to the Special Issue. *Research Policy*, 29(4–5), 535.

Link, A.N., Siegel, D.S., and Bozeman, B. (2007). An empirical analysis of the propensity of academics to engage in informal university technology transfer. *Industrial and Corporate Change*, 16(4), 641–55.

Link, A.N., Siegel, D.S., and Wright, M. (eds) (2015). *The Chicago Handbook of University Technology Transfer and Academic Entrepreneurship*, Chicago: University of Chicago Press.

Scherer, F.S. and Harhoff, D. (2000). Technology policy for a world of skew-distributed outcomes. *Research Policy*, 29(4–5), 559–66.

PART I

KNOWLEDGE TRANSFER

1. The limited transferability of knowledge[1]

Cristiano Antonelli

1.1 INTRODUCTION

Work on technology transfer is increasing rapidly and is providing systematic investigation of the variety of mechanisms enabling agents to share technological and scientific knowledge (Audretsch et al., 2014; Bengoa et al., 2020). This chapter is an attempt to understand the main achievements in this literature with a focus on the determinants and consequences of technology transfer. The aim is to make a contribution to the large literature on technology transfer by proposing an appropriate theoretical framework.

Section 1.2 builds on the Lancastrian notion of utility function as an activity and develops the notion of user competence. Following Arrow's insight, knowledge is compared to standard economic goods and is distinguished by its requirements related to user competence, which dictates the limits related to the transferability of knowledge among economic agents (Arrow, 1962; Antonelli, 2018, 2019).

Section 1.3 focuses on the consequences of the high levels of user competence required to use knowledge, to produce new knowledge and to introduce innovations, and introduces the notion of limited knowledge transferability as a new property of knowledge considered as an economic good. Section 1.4 explores the effects of the limited transferability of knowledge for the range of critical tenets related to the standard economics of knowledge, including level of knowledge appropriability and tradability, organization of the recombinant new knowledge with respect to the choice between internal and external knowledge, and the role of input variety drawing on analyses of Jacobs' increasing returns.

1.2. COMPETENCE TO USE ECONOMIC GOODS

The analysis starts with the utility function developed by Kevin Lancaster (1966: 133):

"the chief technical novelty lies in breaking away from the traditional approach that goods are the direct objects of utility and, instead, supposing that it is the properties or characteristics of the goods from which utility is derived. We assume that consumption is an activity in which goods, singly or in combination, are inputs and in which the output is a collection of characteristics. Utility or preference orderings are assumed to rank collections of characteristics and only to rank collections of goods indirectly through the characteristics that they possess. A meal (treated as a single good) possesses nutritional characteristics but it also possesses aesthetic characteristics, and different meals will possess these characteristics in different relative proportions".

According to Lancaster (1971), consumption refers to the extraction of utility from a good and is, itself, an activity. The Lancastrian consumer is active since consumption requires dedicated activity to benefit from the characteristics of each single good and the bundle of goods that can be purchased and used. This activity requires competence (Bianchi, 1998).

It seems plausible that the argument elaborated by Lancaster could be extended to include the notion that all use of a good requires dedicated activity and specific competence. Application of the Lancastrian utility function then could be stretched to include both the consumption of final goods and the use of intermediary and capital goods.

Little attention is paid in the literature to the competence necessary to use (all kinds of) economic goods. User competence is required for the know-how to use a good. The use of any standard economic good requires some level of competence, to use it and to extract from it the specific services that increase the user's welfare in the case of a final good, or the production of other goods in the case of an intermediary or capital good.

User competence is especially relevant for goods produced by third parties. The production of a good provides the producer with the competence to use it. It seems plausible to assume that only minimum levels of dedicated additional competence are required to use self-produced goods. Competence acquired from learning by doing in the production of the good exerts direct and positive externalities on the competence acquired by learning by using.

The amount of dedicated competence required to use a specific good varies across agents according to their human capital endowment. The larger the agent's endowment of human capital, the easier it will be to acquire (or learn) the specific competence necessary to use a certain good. Agents endowed with high levels of human capital can rely on a larger stock of generic competence and have already learnt to learn. Also, the larger the stock of the agent's generic competence, the lower the absorption costs (Stiglitz, 1987).

The level of competence necessary to use standard economic goods produced by third parties and the level of the related learning efforts vary according to the characteristics, novelty and diffusion of the goods. Use of commodities requires a minimum level of competence; use of durable consumer goods requires higher levels of competence; use of capital goods requires substantial competence.

User competence is acquired by means of efforts dedicated to learning by using. The user's capability to extract the full services provided by a standard good increases with the amount of effort expended on knowing how to use the good, and increases with repeated use.

1.3. USER COMPETENCE AND THE LIMITED TRANSFERABILITY OF KNOWLEDGE

Requirements in terms of user competence have important implications for the transfer of standard economic goods from one party to another. Their transfer from supplier to user may take the form of a transaction or a gift and be coordinated within the firm or mediated by a competent intermediary or via structured interactions between knowledge users and knowledge producers. The assistance of the supplier or the intermediary can affect the customer's use conditions. For instance, the cases of: (i) where the supplier performs dedicated activities to assist the user to acquire the necessary competence; and (ii) where the user elaborates the necessary competence on its own.

In this view, knowledge as an economic good can be regarded as an outlier. Arrow (1969) highlights the problems intrinsic to the "transmission" of knowledge and the important role of the competence required to access the knowledge that is transferred by third parties. Use of knowledge is extremely difficult and requires high levels of competence and substantial learning efforts. Both know-what and know-how are necessary to use knowledge and know-how is associated intrinsically to know-what. Effective use of the know-what of the knowledge transferred by one party to another is enabled only by dedicated efforts to acquire and learn the know-how.

The levels of competence needed to use knowledge are especially high in the case of the essential tacit component of knowledge. Even scientific knowledge and, especially, technological knowledge are characterized by high levels of intrinsic tacitness that can be reduced, but not eliminated. There is a threshold to the progressive reduction of the tacit component of knowledge; there is a fundamental part of knowledge that cannot be codified.

The consequences for the transferability of knowledge of the high levels of user competence required are important. Successful knowledge transfer occurs only if the knowledge receiver is able to fully use the knowledge. That is, the transfer of knowledge will be successful only if the necessary user competence exists. The building of user competence can be the responsibility of the customer side or could involve both producer and customer.

In the first case, the user bears all the costs related to acquisition of the tacit and implicit specifications of its functionality. And absorption costs are relevant. In the second case, the acquisition of know-how is supported by the active participation of the other party and requires dedicated interaction between the two parties. For example, interaction between scholars, teachers and students, and practitioners is necessary to transfer the tacit component of knowledge.

Absorption costs are usually higher than interaction costs. The cost of the interactions required to complete knowledge transfer is both lower than one-party absorption costs and lower than overall absorption costs: interaction is a superior solution from a welfare perspective. However, interaction requires the active involvement of the party transferring the knowledge.

The high level of user competence can reduce the transferability of knowledge drastically to well below the level of transferability of standard economic goods. The transferability of knowledge is lower than the average transferability of standard economic goods.

The major role of the stock of generic competence in the level of user competence triggers a new Matthew effect with non-ergodic effects (Merton, 1968). Agents with large stocks of generic competence learn faster about how to use a good and, especially, about how to use knowledge, compared to agents with limited generic competence. This Matthew effect triggers relevant asymmetries between agents and firms.

The limited transferability of knowledge has important implications for the economics of knowledge.

1.4 EFFECTS OF THE LIMITED TRANSFERABILITY OF KNOWLEDGE

The limited transferability of knowledge has a wide range of effects on the organization of the production and distribution of knowledge as an economic good. The limited transferability of knowledge affects: (i) the level of its appropriability; (ii) knowledge externalities and the basic assumptions of new growth theory; (iii) the tradability of knowledge; (iv) the role of intellectual property rights; (v) the economics of university; and (vi) the recombinant generation of new knowledge with respect to the choice between internal and external knowledge and the role of variety. Let us analyse each in turn.

1.4.1 The Limited Transferability of Knowledge and Its Appropriability

The limited transferability of knowledge increases the actual levels of its appropriability of knowledge substantially and questions Arrovian analysis of the effects of the limited appropriability of knowledge in terms of knowledge market failure and knowledge undersupply (Arrow, 1962).

According to a large literature, knowledge spills into the air; knowledge leakage cannot be eliminated; all knowledge eventually becomes public. The uncontrolled leakage of knowledge limits its appropriability. Third parties can benefit from the economic returns to new knowledge and can participate in the stream of revenue triggered by its production and eventual application to introduce technological innovations.

Spillovers benefit third parties both in the technology production function, where innovations can be imitated and proprietary knowledge used for the production of all other goods, and in the knowledge generation function where proprietary knowledge can be accessed and used as an input to the generation of new knowledge. In both cases, spillover recipients do not bear the costs of the original production of the new knowledge and can outcompete its "inventor". The limited appropriability of knowledge limits the returns that can be appropriated by inventors and triggers the well-known knowledge market failure and undersupply of knowledge.

The appropriability of knowledge and its transferability have been – too often – regarded as specular or complementary: low appropriability implies high transferability and high transferability implies low appropriability. Indeed, high appropriability complements low transferability, but the case for low appropriability cum low transferability does take place and is quite relevant. Appropriability levels fall drastically when a limited number of competent rivals in oligopolistic product markets is able to imitate the new knowledge and yet its transferability is limited by relevant costs of absorption for all other potential users. The imitation of a few competent rivals can undermine drastically the appropriation of knowledge rents and yet the actual transfer of the very same knowledge to all potential users does not take place automatically. Analysis of the limited transferability of knowledge urges reconsideration of the arguments.

The recipients of uncontrolled leaked knowledge have to bear substantial costs related to acquisition of the necessary competence by prospective users. They cannot rely on assistance from its inventors. Absorption costs reduce the opportunistic competitive advantage of spillover users. Mansfield et al. (1981) stressed that imitation costs can reach levels close to the costs of innovation. The cost of absorption activities required to acquire the know-how can be close to the costs of the funded research activities that produced the know-what. In this extreme case, the argument that inventors suffer unfair competition from imitators does not hold. Inventors of knowledge that requires substantial absorption costs can appropriate the full stream of the benefits triggered by their knowledge.

Absorption costs are the cause of barriers to imitative entry: the higher the absorption costs, the higher the barriers to the entry of imitators and, consequently, the higher the levels of the de facto appropriability of the returns from the production of new knowledge and the introduction of innovations.

The opportunistic benefits reaped by the recipients of spillovers should be reconsidered to take account of the difference between the gross marginal product of the spilled knowledge and its absorption costs. The first minus the second identifies the real net opportunistic benefit of spillover recipients.

The pervasive Matthew effect on the limited transferability of knowledge is most relevant in this context. In oligopolistic races, based on a fast sequence of introductions of innovations in given product markets, the limits to knowledge transferability are lower. Competitors command large stocks of generic competence and can learn quickly how to use the new pieces of knowledge introduced by rivals. The barriers to entry stemming from the limited transferability of knowledge are lower, and the de facto appropriability of knowledge is much lower.

1.4.2 The Limited Transferability of Knowledge and Knowledge Externalities

An understanding of the notions of user competence and absorption costs allows an appreciation of the distinction between technical and pecuniary externalities. In Arrovian analysis, spillovers trigger technical externalities that flow in the air and benefit appropriately located potential users. If spillover absorption costs are relevant, but lower than invention costs, then pecuniary rather than technical externalities apply and are enjoyed by the spillover recipients. The levels of these pecuniary benefits depend on each agent's endowment of generic competences and learning capabilities, and the specific conditions of the system in which the agents operate (Antonelli, 2008).

Identification of the limited transferability of knowledge and the relevance of pecuniary as opposed to technical knowledge externalities have significant implications for new growth theory (Aghion and Howitt, 1997). In the original new growth theory framework, there is no variation in the extent to which technological knowledge can be transferred and then received, accessed and used by third parties. All users enjoy the same access conditions to costless knowledge spilling into the atmosphere. And the outcomes of this access are always positive for all third-party users.

New growth theory does not consider the effects of the limited transferability of knowledge and the effort required to absorb knowledge spillovers (Cohen and Levinthal, 1990). It assumes that the returns from spillovers – always – exceed its absorption costs and that the returns do not vary across agents. However, this assumption runs counter to empirical evidence showing the high level of variation in both synchronic and diachronic total factor productivity (TFP) growth – across firms, regions, industries and countries.

Spillover entrepreneurship theory has advanced with the integration of the limited transferability of knowledge in new growth theory, which highlights the crucial role of entrepreneurship in the ability to exploit the potential benefits of knowledge spillovers. This advancement emphasizes the efforts and competence required to take advantage of knowledge spillovers (Acs et al., 2009).

1.4.3 The Limited Transferability of Knowledge and Its Tradability

The most important implication of the limited transferability of knowledge is its tradability. Current analysis of the limited tradability of knowledge is based on the information asymmetries that arise between supplier and customer. Before the supplier reveals the content of the knowledge for sale, the customer bears the risk of a lemon. After the supplier reveals the content of the knowledge, the risk is all the supplier's and the customer can exploit the knowledge without actually buying it.

This interpretation does not take account the burden and effects of the user competence required to use the piece of knowledge on sale. Each piece of knowledge is unique and unlike the case of most standard goods, potential buyers cannot rely on evidence of the successful use of that same piece of knowledge by third parties. In the case of most standard goods, for example, the user of a car can rely on the evidence provided by other successful users of the same product. The user of an advanced computer-assisted design machine has indirect evidence confirming the quality and performances of the type of machinery sold by the same supplier, based on similar, although not identical, capital goods used in other firms.

The prospective user of a new piece of knowledge cannot rely on feedback from other users because, by definition, that piece of knowledge is unique. In this case, the transaction depends on the supplier's reputation. Similarly, the supplier's assistance to the customer to learn how to use the new piece of knowledge is indispensable to overcome the negative effects of the limited transferability of knowledge on its tradability. Transactions-cum-user-producer interactions are necessary to implement the tradability of knowledge. The customer is keen to use the piece of knowledge, but the interactive service complementing the transactions is crucial.

This reveals another important role – that of intellectual property rights. The current patent regime implies public disclosure of the knowledge according to proprietary rights. Public disclosure of the knowledge increases its tradability in two ways: (i) the possessor of the knowledge can reveal all its characteristics before any possible trade without risking the potential customer using it without buying; (ii) the potential buyer can assess the merits of the knowledge and accumulate the necessary user competence before buying.

Secrecy, the alternative to intellectual property rights, is clearly an inferior institutional solution. Only public disclosure of the knowledge makes the trade possible. If knowledge is secret and no intellectual property rights are available, the potential seller risks suffering opportunistic behaviour by the potential buyer if the content is revealed *ex ante*, while the buyer risks opportunistic behaviour from the seller if private disclosure takes place after the exchange contract without the buyer being able to learn the necessary user competence.

The transfer of knowledge enabled by transaction and intellectual property rights differs radically from transfers by means of knowledge spillovers. The latter case is the outcome of unintended and uncontrolled leakage; the former involves both parties. The recipient of knowledge spillovers cannot rely on the assistance of the "inventor" and bears all the cost of the activities required to use it effectively. The seller of a property right has a direct interest in transaction success.

User-producer interactions play a crucial role in knowledge transactions. In fact, knowledge transactions, in most cases, are transactions-cum-interactions. The knowledge seller participates in activities that enable the buyer to use the intellectual property right successfully. Typically, the knowledge transaction is complemented by the provision of knowledge services that reduce the absorption costs.

The role of transactions-cum-interactions in trades of innovative knowledge-intensive goods and, especially, high-tech capital goods, where the assistance of the producers of these advanced technological goods is an indispensable part of the sale, has been highlighted by Lundvall and Von Hippel (Lundvall, 1985; Von Hippel, 1988; Fagerberg, 1995; Lundvall and Lund Vinding, 2004).

The role of transactions-cum-interactions is especially relevant in trades of disembodied knowledge. The cost related to knowledge-user producer interactions is far lower than the cost of the absorption activities which the user alone bears. Knowledge user-producer transaction-cum-interaction benefit both parties in terms of three complementary aspects: (i) the total cost to the user of a property right complemented by user-producer interactions is lower than the cost of the intellectual property right on its own and increases the competitivity of the producers; (ii) users save on absorption costs; (iii) the interactions that parallel and complement the sale of the intellectual property right allow the producer to extract additional information on the functionality of the knowledge which it can use to generate new technological knowledge. Producers have direct access to the positive effects of learning by using in place at customer premises.

An important alternative to the transaction-cum-interaction between knowledge users and producers is the knowledge user's purchase of the knowledge producer. The user simultaneously acquires both the knowledge and the knowledge producer, and the right to use the knowledge as a property right as well as the competence embedded in the producer. This refers to the case of takeover and incorporation of high-tech successful start-ups, dynamics that generate new technological knowledge, ensure technical viability and allow its commercial application. The acquired public company becomes part of the corporation, able to integrate the competence of the "inventors" directly in its production process (Antonelli and Teubal, 2008).

1.4.4 The Limited Transferability of Knowledge and the Recombinant Generation of Knowledge

Recent advances in analysis of the production of knowledge highlight the crucial role of recombination. The generation of knowledge consists of using existing pieces of knowledge and recombining them: new knowledge is the outcome of the recombination of existing knowledge items (Weitzman, 1998; Arthur, 2007, 2009).

Recombinant generation of new knowledge has important consequences in terms of increasing returns: the larger the stock of knowledge, the higher the chances of generating new knowledge. Work in the economics of knowledge production demonstrates the central role played by the variety of the knowledge items that can be used for recombination: the larger the variety of the existing pieces of knowledge to which each agent has access, and wider the scope for their recombination and, hence, the higher the chances that their recombination will generate new knowledge. Increasing returns from the generation of new knowledge are associated not only with the size of the inputs but also – and to a large(r) extent – with their variety. Output increases more than proportionately with the size of the stock of knowledge available to each firm and with its variety (Antonelli et al., 2017, 2022).

The production of knowledge is characterized by Jacobs' increasing returns. Jacobs' increasing returns are peculiar to economies of scope. When economies of scope apply, output increases more than proportionately with the variety of goods manufactured by a given production process with a given amount of inputs. When

Jacobs' increasing returns apply, output increases more than proportionately with the variety of inputs (Jacobs, 1970).

Jane Jacobs identified the positive effects on the growth of cities of the variety of activities localized within each city. These positive effects then became associated with externalities: positive effects external to each co-localized firm. However, Jacobs' increasing returns can be both internal and external to the firm.

It would seem necessary and useful to extend the original Jacobs' externalities notion to include Jacobs' increasing returns. Jacobs' increasing returns can include and take into account the positive effects that arise if the firm is able to combine a variety of activities within its boundaries. We next distinguish between Jacobs' increasing returns internal to firms and Jacobs' increasing returns external to firms. Jacobs' increasing returns are most relevant in the production of knowledge as a recombinant activity.

Increasing returns from the recombinant generation of technological knowledge have significant effects on the organization of knowledge production. The direct engagement of the possessors of the specialized knowledge needed for recombination becomes vital for the organization of the production of knowledge. The variety of internal competences becomes primary to the organization of the teams that perform the research within firms. The literature on knowledge management stresses the importance not only of variety in terms of scientific and technological competence and fields of investigation in which the organization has experience but also in terms of the variety of skills of the researchers involved with special attention to the accumulation of a well-designed mix of skills, human capital and scientific capability of the researchers.

However, the limits of organizations constrain the size of their operations and the coordination of knowledge generation and exploitation. Knowledge generation is characterized by high levels of risk, close to uncertainty, associated with the chances of identifying new pieces of knowledge, the timing of its production, its proximity and its coherence with the initial research objectives. The larger the operation, the larger the risks. The costs of internal coordination limit the size of the organization and define its boundaries and also affect the amount of variety that can be coordinated within the firm's borders. The coordination of variety is even more demanding than the coordination of size (Arrow, 1974).

The trade-off between the limits on coordination and exploitation and the advantages of size and variety in the production of knowledge encourages systematic search for selective access to and use of existing knowledge, external to each agent, but accessible beyond the borders of the organization. The implications of the limited transferability of knowledge are most important for its production.

The pervasive Matthew effect on the limited transferability of knowledge is most relevant in this context. Firms endowed with a large stock of generic competence are better able to overcome the limited transferability of knowledge and can rely on external knowledge, at lower cost, compared to firms with small stocks of generic user competence.

The Matthew effect triggers specific increasing returns. The larger the stock of generic competence, the lower the costs of external knowledge and, hence, the larger the benefits derived from the division of labour and the specialization, which, in turn, trigger lower knowledge costs and higher rates of increase in the stock of knowledge.

Access to external knowledge can be enabled by its purchase or by means of structured knowledge interactions. As a large literature suggests, structured interaction with external sources of knowledge that can be used as inputs to knowledge generation becomes the basic alternative to their internal coordination. Knowledge interactions allow access to external knowledge, reducing the burden on the internal stock of knowledge and the absorption costs, but require the active participation of the parties involved. It is necessary to engage the providers of external knowledge in the undertaking. A variety of mechanisms is available.

Access to the competence of specialized inventors and its integration in the internal production of new knowledge is far easier among co-localized firms in regional knowledge-intensive clusters. Researcher mobility favours reciprocal access to tacit knowledge and increases the range of competences contributing to the knowledge generation process. Numerous studies confirm the selective localization in regional clusters, which increases access to external knowledge, as one of the main mechanisms to increase the inclusion of variety and, hence, the implementation of Jacobs' increasing returns.

Cooperation between two firms enables each to share its internal knowledge and facilitates reciprocal access to each party's tacit knowledge. Here, the distinction between horizontal and vertical cooperation is crucial.

In the case of horizontal cooperation between firms active in the same product market or in a product market with low barriers to mobility, the opportunity costs of the missing exclusivity of the property rights on the output of the shared research activities limit the scope of application of cooperative research, in order to increase variety, but reduce the costs of absorbing external knowledge.

However, if cooperation is among firms that participate in the same but different parts of the value chain, collaboration to generate technological knowledge and introduce innovations yields major benefits. Firms promote platforms to organize complementary advances and activate important transaction-cum-interaction flows between the users and the producers of goods that contribute to the same final product.

There is a large stream of work on the structured interactions between firms and universities as effective mechanisms allowing exploitation of Jacobs' increasing returns in the production of knowledge. The interaction between business research and academic research is based on the transaction-cum-interaction mechanism. Firms buy academic services that are performed jointly by firm and academic researchers.

The success of the transaction-cum-interaction mechanism applied to business and academia is positive not just for the firm. The very foundations of the economic rationale for academic research have been subject to increasing criticism. This critique was based on acknowledgment of the limits to the classic interactions between universities and firms, triggered by increasing evidence of the problems related to the limited transferability of knowledge.

The limited appropriability of knowledge was touted as the main economic rationale of the university as an economic institution. The university would help the system to overcome the market failure stemming from the limited appropriability of knowledge, providing incentives for scholars to generate knowledge and make it available to the system through publication. The "publish or perish" imperative combines the private incentives to scholars to obtain tenure with the public incentive to provide firms with the output of basic research. The limited transferability of knowledge undermined the working of this mechanism; dissemination of publications is not enough to outweigh the absorption costs and the hiring of doctoral graduates is not sufficient to promote the necessary knowledge interactions.

The active, direct and contemporaneous involvement of academic research in recombinant generation of new knowledge by firms has become the new central organizational mechanism. The sequence of academic research, publication in scientific journals, absorption by highly educated managers and use in applied and development research has become blurred and has been substituted by direct horizontal division of labour between active academic research laboratories and firms.

1.5 CONCLUSIONS

The analysis of knowledge as an economic good is especially rich since it captures the specific properties of knowledge and its implications for the economic organization of its production, exploitation and use. Much attention has been paid to the limited appropriability, exhaustibility and tradability of knowledge, but its limited transferability has attracted less research.

Analysis of the limited transferability of knowledge is based on the extension and implementation of the Lancastrian analysis of the utility function as an activity that suggests the need for dedicated competence to use all economic goods and, especially, the economic good of knowledge.

The limited transferability of knowledge is caused by the high levels of competence required for its use. The costs of absorbing knowledge produced elsewhere by third parties are especially high, compared to the knowledge related to standard economic goods.

The transfer of knowledge is necessary to implement the recombinant production of new knowledge and is the base of Jacobs' increasing returns, according to which the amount of knowledge increases more than proportionately with the increase in the size of the stock of knowledge that can be accessed and its composition in terms of the variety of the competences involved.

The limited transferability of knowledge reduces the negative effects of both its limited appropriability and tradability, and feeds the search for specific mechanisms to enable interaction between knowledge producers and knowledge users.

The implementation of dedicated transaction-cum-interaction mechanisms seems to qualify the development and dissemination of knowledge in the system and highlights the central role of industry-university cooperation in the production of new knowledge.

Analysis of the appropriability trade-off between the negative effects of the limited appropriability of knowledge in terms of lack of incentives for its production and the consequent – and well-known – knowledge market failure and the positive effects of knowledge spillover on the overall productivity at system level has been mother's milk to the economics of knowledge.

Limited transferability is the best antidote to the limited appropriability of knowledge. However, at the same time, it is clear that the increased transferability of knowledge has important positive effects in terms of reducing the firm's knowledge absorption and coordination costs, and increasing the scope and viability of increasing returns from recombinant knowledge generation. Analysis of the limited transferability of knowledge combined with an appreciation of the limits of organization in the production and exploitation of knowledge and the increasing returns to the size and variety of the stock of knowledge that can be accessed and used in the generation of knowledge reveals a new and crucial trade-off for the economics of knowledge.

From a societal viewpoint, the augmented transferability of knowledge made possible by knowledge interactions – for given levels of incentives to generate new knowledge – allows overall increased output from the resources invested in the production of knowledge. However, knowledge interactions require the active participation of all the parties involved. Due to the limitations imposed by knowledge appropriability, profit maximization limits the availability of individual agents to engage in dedicated knowledge interactions to support the transfer of their knowledge to other parties.

The new knowledge transferability trade-off can be identified. At system level, the sum of the benefits from the generation of the new knowledge produced by inventors that try to retain control over their proprietary knowledge, including the positive effects of productivity for consumers, is smaller than the benefits that would accrue to agents able to transfer and share their proprietary knowledge while also appropriating a fair return for it.

The dynamic efficiency of the system rests on the solution to this trade-off through the implementation of organizational modes of knowledge sharing and interaction protocols and mechanisms able to combine appropriation incentives with transferability mechanisms.

NOTE

1. This chapter is based on the PRIN 20177J2LS9 research project which, along with the Università di Torino and the Collegio Carlo Alberto, provided funding and support.

REFERENCES

Acs, Z.J., Braunerhjelm, P., Audretsch, D.B. et al. (2009). The knowledge spillover theory of entrepreneurship. *Small Business Economics*, 32, 15–30.

Aghion, P., Howitt, P.W. (1997). *Endogenous Growth Theory*, Cambridge: MIT Press.

Antonelli, C. (2008). Pecuniary knowledge externalities: The convergence of directed technological change and the emergence of innovation systems. *Industrial and Corporate Change*, 17, 1049–70.

Antonelli, C. (2018). Knowledge properties and economic policy: A new look. *Science and Public Policy*, 45(2), 151–8.

Antonelli, C. (2019). Knowledge as an economic good: Exhaustibility vs appropriability? *Journal of Technology Transfer*, 44, 647–58.

Antonelli, C., Teubal, M. (2008). Knowledge-intensive property rights and the evolution of venture capitalism. *Journal of Institutional Economics*, 4, 163–82.

Antonelli, C., Crespi, F., Mongeau Ospina, C.A., Scellato, G. (2017). Knowledge composition, Jacobs externalities and innovation performance in European regions. *Regional Studies*, 51(11) 1708–20.

Antonelli, C., Crespi, F., Quatraro, F. (2022). Knowledge complexity and the mechanisms of knowledge generation and exploitation: The European evidence. *Research Policy*, forthcoming.

Arrow, K.J. (1962). Economic welfare and the allocation of resources for invention. In Nelson, R.R. (ed.), *The Rate and Direction of Inventive Activity: Economic and Social Factors*, Princeton: Princeton University Press for NBER, pp. 609–25.

Arrow, K.J. (1969). Classificatory notes on the production and transmission of technical knowledge. *American Economic Review*, 59, 29–35.

Arrow, K.J. (1974). *The Limits of Organization*, New York: W.W. Norton.

Arthur, W. B. (2007). The structure of invention. *Research Policy*, 36(2), 274–87.

Arthur, W. B. (2009). *The Nature of Technology: What It Is and How It Evolves*, New York: Free Press.

Audretsch, D.B., Lehmann, E.E., Wright, M. (2014). Technology transfer in a global economy. *Journal of Technology Transfer*, 39, 301–12.

Bengoa, A., Maseda, A., Iturralde, T. et al. (2020). A bibliometric review of the technology transfer literature. *Journal of Technology Transfe*r. https://doi.org/10.1007/s10961–019–09774–5

Bianchi, M. (ed.) (1998). *The Active Consumer. Novelty and Surprise in Consumer Choice*, London, Routledge.

Cohen, W.M., Levinthal, D.A. (1990). Absorptive capacity: A new perspective on learning and innovation. *Administrative Science Quarterly*, 35(1), 128–52.

Fagerberg, J. (1995). User-producer interaction, learning and comparative advantage. *Cambridge Journal of Economics*, 19(1), 243–56.

Jacobs, J. (1970). *The Economy of Cities*, New York: Vintage Books.

Lancaster, K.J. (1966). A new approach to consumer theory. *Journal of Political Economy*, 74(2), 132–57.

Lancaster, K.J. (1971). *Consumer Demand: A New Approach*, New York: Columbia University Press.

Lundvall, B-A. (1985). *Product Innovation and User-Producer Interactions*, Aalborg: Aalborg University Press.

Lundvall, B-A., Lund Vinding, A. (2004). Product innovation and economic theory – user-producer interaction in the learning economy. In Christenses, J.L., Lundvall, B-A. (eds), *Product Innovation, Interactive Learning and Economic Performance* (*Research on Technological Innovation, Management and Policy, Vol. 8*), Bingley: Emerald Group Publishing, pp. 101–28.

Mansfield, E., Schwartz, M., Wagner, S. (1981). Imitation costs and patents: An empirical study. *Economic Journal*, 91(364), 907–18.
Merton, R.K. (1968). The Matthew effect in science. *Science*, 159(3810), 56–63.
Stiglitz, J.E. (1987). Learning to learn, localized learning and technological progress. In Dasgupta, P., Stoneman, P. (eds), *Economic Policy and Technological Performance*, Cambridge: Cambridge University Press, pp.125–53.
Von Hippel, E. (1988). *The Sources of Innovation*, Oxford: Oxford University Press.
Weitzman, M.L. (1998). Recombinant growth. *Quarterly Journal of Economics*, 113(2), 331–60.

2. The impact of knowledge transfer on innovation: exploring the cross-fertilization of basic and applied research[1]

Dennis P. Leyden and Matthias Menter

2.1 INTRODUCTION

Despite the general understanding that knowledge and innovation serve as key drivers of economic growth, innovation policies that are meant to stimulate the creation and commercialization of new ideas and inventions, thus boosting economic prosperity, have not met expectations – especially in recent years. In the context of current productivity slowdowns, scholars like Gordon (2016) even question whether we have reached the end of an exceptional growth period in economic history, calling for new solutions to return to former growth paths. The relationships between knowledge, innovation, and economic growth are indeed complex and difficult to conceptualize. The linear model of innovation has been a first attempt to explain the different phases of an innovation process, emphasizing the importance of scientific research as the basis of any innovation, almost neglecting the role of later-stage actors in the innovation process. Although this model has been useful to understand the relationship between science and technology and has contributed to the development of innovation as a policy field, Hall et al. (2001) as well as Medda et al. (2006) suggest that the continuous application of this oversimplified concept might at least partially explain the failure of today's innovation policies.

Whereas the linear model of innovation posits that knowledge created through basic research serves as the foundation of applied research and is ultimately transformed into new products or technologies, Leyden and Menter (2018) argue that a more sophisticated model of innovation is needed that takes the potential cross-fertilization of basic and applied research into account. Hence, they propose to break up the unidirectional sequence of the linear model of innovation and consider the interactions between basic and applied research. This argument is further supported by Leyden and Link (2015) who state that a mismatch between actors conducting basic research and actors conducting applied research exists that limits the flow and absorption of knowledge and thus the realization of the full potential of regional knowledge creation and exploitation. From a policy perspective, fuller models are needed that, on the one hand, appreciate the complexity of the research process and, on the other hand, consider the determinants affecting the translation process from research toward innovation in order to effectively shape innovation processes.

The purpose of this chapter is to conceptualize a theoretical model of regional innovation output, taking into account the knowledge flows from basic to applied research and vice versa. We thereby emphasize the role of the government as an enabler and facilitator of knowledge creation and diffusion. Using Leyden and Menter's (2018) two-sector model of the innovation process, we model regional innovation output as the result of regional basic and applied research activities. Regional innovation output is directly affected by the amount of applied knowledge in the region. However, because applied knowledge is itself a function of the region's basic research activity and the region's applied research activity, and because both types of research activity are themselves functions of each other, the model is developed in two steps. In the first step, the interdependence of basic and applied research is developed. In the second step, the region's innovation output as a function of this complex mix of research activity is developed.

From a policy perspective, the optimal allocation of resources to basic and applied research is highly relevant, yet is still an open issue. Ideally, basic and applied research funding would act synergistically in the production of regional innovation output. This chapter contributes to our understanding of the optimal mix of basic research funding and applied research funding and provides the basis for empirical investigations. Considering the knowledge cross-fertilization of basic and applied research, we develop a theoretical model of regional innovation output that disentangles the underlying mechanisms of knowledge production and innovation. It shall thus serve as the starting point for further investigations broadening our understanding of the innovation process that makes best use of basic and applied research resources.

The chapter is structured as follows. The next section focuses on the knowledge cross-fertilization of basic and applied research and constitutes the first step of our new innovation model. The third section models regional innovation output as a function of basic and applied research and constitutes the second step of our new innovation model. The fourth section discusses implications for science and innovation policy. A final section concludes.

2.2 KNOWLEDGE CROSS-FERTILIZATION OF BASIC AND APPLIED RESEARCH

Focusing on the process of basic research, the quantity of basic knowledge in any given year, B_t, can be modeled as the result of the prior year's basic research process that in turn is a positive, convex function of prior existing basic knowledge, LB_t, prior existing applied knowledge, LA_t, and a prior vector of resources, LX_t^B, that were devoted to the basic research process.[2] However, while basic researchers will have full access to prior existing basic knowledge, LB_t, they are unlikely to have full access to all prior applied knowledge, LA_t, because of institutional and cultural constraints that inhibit the flow of applied knowledge from applied researchers to basic researchers (Adams, 2006; Medda et al., 2006).[3] Letting the fraction of total pre-existing applied knowledge, LA_t, that basic researchers have access to be some *a*, with $0 < a < 1$, we can define the quantity of basic knowledge as:

$$B_t = B\left(LB_t, aLA_t, LX_t^B\right) \tag{2.1}$$

such that:

$$\frac{\partial B_t}{\partial LB_t} > 0, \frac{\partial^2 B_t}{\partial LB_t^2} < 0, \frac{\partial B_t}{\partial aLA_t} > 0, \frac{\partial^2 B_t}{\partial aLA_t^2} < 0, \frac{\partial B_t}{\partial LX_t^B} > 0, \frac{\partial^2 B_t}{\partial LX_t^{B2}} < 0$$
(2.2)

It is interesting to note that the effect of lagged applied knowledge, LA_t, on basic knowledge, B_t, will also be positive and a positive function of a:

$$\frac{\partial B_t}{\partial LA_t} = \frac{\partial B_t}{\partial aLA_t} \cdot \frac{\partial aLA_t}{\partial LA_t} = \frac{\partial B_t}{\partial alA_t} \cdot a > 0 \tag{2.3}$$

and that $\frac{\partial^2 B_t}{\partial LA_t^2}$ will be negative, and increasingly so with increases in a:

$$\frac{\partial^2 B_t}{\partial LA_t^2} = \frac{\partial^2 B_t}{\partial aLA_t^2} \cdot \left(\frac{\partial aLA_t}{\partial LA_t}\right)^2 = \frac{\partial^2 B_t}{\partial aLA_t^2} \cdot a^2 < 0 \tag{2.4}$$

Likewise, the quantity of applied knowledge in any given year, A_t, can be modeled as the result of the prior year's applied research process that in turn is a positive, convex function of prior existing applied knowledge, LA_t, prior existing basic knowledge, LB_t, and a prior vector of resources, LX_t^A, that were devoted to the applied research process. However, like the basic research process, applied researchers are unlikely to have full access to all prior basic knowledge, LB_t, because of institutional and cultural constraints that inhibit the flow of basic knowledge from basic researchers to applied researchers. Boardman and Bozeman (2006) find the nature of knowledge transfers between firms to be different from transfers from universities to firms, and Medda et al. (2006), for example, in their examination of the flow of knowledge from basic researchers to applied researchers and from applied researchers to basic researchers find that the flow of knowledge from universities to firms (i.e., basic knowledge impacting applied researchers) has a less direct impact and is more likely to occur when appropriability conditions are weak and outcomes are not direct or immediate.[4]

Letting the fraction of total pre-existing basic knowledge, LB_t, that applied researchers have access to be some b, with $0 < b < 1$, we can define the quantity of applied knowledge as:

$$A_t = A\left(LA_t, bLB_t, LX_t^A\right) \tag{2.5}$$

such that:

$$\frac{\partial A_t}{\partial LA_t} > 0, \frac{\partial^2 A_t}{\partial LA_t^2} < 0, \frac{\partial A_t}{\partial bLB_t} > 0, \frac{\partial^2 A_t}{\partial bLB_t^2} < 0, \frac{\partial A_t}{\partial LX_t^A} > 0, \frac{\partial^2 A_t}{\partial LX_t^{A2}} < 0 \tag{2.6}$$

It is interesting to note that the effect of lagged basic knowledge, LB_t, on applied knowledge, A_t, will also be positive and a positive function of b:

$$\frac{\partial A_t}{\partial LB_t} = \frac{\partial A_t}{\partial bLB_t} \cdot \frac{\partial bLB_t}{\partial LB_t} = \frac{\partial A_t}{\partial bLB_t} \cdot b > 0 \tag{2.7}$$

and that $\frac{\partial^2 A_t}{\partial LB_t^2}$ will be negative, and increasingly so with increases in b:

$$\frac{\partial^2 A_t}{\partial LB_t^2} = \frac{\partial^2 A_t}{\partial bLB_t^2} \cdot \left(\frac{\partial bLB_t}{\partial LB_t}\right)^2 = \frac{\partial^2 A_t}{\partial bLB_t^2} \cdot b^2 < 0 \tag{2.8}$$

Schematically, this innovation process can be illustrated by Figure 2.1.

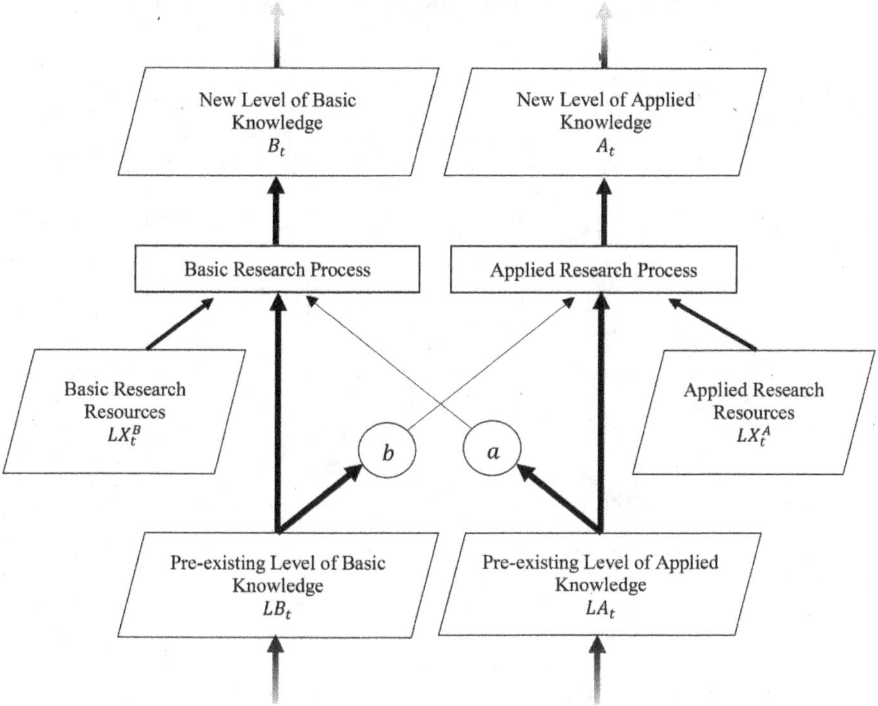

Figure 2.1 *Knowledge cross-fertilization between basic and applied research*

2.3 TOWARD A THEORETICAL MODEL OF REGIONAL INNOVATION OUTPUT

Given the above described basic and applied research structure, regional innovation output at a given time, y_t, can be modeled as a positive, convex function of the state of applied knowledge at the same time:

$$y_t = f\left(A_t, Z_t\right) \ni \frac{\partial f}{\partial A_t} > 0, \frac{\partial^2 f}{\partial A_t^2} < 0 \tag{2.9}$$

with the level of applied knowledge being a function of the interaction between applied research and basic research activity defined by Equations (2.1)–(2.8) above and with the vector Z_t representing aspects of the regional economic environment that mediate the effects of applied knowledge on regional innovation output. Figure 2.2 provides an illustration of this overall model with the greyed-out material indicating inputs and processes that take place too late to affect regional innovation output, y_t.

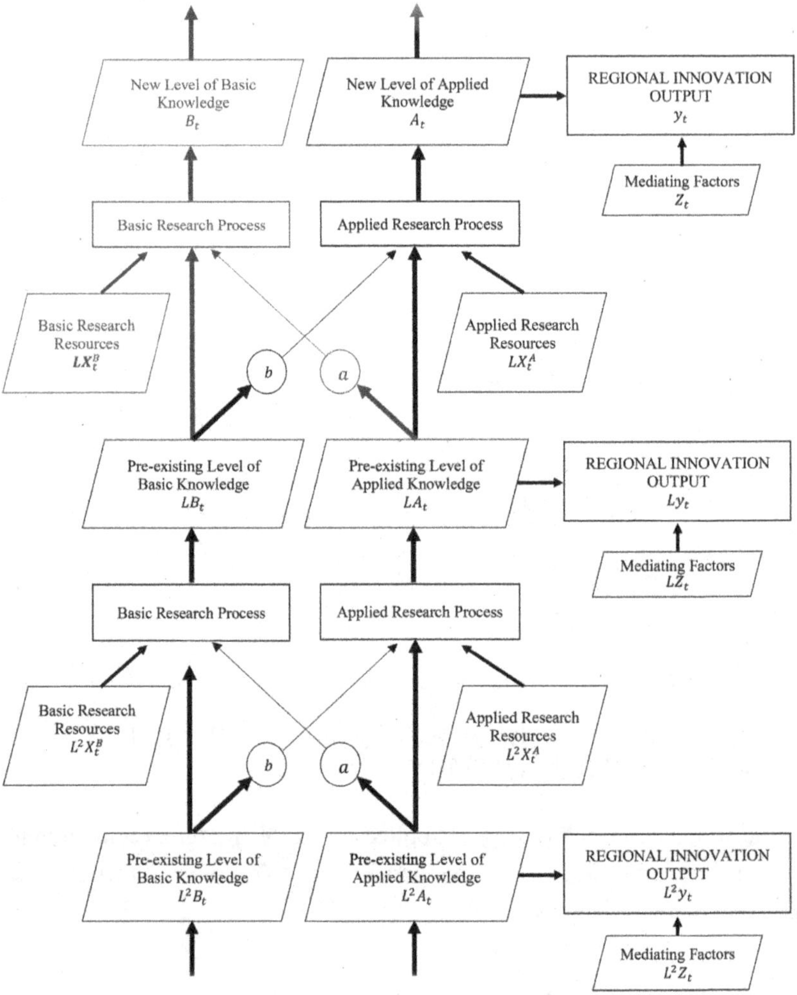

Figure 2.2 A theoretical model of regional innovation output

As Figure 2.2 suggests, the level of applied knowledge, is a function of all past applied research resource input vectors, $\{L^i X_t^A | i = 1,2,3,\ldots\}$, and all past basic research input vectors save the most recent, $\{L^i X_t^B | i = 2,3,4,\ldots\}$. This suggests that an equivalent to Equation (2.9) exists with the set of past research resource input vectors substituting for A_t. Of course, an infinite sequence of past resource input vectors will not be available. But noting that Ly_t is a function of all past applied research resource input vectors save the most recent, $\left\{L^i X_t^A | i = 2,3,4,\ldots\right\}$, and all past basic research input vectors save the two most recent, $\left\{L^i X_t^B | i = 3,4,5,\ldots\right\}$, allows us to rewrite Equation (2.9) as:

$$y_t = g\left(Ly_t, LX_t^A, L^2 X_t^B, Z_t\right) \tag{2.10}$$

such that:

$$\frac{\partial g}{\partial Ly_t} > 0, \frac{\partial^2 g}{\partial Ly_t^2} < 0, \frac{\partial g}{\partial LX_t^A} > 0, \frac{\partial^2 g}{\partial LX_t^{A2}} < 0, \frac{\partial g}{\partial L^2 X_t^B} > 0, \frac{\partial^2 g}{\partial L^2 X_t^{B2}} < 0 \tag{2.11}$$

Based on the logic of Equations (2.7) and (2.8), note that $\frac{\partial g}{\partial L^2 X_t^B}$ will be an increasing function of *b*.[5]

The marginal productivity of basic research funding and of applied research funding are of particular interest because of their connection to policy variables. Because Equation (2.10) is derived from Equation (2.9), we can write the marginal product of applied research funding in producing regional innovation output as:

$$\frac{\partial g}{\partial LX_t^A} = \frac{\partial f}{\partial A_t} \cdot \frac{\partial A_t}{\partial LX_t^A} > 0 \tag{2.12}$$

That is, the marginal product of applied research funding in producing regional innovation output will be the product of the marginal productivity of applied research in producing regional innovation output and the marginal product of applied research funding in producing applied knowledge. Likewise, we can write the marginal product of basic research funding in producing regional innovation output as:

$$\frac{\partial g}{\partial L^2 X_t^B} = \frac{\partial f}{\partial A_t} \cdot \frac{\partial A_t}{\partial bLB_t} \cdot b \cdot \frac{\partial LB_t}{\partial L^2 X_t^B} > 0 \tag{2.13}$$

where the marginal product of basic research funding in producing regional innovation output is the product of the marginal productivity of applied research in producing regional innovation output, the marginal product of basic knowledge in producing applied knowledge, and the marginal product of basic research funding in producing basic knowledge. Note that this marginal product of basic research funding is a positive function of *b*.

The mutual dependence of basic research and applied research on each other suggests that there may be interactions between those different methods of funding the two types of research. To explore the interactions between basic research funding, $L^2X_t^B$, and applied research funding, LX_t^A, consider the cross partial derivative of regional output with respect to applied research funding and basic research funding:[6]

$$\frac{\partial^2 g}{\partial L^2 X_t^B \, \partial L X_t^A} = \frac{\partial^2 f}{\partial A_t^2} \cdot \frac{\partial A_t}{\partial bLB_t} \cdot b \cdot \frac{\partial LB_t}{\partial L^2 X_t^B} \cdot \frac{\partial A_t}{\partial LX_t^A} + \frac{\partial f}{\partial A_t} \cdot \frac{\partial^2 A_t}{\partial bLB_t \partial LX_t^A} \cdot b \cdot \frac{\partial LB_t}{\partial L^2 X_t^B} \tag{2.14}$$

Based on the model in Equations (2.1)–(2.9) above, the first term of Equation (2.14) is negative. However, the second term cannot be as easily signed because of the indeterminately signed cross partial derivative of applied knowledge with respect to basic knowledge and applied research funding, $\frac{\partial^2 A_t}{\partial bLB_t \partial LX_t^A}$. If that cross partial derivative is negative, that is, if greater basic knowledge results in a lower marginal productivity of applied research funding in producing applied knowledge, then the second term in Equation (2.14) will be negative and decreasing in *b*, and thus applied research funding and basic research funding will exhibit typical convexities in the production of regional innovation output without any observable synergies. However, if that cross partial derivative is positive, then Equation (2.14) could be positive or negative. If Equation (2.14) is positive, then applied and basic research funding have the remarkable character that an increase in one will increase the marginal productivity of the other, thus acting synergistically in the production of regional innovation output with that synergy increasing in *b*. Regardless of whether synergies are present, however, an increase in basic research funding will increase total regional innovation output. The issue is simply whether the two types of research funding interact conventionally, that is, non-synergistically, or, instead, essentially "supercharge" each other synergistically in the production of regional innovation output. Figures 2.3 and 2.4 provide illustrations of the production function for regional innovation output for the case where applied research funding and basic research funding are non-synergistic and for the case where they are synergistic.

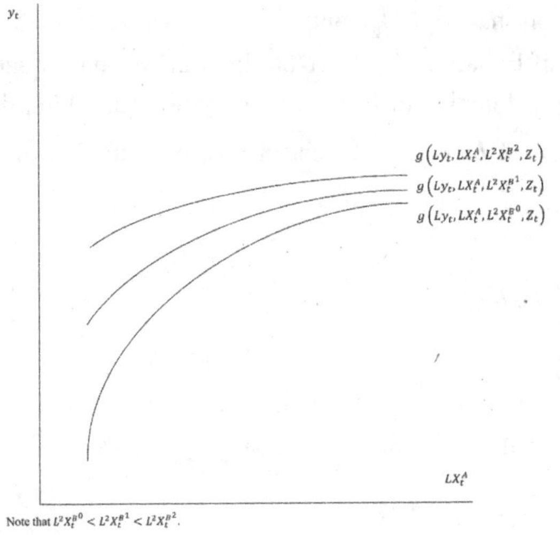

Figure 2.3 *Regional innovation output production with non-synergistic funding inputs*

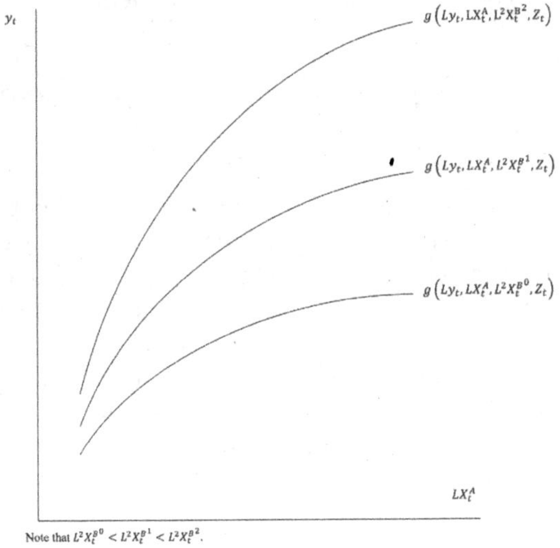

Figure 2.4 *Regional innovation output production with synergistic funding inputs*

Given this discussion, a linearized version of Equation (2.10) can be written:

$$y_t = \beta_1 Ly_t + \beta_2 LX_t^A + \beta_3 L^2 X_t^B + \beta_4 LX_t^A L^2 X_t^B + \beta_5 Z_t + \epsilon_t \tag{2.15}$$

with the expectation that β_1, β_2, and β_3 are positive and that the sign of β_4 is equal to the sign of Equation (2.14).[7] Note in addition that Equation (2.13) implies that β_3 is a positive function of *b*, and that Equation (2.14) implies that β_4 will be a negative function of *b* if $\beta_4 < 0$ and a positive function of *b* if $\beta_4 > 0$. The marginal effect of LX_t^A on y_t will be:

$$\frac{\partial y_t}{\partial LX_t^A} = \beta_2 + \beta_4 L^2 X_t^B \tag{2.16}$$

which, given the sign on Equation (2.12), we anticipate having a positive sign, and the marginal impact of $L^2 X_t^B$ on y_t will be:

$$\frac{\partial y_t}{\partial L^2 X_t^B} = \beta_3 + \beta_4 LX_t^A \tag{2.17}$$

which, given the sign on Equation (2.13), we also anticipate will have a positive sign.

2.4 IMPLICATIONS FOR SCIENCE AND INNOVATION POLICY

Public policy can affect the outcome of the above innovation process in two ways: (1) by affecting the allocation of resources that go to basic research and to applied research, and (2) by affecting *a*, the fraction of total pre-existing applied knowledge that basic researchers have access to, and *b*, the fraction of total pre-existing basic knowledge that applied researchers have access to.

Consider first the allocation of resources to basic research and to applied research. Following Leyden and Menter (2018), the optimal allocation of resources will be where the marginal impact of an increase in resources on regional innovation output is the same for resources devoted to applied research and resources devoted to basic research:

$$\frac{\partial g}{\partial LX_t^A} = \frac{\partial g}{\partial L^2 X_t^B} \tag{2.18}$$

Given the terms in Equation (2.18) are defined by Equations (2.12) and (2.13), this condition can be reduced to the requirement that:

$$\frac{\partial A_t}{\partial LX_t^A} = \frac{\partial A_t}{\partial LB_t} \cdot \frac{\partial LB_t}{\partial L^2 X_t^B} \qquad (2.19)$$

which, in terms of the linear model, is equivalent to the condition:

$$\beta_2 + \beta_4 L^2 X_t^B = \beta_3 + \beta_4 LX_t^A \qquad (2.20)$$

Hence, if $\beta_2 + \beta_4 L^2 X_t^B > \beta_3 + \beta_4 LX_t^A$, public policy should focus on shifting resources toward applied research and away from basic research. Likewise, if $\beta_2 + \beta_4 L^2 X_t^B < \beta_3 + \beta_4 LX_t^A$, public policy should focus on shifting resources toward basic research and away from applied research. Figure 2.5 provides a graphical representation of this analysis for a given sum of resources devoted to applied and basic research, that is, for some total amount of resources X_0 such that:

$$LX_t^A + L^2 X_t^B = X_0 \qquad (2.21)$$

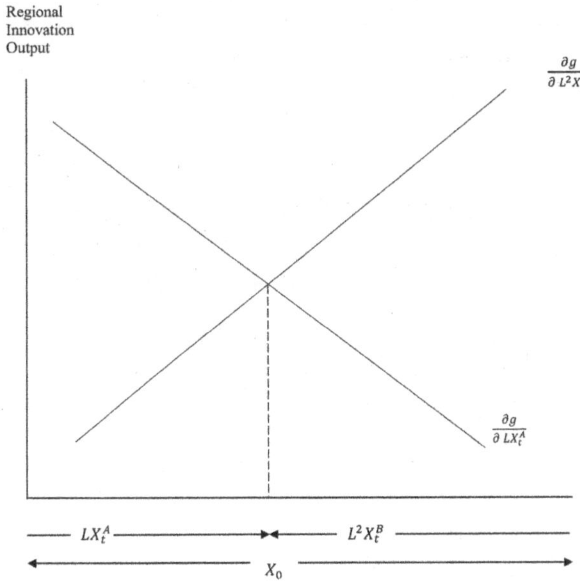

Figure 2.5 *Optimal mix of basic research funding and applied research funding*

There are a variety of mechanisms that could be employed to effect such a policy, ranging from direct governmental grants to basic research and applied research to more indirect mechanisms including tax policy or other expenditure policies to manipulate other organizations to alter their support for basic research and applied research.[8]

Finally, consider public policy directed at altering *a*, the fraction of total pre-existing applied knowledge that basic researchers have access to, and *b*, the fraction of total pre-existing basic knowledge that applied researchers have access to. Given the structure of the model above, it is clear that an increase in *a* or *b* will result in greater regional innovation activity through the increase in the flow of basic knowledge to applied researchers and the increase in the flow of applied knowledge to basic researchers.

There are a variety of policies that might be able to increase the values of *a* and/or *b* including, for example, policies intended on changing institutional arrangements among research-focused organizations,[9] a reallocation of property rights to research results,[10] or financial incentives for greater transfer of knowledge. However, all such policies are likely to be costly and subject to diminishing returns. As a result, the efficient levels of such policy actions are difficult to characterize a priori in a way that might provide guidance to policy makers. Instead, such policies are likely to require a learning-by-doing approach that also includes the learning from failures. Dutz et al. (2014: 3) thus recommend to boost innovation by engaging in experimentation through "designing a portfolio of policies to solve problems step-by-step; monitoring and evaluating intermediate outcomes as rapidly as possible; and constant learning, feedback and adjustment."

2.5 CONCLUSION

Governments play a crucial role in enabling knowledge and technology transfer between regional actors, thus stimulating regional innovation outcomes (Cunningham et al., 2019, 2021; Kochenkova et al., 2016). Public policy measures can thereby exert influence through various means. On the one hand, governments may stimulate knowledge creation and diffusion through financial subsidies, that is, funding of both basic and applied research activities with the idea to provide incentives for collaboration and to reduce predominant gaps impeding efficient interactions between basic and applied research. Besides those direct policy interventions, governments may, on the other hand, more indirectly shape knowledge transfer through creating fruitful environments that facilitate the flow of knowledge and ultimately enhance knowledge cross-fertilization.

Irrespective of the type of policy measure, policy makers need to balance their support as neither a sole focus on basic research nor on applied research seems to be beneficial to achieve the optimal outcome and leverage the full regional innovation potential. Considering the heterogeneity of regional innovation actors and systems is thereby crucial. In their investigation of the German research landscape, Graf and

Menter (2022) show that public research institutions vary significantly in their scientific orientation, ranging from a focus on basic research (e.g., Max Planck institutes) to a focus on applied research (e.g., Fraunhofer institutes). Hence, context matters and one-size-fits-all approaches are likely to fail. Policy makers need to find solutions that take account of and enhance the individual strengths of regional innovation actors as well as enable framework conditions that are conducive for knowledge creation, diffusion, and exploitation.

Future research should empirically assess which policy measures promise the greatest success with regard to regional innovation outcomes. We need a better understanding of the underlying mechanisms of knowledge production and innovation in different contexts as well as the preconditions to leverage synergies and facilitate the cross-fertilization of basic and applied research. Ultimately, it is not about favoring one type of research over the other, but to better coordinate and integrate both basic and applied research activities. Therefore, more guidance for policy makers is needed.

NOTES

1. JEL classification: O31–O38.
2. L is the lag operator. Thus, for example, $Lx_t = x_{t-1}$, and $L^2 x_t = x_{t-2}$.
3. The degree of knowledge transfer may also vary by knowledge discipline. Thus, for example, Audretsch et al. (2004) find that while geographic proximity generally plays a role in the ability to transfer knowledge, the importance of geographic proximity is much lower for researchers in the natural sciences than for researchers in the social sciences. This incomplete process of knowledge transfer is reminiscent of the role that the knowledge filter plays in the knowledge spillover theory of entrepreneurship (Acs et al., 2013).
4. Reinforces earlier work by Hall et al. (2001).
5. Note that the effect of a is embodied in the value of y_t.
6. Equation (2.14) can be interpreted as either (1) the effect of a rise in greater basic research funding on the marginal product of applied research funding or as (2) the effect of a rise in greater applied research funding on the marginal product of basic research funding.
7. One advantage of this linear model is that it provides the potential for determining empirically the sign of β_4 and therefore whether basic research funding and applied research funding interact synergistically.
8. In the United States, examples include the R&E Tax Credit of 1981, the Small Business Innovation Development Act of 1982, and the Omnibus Trade and Competitiveness Act of 1988. For an analysis of these policies and their effects, see Leyden and Link (2015).
9. The National Cooperative Research Act of 1984 in the United States is an example of such a change in institutional arrangements. For an analysis of this policy and its effects, see Leyden and Link (2015).
10. The Bayh-Dole Act of 1980 and the Stevenson-Wydler Act of 1980 in the United States are examples of such a reallocation. For an analysis of these policies and their effects, see Leyden and Link (2015).

REFERENCES

Acs, Z.J., Audretsch, D.B., and Lehmann, E.E. 2013. "The knowledge spillover theory of entrepreneurship." *Small Business Economics* 41 (4): 757–74.

Adams, J.D. 2006. "Learning, internal research, and spillovers." *Economics of Innovation and New Technology* 15 (1): 5–36.

Audretsch, D.B., Lehmann, E.E., and Warning, S. 2004. "University spillovers: Does the kind of science matter?" *Industry and Innovation* 11 (3): 193–206.

Boardman, C., and Bozeman, B. 2006. "Implementing a 'bottom-up,' multi-sector research collaboration: The case of the Texas Air Quality Study." *Economics of Innovation and New Technology* 15 (1): 51–69.

Cunningham, J.A., Lehmann, E.E., Menter, M., and Seitz, N. 2019. "The impact of university focused technology transfer policies on regional innovation and entrepreneurship." *The Journal of Technology Transfer* 44 (5): 1451–75.

Cunningham, J.A., Lehmann, E.E., Menter, M., and Seitz, N. 2021: "Regional innovation, entrepreneurship and the reform of the professor's privilege in Germany." In Guerrero, M., and Urbano, D. (eds), *Technology Transfer and Entrepreneurial Innovations* (pp. 175–205). New York: Springer.

Dutz, M.A., Kuznetsov, Y., Lasagabaster, E., and Pilat, D. 2014. *Making Innovation Policy Work: Learning from Experimentation*. Paris and New York: OECD and The World Bank.

Gordon, R.J. 2016. *The Rise and Fall of American Growth: The US Standard of Living since the Civil War*. Princeton, NJ: Princeton University Press.

Graf, H., and Menter, M. 2022. "Public research and the quality of inventions: The role and impact of entrepreneurial universities and regional network embeddedness." *Small Business Economics* 58 (2): 1187–204.

Hall, B.H., Link, A.N., and Scott, J.T. 2001. "Barriers inhibiting industry from partnering with universities: Evidence from the advanced technology program." *The Journal of Technology Transfer* 26 (1): 87–98.

Kochenkova, A., Grimaldi, R., and Munari, F. 2016. "Public policy measures in support of knowledge transfer activities: A review of academic literature." *The Journal of Technology Transfer* 41 (3): 407–29.

Leyden, D.P., and Link, A.N. 2015. *Public Sector Entrepreneurship: US Technology and Innovation Policy*. New York: Oxford University Press.

Leyden, D.P., and Menter, M. 2018. "The legacy and promise of Vannevar Bush: Rethinking the model of innovation and the role of public policy." *Economics of Innovation and New Technology* 27 (3): 225–42.

Medda, G., Piga, C., and Siegel, D.S. 2006. "Assessing the returns to collaborative research: Firm-level evidence from Italy." *Economics of Innovation and New Technology* 15 (1): 37–50.

3. The role of public finance in knowledge transfer and innovation

David B. Audretsch and Maksim Belitski

3.1 INTRODUCTION

While reaching the factors of innovation and productivity, the analysis of a production productivity frontier (PPF) may be the ultimate and most desired objective for firm managers to understand firm growth's frontiers and boundaries better. From the public policy perspective, approaching PPF is important to challenge incumbents and put them through competitive pressure, achieving wider economic benefits for the industry and regions (Marshall, 1890; Jaffe et al., 1993; Jaffe and Lerner, 2001; Audretsch and Lehmann, 2005; Link et al., 2007; Cappelli et al., 2014).

The promotion of factors that will enhance the firm's innovation and productivity, allowing an entrepreneurial firm to catch up with the top performers in the industry regionally, nationally and internationally, has remained a key agenda for the regional and national innovation policy in both developed and developing countries (Audretsch et al., 2021). Two disparate strands of literature have identified two very different ways to use knowledge for innovation and productivity. The first is direct within the firm through R&D investment and technology improvement. The second is indirect through accessing external knowledge collaboration with third-party firms.

This study overcomes the limitations in prior research on policy for innovative entrepreneurship by employing large, unbalanced panel data and econometric techniques to explain how public finance can facilitate the effect of knowledge collaboration on innovation for firms at different productivity levels. Finance can help firms to overcome their limited capabilities while approaching the PPF and continue to innovate and may be most important for scaling up and high-productive firms. Our data on microfoundations include 17,859 innovative firms in the United Kingdom (UK) during 2002–14. Average firm age in both samples varies between 10 and 11 years since establishment, while all firms in a sample exert innovation effort and costs (R&D and other technology expenditure, legal and strategic protection of innovation, process, product and organizational innovation, exploration and exploitation, knowledge collaboration on innovation).

We find that various combinations of external knowledge facilitate innovation as well as access to public finance. The effect is stronger for the most productive firms as firms approach their productivity frontier. While an increase in external knowledge bundles may limit innovation outcomes and, particularly, at PPF, public finance support programs positively mitigate the diminishing returns to external knowledge on innovation output.

This study addresses a call in knowledge transfer and innovation literature on heterogeneity in firms and financial support (Chung, 2001; David et al., 2013; Colombo et al., 2016).

In the next section, we develop our theoretical framework and formulate our main hypotheses. From this, we discuss data and methodology in Section 3.3. We summarize the results and discuss them in Section 3.4. Section 3.5 offers policy implications and conclusions.

3.2 THEORETICAL FRAMEWORK

More recently, entrepreneurship and innovation scholars have become interested in the synergies between various forms of knowledge collaboration (Cassiman and Valentini, 2016; Audretsch and Belitski, 2020a) and how environmental characteristics can facilitate these synergies (Autio et al., 2014, 2018). This is highlighted by increasing references to contingent and configurational effects in developing a firm's innovation. As noted, innovators are subject to significant financial resource constraints, both at the origin of PPF and at the border of PPF, which compels them to seek external support from private and public foundations, such as universities, the government, external suppliers, competitors, consultants and others to access knowledge, funds, mentorship and incubation. This can lead to two effects. Firstly, a firm is more likely to signal external investors on "investment readiness" and high competitive advantage while approaching the PPF as external investors preoccupied with the potential for business growth (Mason and Kwok, 2010). Secondly, while approaching the PPF, a firm is more likely to obtain public financial support for innovation activities from the following local, regional and national governments (e.g. Innovate UK and local enterprise partnerships initiatives in the UK), which includes financial support via tax credits or deductions, grants, subsidized loans and equity investments and excludes research and other innovation activities conducted entirely for the public sector under contract. This direct financial support enables the reduction of innovation costs (via tax and R&D credits or deductions) as well as the obtaining of lump-sum finance on the development of specific innovation (grants, subsidized loans) and important consultancy and mentorship provided along with financial aid.

In the form of external finance, these two channels of financial support incentivize innovators to organize their business operations strategically to attract external investors and public financial support. External funding is an important feature of an innovator's strategic environment (De Bettignies and Brander, 2007; Cumming et al., 2017) at any firm's growth and productivity stage. We develop these ideas by hypothesizing that firms that secured external funding (public and private) are more likely to finance innovation and deploy open innovation by intensifying external knowledge collaborations. In other words, access to public finance moderates the relationship between knowledge collaboration and innovation.

Knowledge collaboration is costly and firms on their own will face resource constraints when interacting with national and international partners. Hence, without sufficient funding of innovation, firm managers are likely to underuse the potential of knowledge collaborations across different geographical dimensions. In other words, firms with public finance access are likely to invest more intensively in knowledge collaboration than those without such public support and those who face significant financial barriers to innovate (Mason and Kwok, 2010).

Building our argument hierarchically, it is important to control for the configurational effects (knowledge collaboration-public finance) embedded in the simultaneous interaction of knowledge acquisition and the ability of public funding to support innovation. This allows us to capture the complex relationship between knowledge inputs and financial constraints for firms with different productivity levels and understand who can benefit most from the availability of resources for innovation – the least or most productive firms. Given that government financial support usually targets the least productive, early-stage and more promising and fast-growing firms, we may anticipate that firms in the origin of PPF will benefit more from public finance. At the same time, financial constraints are most appealing at the origins of PPF when resources are tight and insufficient to invest in R&D, knowledge collaboration and support would be most welcome at this stage (Audretsch and Belitski, 2020b). Our theoretical imperative is based on the notion that complementarity exists between knowledge collaboration and public financial support (Gibson and Birkinshaw, 2004; Chesbrough et al., 2006; West et al., 2014; Colombo et al., 2016). We hypothesize:

H1: Constraints to access finance exerts a negative effect on firm innovation.

H2: Access to public financial support exerts a positive effect on firm innovation, intensifying knowledge collaboration on innovation.

3.3 DATA AND METHOD

3.3.1 Sample

To test our hypotheses, we used six pooled cross-sectional datasets from the Business Structure Database known as the Business Register and the UK Innovation Survey (UKIS) from the period 2002–14. Although two datasets were pooled together and constructed from two different sources, they are matchable. Firstly, we collected and matched six consecutive UKIS waves (UKIS 4 2002–04, UKIS 5 2004–06, UKIS 6 2006–08, UKIS 7 2008–10, UKIS 8 2010–12 and UKIS 9 2012–14) each conducted every second year by the Office of National Statistics (ONS); the UK was included in this study on behalf of the Department for Business Innovation and Skills (BIS).

Secondly, we matched the Business Structure Database (BSD) data for years 2002, 2004, 2006, 2008, 2010 and 2012 to correspondent Community Innovation Survey (CIS) survey waves with the data from BSD taken for the initial year of UKIS period. The BSD is a version of the Inter-Departmental Business Register for research use, taking full account of changes in firm legal status, ownership (foreign or national firm), alliance information (firm belongs to a larger enterprise network), export, turnover, employment, industry at 5-digit level and a firm location by the postcode. The BSD is the key sampling frame for UK business statistics and is maintained and developed by the Business Registers Unit (BRU) within the ONS. The construction of the data has used specifically Value Added Tax (VAT) businesses and Company Registration (for businesses that wish to operate with limited liability).

Given the availability of data and our research question, we use two samples with two dependent variables – innovation sales and product innovation propensity to test our hypotheses. Each sample was split into subsamples by a level of productivity in percentiles: 0–10%, 20–30%, 40–50%, 60–70%, 70–80% and 90–100%. Having cleaned our sample for missing values, outliers and as a result of a match we end up with 17,859 firms and 21,702 observations.

Most of the firms in innovation sales and product innovation samples come from high-tech manufacturing (15.1 percent and 19.44 percent, respectively), construction (9.9 percent and 10.2 percent, respectively), wholesale and retail trade (16.8 percent and 16.0 percent, respectively), real estate and business activities (14.4 percent and 12.3 percent, respectively) as well as public services (including healthcare and defense) (11.1 percent and 10.1 percent, respectively). Sectors where firms from both samples are underrepresented are mining and quarrying (<1 percent), utility electricity (<1 percent) education (approximately 1 percent) (see Appendix 3.A1).

Most of the firms in innovation sales and product innovation samples are located in the South East of England (10.91 percent and 10.88 percent, respectively), London (9.51 percent and 9.72 percent, respectively), the North West (9.20 percent and 9.08 percent, respectively) and Eastern England (8.97 percent and 9.09 per cent, respectively). Wales (<6 percent), Scotland (<9 percent) and Northern Ireland (<8 percent) are least represented in both samples. The industrial and geographical composition of firms does not change across multiple samples which illustrates that the reduced sample is representative of the original one (89,518 observations) (see Appendices 3.A1 and 3.A2).

The major differences were observed across survey waves 2002–14. Most of the observations in our sample come from the first UKIS4 round (2002–04) – 57.8 percent in the (innovative sales) sample, while it is only 18.4 percent in the original sample. Although there is a symmetric distribution of firms in both samples and after the UK 2002–04 wave, we find that the wave 2010–12 (post-crises) is least representative for both samples (16.1 percent and 6.1 percent, respectively). It is likely that the share of new product innovators (Colombelli et al., 2016) as well as the number of firms who responded to a survey dropped significantly in post-crises period (see Appendix 3.A2). Description and sources of variables included in this study are in

Table 3.1, while their descriptive statistics of variables are available in the original sample (89,518 observations) and innovation sales sample (21,702 observations).

Table 3.1 Description of variables

Variable (source)	Definition
Productivity	Difference between firm's labor productivity and average labor productivity (sales per employee) by 3-digit SIC industry using a full sample of firms from the Business Registry by each year. Based on productivity, variable percentile subsamples were created.
Innovation (UKIS)	% of firm's total turnover from goods and services that were new to the market
Product innovator (UKIS)	Binary variable =1 if firm reports positive firm's turnover from goods and services that were new to the market or new to the firm, zero otherwise
Age (BSD)	Age of a firm (years since the establishment), in logs
Employment (BSD)	Number of full-time employees, in logarithms
High-tech manufacturing (UKIS)	Binary variable =1 if SIC2007 (2 digit): 21, 26, 30, 31, zero otherwise
Med-tech manufacturing (UKIS)	Binary variable =1 if SIC2007 (2 digit): 20, 22–25, 27–29, 32, zero otherwise
Finance cost	Sum of scores (0 to 3) of the factors which are the main constraints to innovation: direct innovation cost too high; cost of finance; availability of finance (rescaled between zero and one)
Government funding[a]	Binary variable =1 if firm receives public (UK regional and national government) financial support (if any) for innovation activities such as financial support via tax credits or deductions, grants, subsidized loans and loan guarantees and excludes R&D and other innovation activities conducted entirely for the public sector under contract, zero otherwise
Risk (UKIS)	Firm has experienced constraining innovation activities such as excessive perceived economic risks (zero – not experienced, 3 – high)
Technology (UKIS)	Firm has experienced constraining innovation activities such as lack of information on technology (zero – not experienced, 3 – high)
Scientist (UKIS)	The proportion of employees that hold a degree or higher qualification in science and engineering at BA/BSc, MA/PhD, PGCE levels
Exporter (UKIS)	Binary variable =1 if a firm sells its products in foreign markets, zero otherwise
Survival (BSD)	Binary variable =1 if a firm survived as an independent unit or as a part of a group until year 2017, zero otherwise
Herfindahl Index	Herfindahl Index calculated using concentration in employment by 2 SIC digit industry
Foreign (BSD)	Binary variable =1 if a firm has headquarters abroad, zero otherwise
Subsidiaries	Number of firm's subsidiaries and local units, in logarithms
Cooperation regional (UKIS)	Binary variable =1 if firm cooperates on innovation regionally within enterprise group, suppliers; clients or customers; competitors; consultants, commercial labs, private R&D institutes; universities; government and public research institutes, zero otherwise
Cooperation national (UKIS)	Binary variable =1 if firm cooperates on innovation nationally within enterprise group, suppliers; clients or customers; competitors; consultants, commercial labs, private R&D institutes; universities; government and public research institutes, zero otherwise
Cooperation international (UKIS)	Binary variable =1 if firm cooperates on innovation in Europe and other world within enterprise group, suppliers; clients or customers; competitors; consultants, commercial labs, private R&D institutes; universities; government and public research institutes, zero otherwise

Variable (source)	Definition
R&D intensity (UKIS)	The amount of expenditure for internal Research and Development (000s), to total sales (000s pound sterling)
Software (UKIS)	The amount of expenditure for purchasing advanced machinery, equipment and software (000s) to total sales (000s pound sterling)
Appropriability (UKIS)	Sum of scores of the effectiveness of the following methods for protecting new products and processes: secrecy, complexity of goods and services, lead time advantages, patenting, design, copyright, trademarks, lead, complexity (rescaled between zero and one)

Note: [a] The public sector includes government-owned organizations such as local, regional and national administrations and agencies, schools, hospitals and government providers of services such as security, transport, housing and energy, etc. UK central government's agencies or funding bodies include, for example, Innovate UK, formerly known as TSB, Department of Business Innovation and Skills.
Source: Department for Business, Innovation and Skills, Office for National Statistics, Northern Ireland. Department of Enterprise, Trade and Investment (2018). UK Innovation Survey, 1994–2016: Secure Access [data collection]. 6th Edition. UK Data Service. SN: 6699, http://doi.org/10.5255/UKDA-SN-6699–6.
Office for National Statistics. (2017). *Business Structure Database, 1997–2017: Secure Access* [data collection]. *9th Edition.* UK Data Service. SN: 6697, http://doi.org/10.5255/UKDA-SN-6697–9.
Further citation: UK Innovation Survey, 1994–2016; Business Structure Database, 1997–2017.

3.3.2 Variables

Our dependent variable (DV) is the commercial success of the innovation (innovative sales) [0–100]. A turnover-based measure enables us to integrate these innovations' highly variable commercial value (Negassi, 2004; Laursen and Salter, 2014). Given the potential endogeneity between collaboration variables and innovation and descriptive statistics across all samples, we perform IV Tobit estimation (Wooldridge, 2009).

We use a firm's labor productivity to distinguish subsamples for innovation. A firm's labor productivity is measured as a difference between firm's sales per employee and the industry average sales per employee (by 3-digit SIC). This is appealing for several reasons. Industry competitors are most likely to face similar conditions and experience common shocks to performance. By comparing a firm's performance to its 3-digit SIC industry peers' performances, it is likely to be the closest approximation of potential joint product and joint competitors, the experience of common industry shocks, maintaining a strong baseline of comparability. Interestingly, in their study of relative performance evaluation, Gibbons and Murphy (1990) do not find evidence of the use of relative performance evaluation at the industry level. They find more evidence of relative performance evaluation at the level of the stock market as a whole.

Our explanatory variables for external knowledge are knowledge collaboration (Audretsch and Belitski, 2019). In this study, regional knowledge collaboration is considered with an establishment located within a 100 miles area for several reasons. Knowledge collaboration nationally takes place within the UK geographical boundaries, including Northern Ireland. Collaboration with international partners includes contacts with the European Union (EU) and other non-EU countries (van Beers and Zand, 2014; Choi and Contractor, 2019).

Control variables are described in Table 3.1.

3.3.3 Method

3.3.3.1 Innovation production function (EU)

We estimate the innovation production function using a random-effects Tobit model with a dependent variable y_i (firm's innovation) and an endogenous variable m_i (knowledge collaboration):

$$y_{it} = \beta_0 + \beta_1 x_{it} + \beta_2 r_{it} + \omega_1 m_{it} + \beta_3 m_{it} * r_{it} + \lambda_t + \tau_s + \psi_j + u_{it} \qquad (3.1)$$

We can also call it structural equation to emphasize that we are interested in β_i and that the equation to be measured as causal. Variable r_{it} indicates availability of financial capital for a firm or public financing; m_{it} is explanatory variables of firm's innovation – knowledge collaboration, and u_{it} is an error term. x_{it} are exogenous control variables not correlated with u_{it}, while m_{it} is likely to be correlated with u_{it} (Wooldridge, 2009: 517). λ_t *and* τ_s are time and industry fixed effects, ψ_j is city fixed effects where firm is located. m_{it} represents four variables.

3.3.3.2 Endogeneity issue

A firm decides whether to collaborate on knowledge strategically, and firms with high levels of innovation output are more likely to source knowledge externally (West et al., 2014). This raises a possible endogeneity issue. In order to analyze the relationship between external knowledge collaboration and innovation we estimate an innovation production function in two steps (Wooldridge, 2009) to correct for endogeneity. We split productivity variable (ρ_i) across ten percentiles and estimate selectively seven innovation, production functions for 10%, 30%, 50%, 70%, 80%, 90% and 100% percentiles of firm productivity. We estimate innovation production function correcting for endogeneity and as a robustness check we estimate it without correcting for endogeneity with the original values of knowledge collaboration.

Table 3.2 reports the second-stage IV Tobit estimation with $\widehat{\varphi_i}$ and x_i as explanatory variables.

3.4 RESULTS AND DISCUSSION

3.4.1 Access to Public Financial Support for Innovation

We follow Wiklund and Shepherd (2005), who used the regression coefficient estimates to plot the total effects of external finance on small and medium-sized enterprises' (SMEs) propensity to export for a range of values. Table 3.2 regressions (1–6) demonstrate the role that constraints in accessing finance have on innovation for highly productive firms supporting H1.

Table 3.2 Results of IV Tobit estimation for innovation output at different level of productivity percentiles (stage two). Dependent variable: innovation is taken in % and varies from 0–100 (N = 21,702 observations)

Specification	(1)	(2)	(3)	(4)	(5)	(6)
Percentile of productivity	10%	30%	50%	70%	80%	100%
Age	−0.50	−0.14	−0.28	0.56	−0.83	−0.47
	(0.57)	(0.44)	(0.50)	(0.40)	(0.43)	(0.52)
Age squared	0.01	0.02	0.02	0.01	0.01	0.01
	(0.01)	(0.01)	(0.01)	(0.01)	(0.01)	(0.01)
Employment	−2.02	0.78	−0.12	−0.80	−0.93	0.59
	(1.3)	(1)	(1.1)	(.82)	(.81)	(.94)
High-tech manufacturing	−15.98	−1.97	−6.123	2.069	−12.05	8.50
	(15)	(9.8)	(17)	(16)	(24)	(14)
Med-tech manufacturing	0.37	−6.04	5.077	0.02	0.95	−4.29
	(8.3)	(4.3)	(3.7)	(3.3)	(3.2)	(5.6)
Finance cost (H1)	20.43	−1.17	2.58	−1.13**	−0.62**	−0.67**
	(10)	(4.5)	(4.6)	(0.5)	(0.2)	(0.2)
Government funding	14.70*	12.46**	0.040	3.21**	6.90***	8.66***
	(6.7)	(4.6)	(5.1)	(1.9)	(2.3)	(2.1)
Risk	−2.31	4.01**	1.87	2.13*	1.89	3.41*
	(1.7)	(1.2)	(1.3)	(1.0)	(1.1)	(1.4)
Technology	−1.41	0.21	−0.72	2.49*	2.04	0.37
	(2.0)	(1.4)	(1.4)	(1.1)	(1.3)	(1.6)
Scientist	0.30***	0.08	0.04	0.09*	0.17**	0.01
	(0.08)	(0.05)	(0.05)	(0.04)	(0.04)	(0.07)
Exporter	12.94***	5.81*	7.49**	6.70***	7.23***	9.69***
	(3.0)	(2.3)	(2.4)	(1.9)	(2.0)	(2.5)
Survival	−2.13	5.68*	-1.37	1.60	1.75	1.85
	(3.1)	(2.2)	(2.3)	(1.7)	(1.8)	(2.3)
Herfindahl Index	−28.46	9.471	−15.37	−8.510	33.02*	26.11
	(17)	(21)	(23)	(21)	(16)	(17)
Foreign	−5.75	0.08	1.23	−4.09*	−0.81	−4.43
	(3.8)	(2.5)	(2.5)	(2)	(2.2)	(3.1)
Subsidiaries	1.05	−2.65	0.35	−1.00	−0.87	−1.13
	(2)	(1.8)	(1.6)	(1.2)	(1.2)	(1.3)

Specification	(1)	(2)	(3)	(4)	(5)	(6)
Collaboration regional β_2	11.25**	5.51*	6.04*	7.39**	3.82	4.45
	(3.7)	(2.5)	(3)	(2.3)	(2.4)	(3.1)
Collaboration national β_3	9.36*	12.00***	12.42***	10.72***	13.03***	7.41*
	(3.7)	(2.8)	(3.1)	(2.6)	(2.6)	(3.2)
Collaboration international β_4	−5.12	4.58	3.94	0.37	0.56	1.84
	(4.3)	(3.1)	(3.5)	(2.8)	(2.8)	(3.4)
R&D intensity	81.75***	38.00**	94.55***	45.39**	48.76**	29.66
	(21)	(14)	(19)	(17)	(19)	(25)
Software	44.86*	−0.829	43.55*	32.56	21.26	20.62
	(22)	(18)	(21)	(18)	(19)	(28)
Appropriability	45.95***	49.19***	40.18***	41.26***	38.37***	46.01***
	(11)	(8.4)	(7.6)	(5.6)	(5.8)	(7.4)
Constant	−23.72*	44.35***	−55.21***	−52.48***	−29.37***	−47.84***
	(12)	(10)	(13)	(13)	(7.9)	(11)
Variance of error term	720.84**	613.84**	708.94**	508.71**	562.41**	704.51**
	(67)	(48)	(59)	(38)	(41)	(60)
Industry, year and city-region fixed effects	Yes	Yes	Yes	Yes	Yes	Yes
Number of observations	1133	1801	1972	2270	2201	1557
Chi^2	603.58	819.50	672.93	768.44	784.12	487.88
Left-censored	861	1411	1610	1832	1751	1218
Log-likelihood	−1477.63	−2094.88	−2041.35	−2376.68	−2449.71	−1891.86
Pseudo R^2	0.16	0.16	0.14	0.13	0.13	0.11

Note: Reference category for legal status is company (limited liability company), industry (mining), city-region (Newcastle). Instead of industry dummies in this estimation, employment (in logs) is used. Robust standard errors are in parentheses. The coefficients of the Tobit regressions are the marginal effect of the independent variable on the probability of incoming spillover, knowledge collaboration, *ceteris paribus*. For dummy variables, it is the effect of a discrete change from 0 to 1. Significance level: * $p < 0.05$; ** $p < 0.01$, *** $p < 0.001$.
Source: UK Innovation Survey, 1994–2016; Business Structure Database, 1997–2017.

One-unit change in the difficulty to access finance reduces innovation by 0.67 to 1.13 percent in economic terms. We can also note that the direct effect of government financing is positive and statistically significant. Access to government funding increases innovation sales from 3.21 to 8.66 percent. Interestingly, that the effect is stronger for the most productive firms means that firms with high productivity will benefit more from government funding.

Table 3.3 regressions (1–6) report the moderation effects between knowledge collaborations and innovation. Receiving public finance support means various forms of mentoring and financial aid from the local or national UK government and other public institutions aiming to support innovation. The interactions are positive again for firms with 70–100 percentile of productivity. This means that firms in the PPF will benefit more from public support and will be able to translate their knowledge collaboration regionally and internationally into innovation better. This finding supports H2.

Table 3.3 *Results of IV Tobit estimation for innovation output at different level of productivity percentiles (stage two). Dependent variable: innovation is taken in % and varies from 0–100 (N overall = 21,702)*

Specification	(1)	(2)	(3)	(4)	(5)	(6)
Percentile of productivity	10%	30%	50%	70%	80%	100%
Age	−0.35	−0.07	−0.31	0.65	−0.74	−0.37
	(0.56)	(0.44)	(0.5)	(0.41)	(0.43)	(0.52)
Age squared	0.01	−0.01	−0.01	−0.02	0.01	0.01
	(0.01)	(0.01)	(0.01)	(0.01)	(0.01)	(0.01)
Employment	−2.63*	1.04	−0.20	−0.98	−0.88	0.49
	(1.3)	(1.0)	(1.1)	(0.83)	(0.81)	(0.93)
High-tech manufacturing	−16.94	−4.17	−9.57	3.49	−11.36	12.14
	(15)	(9.9)	(18)	(16)	(23)	(14)
Med-tech manufacturing	0.02	−6.12	5.45	0.59	1.64	−5.16
	(8.4)	(4.3)	(3.7)	(3.3)	(3.2)	(5.6)
Finance cost	19.42**	−0.09	2.51	−1.00	1.20	−3.61
	(6)	(4.5)	(4.7)	(3.7)	(4.1)	(5.2)
Government funding	11.89	9.31	5.30	14.01**	5.51**	14.36**
	(16)	(16)	(15)	(8.5)	(1.8)	(5.2)
Risk	−2.62	3.99**	1.74	1.99	1.79	3.27*
	(1.7)	(1.2)	(1.3)	(1)	(1.1)	(1.4)
Technology	−1.34	−0.27	−0.78	2.52*	2.29	−0.17
	(1.9)	(1.4)	(1.4)	(1.1)	(1.3)	(1.6)
Scientist	0.28**	0.06	0.03	0.08	0.19***	−0.01
	(0.08)	(0.05)	(0.05)	(0.04)	(0.04)	(0.07)
Exporter	11.67**	5.87*	7.39**	6.24**	7.10***	9.89***
	(3)	(2.3)	(2.4)	(1.9)	(2.0)	(2.5)
Survival	−0.36	−2.02	5.50*	−1.63	1.77	3.34
	(2.9)	(2.2)	(2.3)	(1.7)	(1.8)	(2.3)
Herfindahl Index	−26.27	12.33	−16.56	−11.38	32.95*	20.66
	(17)	(21)	(23)	(21)	(16)	(17)
Foreign	−5.32	0.75	0.81	−3.85	−0.93	−4.39
	(3.7)	(2.5)	(2.5)	(2)	(2.2)	(3.1)
Subsidiaries	1.00	−3.02	0.52	−1.13	−0.51	−0.96
	(2)	(1.8)	(1.6)	(1.3)	(1.2)	(1.3)
Collaboration regional β_2	18.61**	2.87	1.53	9.31	7.40	4.99
	(7)	(5.6)	(6.9)	(5.2)	(5.8)	(7.2)
Collaboration national β_3	17.45*	21.63***	21.02***	19.39***	17.14**	17.24**
	(6.8)	(5.9)	(5.7)	(5)	(5.4)	(6.1)

Specification	(1)	(2)	(3)	(4)	(5)	(6)
Collaboration international	−16.30	13.87	11.17	9.693	10.55	16.46*
β_4	(10)	(7.5)	(7.2)	(6.7)	(6.2)	(6.5)
R&D intensity	99.08**	42.13**	96.69***	53.91**	41.51*	45.19
	(21)	(15)	(19)	(17)	(19)	(25)
Software	33.68*	1.864	42.72*	31.72*	23.46	9.54
	(17)	(18)	(20)	(15)	(19)	(28)
Appropriability	41.63**	49.94***	39.14***	40.51***	38.84***	42.74***
	(11)	(8.4)	(7.6)	(5.6)	(5.8)	(7.3)
Collaboration regional x Government funding (H2)	22.19 (22)	1.501 (18)	−27.11 (38)	2.41** (0.5)	6.16** (3.25)	25.75** (11)
Collaboration national x Government funding (H2)	7.50 (24)	−0.94 (19)	−20.63 (37)	72.11* (32.2)	2.91** (1.2)	89.49** (40.1)
Collaboration international x Government funding (H2)	19.40 (25)	4.98 (21)	3.03 (39)	41.44** (21)	25.74** (12.3)	142.61** (35.5)
Industry, year and city-region fixed effects	Yes	Yes	Yes	Yes	Yes	Yes
Constant	−19.39	−47.49***	−55.12***	−54.24***	−32.74***	−51.59***
	(11)	(10)	(13)	(14)	(8.0)	(11)
Variance of error term	674.4***	598.4***	694.7***	502.6***	547.9***	674.1***
	(62)	(47)	(58)	(38)	(40)	(57)
Number of observations	1133	1801	1972	2270	2201	1557
Chi^2	630.94	839.73	687.39	789.70	808.76	519.02
Left-censored	861	1411	1610	1832	1751	1218
Log-likelihood	−1463.94	−2084.75	−2034.12	−2366.06	−2437.39	−1876.28
Pseudo R^2	0.17	0.16	0.14	0.14	0.14	0.12

Note: Reference category for legal status is company (limited liability company), industry (mining), city-region (Newcastle). Instead of industry dummies in this estimation, employment (in logs) is used. Robust standard errors are in parentheses. The coefficients of the second stage estimation (Tobit) regressions are the marginal effect of the independent variable on the probability of incoming spillover, knowledge collaboration, *ceteris paribus*. For dummy variables, it is the effect of a discrete change from 0 to 1. Significance level: * $p < 0.05$; ** $p < 0.01$, *** $p < 0.001$.
Source: UK Innovation Survey, 1994–2016; Business Structure Database, 1997–2017.

The nature innovation output is more likely when firms actively search for knowledge collaboration internationally and obtain public investment from the UK local or national government (Table 3.3). This is an important result that does not emerge from a simple assessment of interaction effects. We also find that most productive firms with high financial constraints to innovation will innovate less. Public financial support is important as it may achieve greater complementarities between international and regional knowledge collaborations and the government.

3.4.2 Other Determinants of Innovation

Next we discuss other determinants of innovation using Tables 3.2 and 3.3. Once controlling for firm age and size, we find that innovation sales do not depend on firm size and age, which remain insignificant for various levels of productivity.

There is no difference in innovation sales between high- and medium-tech manufacturing firms and non-manufacturing firms, including low-tech manufacturing. Although the results seem surprising, all industries are becoming technologically and digitally savvy, as well as product innovation taking time, which provides essential competitive advantage to other firms in non-manufacturing industries to transform their knowledge inputs into innovation quicker.

The cost of technology and risk sharing are important drivers of innovation. On the contrary, firms which perceive the risk of innovation and technology availability as a barrier to innovate are not likely to have higher innovation outputs. Viewed from the innovation perspective, this result is not so surprising. All innovators deal with higher uncertainty and risk as well as available technology to create new products. These indicators also perceive those factors which may not directly the capture financial health of a firm, rather than market risks. Human capital is important for innovation; firms that had a higher share of employees with university degrees had higher innovation sales. A positive effect was found for both least innovative firms and most innovative firms.

Firms that export to Europe and the rest of the world have on average 5–12 percent higher innovation sales than firms that do not innovate; interestingly, the effect remains positive and significant for firms with low, medium and high levels of productivity (Tables 3.2 and 3.3). A binary variable of survival that picks up firms that survived or not until 2017 is positive but insignificant. The result demonstrates that innovation and higher productivity are not a guarantee for survival. Competition in the industry proxied by the Herfindahl Index is not associated with differences in innovation sales at different levels of productivity.

Finally, firms which invest in software and advanced technology are likely to have improved technological competences and better leverage the physical distance when collaborating (Cassiman and Veugelers, 2002, 2006). It is likely that the least productive firms will benefit doubly from investment in internal R&D than the most productive firms in the 80th percentile (approaching the PPF). The effect of appropriability on innovation sales is positive and statistically significant across various levels of productivity, including the 10th and 100th percentiles (Table 3.3, regressions 1–6).

3.5 CONCLUSION

There are several important findings and implications in this chapter. Firstly, returns to a portfolio of external knowledge collaborations are conditional on the type of knowledge collaboration partner and its productivity.

Secondly, public finance and programs facilitate innovation for firms that exploit external knowledge portfolios and are most productive. This demonstrates the importance and higher degree of special innovation support programs' applicability, rather than making finance easily available to firms with various levels of productivity.

Our findings support public finance's role, rather than access to finance or reducing the cost of finance, in exerting positive complementarities between knowledge collaborations and public finance. Our findings also demonstrated that this cost of finance matters for innovators and that systemic financial support created through the government mechanisms for innovation policy is important.

In particular, public finance helped the most productive firms in taking the lead in innovation and integrating external knowledge in the organization.

Given the major limitation of this study (small panel element) and the cross-sectional character of data over 12 years, we cannot directly enforce the causality. This means we do not know whether predictive firms access finance easier or it is financial support that leads to greater productivity.

3.5.1 Managerial and Policy Implications

Using PPF, we unpack the role that the firm's productivity and public finance may exert on the firm's ability to use, absorb and appropriate the results of innovation, and exploit various bundles of collaborations to achieve higher innovation output. At the same time, our results demonstrate that availability and cost of financial capital and the cost of investment in innovation have little to explain the joint negative effect of collaborations on innovation output and sales. We find that firms that accessed UK regional and national government financial support and mentorship and were able to apply R&D and other tax exceptions were able to mitigate the negative effect and turn knowledge collaborations to innovation output. This is an important finding, which demonstrates that finance cost and access per se do not change the diminishing marginal returns to collaboration; rather it is institutionalized local and national support which helps.

Public financial support in the UK includes both tax exceptions mechanisms, as well as ad hoc mentoring, collaboration with universities via Innovate UK grants, innovation vouchers and other forms of boosting innovation as an institutional system of support (local enterprise partnerships) and national programs (e.g. Innovate UK, UK Trade and Investment department schemes, etc.) which offer both financial and non-financial support (Innovate UK, 2019).

Our results indicate that financial support for innovation instruments in the UK deliver on their purpose and enable firms to mitigate resource limitations and in bestowing collaboration. While our main focus was not on any specific industry, future industry-specific studies (e.g. information and communication, knowledge-intense business services, creative, key enabling technologies and R&D) are welcome.

In doing this, scholars can draw managers' and policy makers' attention to how changes in knowledge partnerships and their complementarities change innovation propensity and commercialization of new products. This will identify a necessary condition for innovation while positioning firms at a very specific point within PPF. For example, in light of Brexit and an increase in accessibility of finance via local enterprise partnerships and UK Trade and Investment (UKTI), which fosters internationalization of innovators, it is important to evidence whether policy instruments will have similar effects across different sectors, for example, those which may be most affected by the UK leaving the European Union common market.

REFERENCES

Audretsch, D. B., and Belitski, M. (2020a). The limits to collaboration across four of the most innovative UK industries. *British Journal of Management*, *31*(4), 830–55.

Audretsch, D. B., and Belitski, M. (2020b). The role of R&D and knowledge spillovers in innovation and productivity. *European Economic Review*, *123*, 103391.

Audretsch, D.B., and Lehmann, E.E. (2005). Does the knowledge spillover theory of entrepreneurship hold for regions? *Research Policy*, 34(8), 1191–202.

Audretsch, D.B., Belitski, M., and Caiazza, R. (2021). Start-ups, innovation and knowledge spillovers. *The Journal of Technology Transfer*, *46*(6), 1995–2016.

Autio, E., Kenney, M., Mustar, P., Siegel, D., and Wright, M. (2014). Entrepreneurial innovation: The importance of context. *Research Policy*, 43(7), 1097–108.

Autio, E., Nambisan, S., Thomas, L.D.W., and Wright, M. (2018). Digital affordances, spatial affordances, and the genesis of entrepreneurial ecosystems. *Strategic Entrepreneurship Journal*, 12(1), 72–95.

Cappelli, R., Czarnitzki, D., and Kraft, K. (2014). Sources of spillovers for imitation and innovation. *Research Policy*, 43, 115–20.

Cassiman, B., and Valentini, G. (2016). Open innovation: Are inbound and outbound knowledge flows really complementary? *Strategic Management Journal*, 37, 1034–46.

Cassiman, B., and Veugelers, R. (2002). R&D cooperation and spillovers: Some empirical evidence from Belgium. *American Economic Review*, 92(4), 1169–84.

Cassiman, B., and Veugelers, R. (2006). In search of complementarity in innovation strategy: Internal R&D and external knowledge acquisition. *Management Science*, 52, 68–82.

Chesbrough, H., Vanhaverbeke, W., and West, J. (2006). *Open Innovation: Researching a New Paradigm*. Oxford: Oxford University Press.

Choi, J., and Contractor, F.J. (2019). Improving the progress of research & development (R&D) projects by selecting an optimal alliance structure and partner type. *British Journal of Management*, 30(4), 791–809.

Chung, C. (2001). Markets, culture and institutions: The emergence of large business groups in Taiwan, 1950s–1970s. *Journal of Management Studies*, 38(5), 719–45.

Colombelli, A., Krafft, J., and Vivarelli, M. (2016). To be born is not enough: The key role of innovative start-ups. *Small Business Economics*, 47(2), 277–91.

Colombo, M.G., Cumming, D.J., and Vismara, S. (2016). Governmental venture capital for innovative young firms. *Journal of Technology Transfer*, 41(1), 10–24.

Cumming, D., Filatotchev, I., Knill, A., Reeb, D.M., and Senbet, L. (2017). Law, finance, and the international mobility of corporate governance. *Journal of International Business Studies*, 48(2), 123–47.

David, R.J., Sine, W.D., and Haveman, H.A. (2013). Seizing opportunity in emerging fields: How institutional entrepreneurs legitimated the professional form of management consulting. *Organization Science*, 24(2), 356–77.

De Bettignies, J-E., and Brander, J.A. (2007). Financing entrepreneurship: Bank finance versus venture capital. *Journal of Business Venturing*, 22(6), 808–32.
Gibbons, R., and Murphy, K.J. (1990). Relative performance evaluation for chief executive officers. *ILR Review*, 43(3), 30-S.
Gibson, C.B., and Birkinshaw, J. (2004). The antecedents, consequences, and mediating role of organizational ambidexterity. *Academy of Management Journal*, 47, 209–26.
Innovate UK (2019) UK Innovators support. Available at: https://www.gov.uk/government/organisations/innovate-uk. Accessed 31 May 2019.
Hall, B.H., and Sena, V. (2017). Appropriability mechanisms, innovation, and productivity: Evidence from the UK. *Economics of Innovation and New Technology*, 26(1–2), 42–62.
Jaffe, A.B., and Lerner, J. (2001). Reinventing public R&D: Patent policy and the commercialization of national laboratory technologies. *RAND Journal of Economics*, 32, 167–98.
Jaffe, A.B., Trajtenberg, M., and Henderson, R. (1993). Geographic localization of knowledge spillovers as evidenced by patent citations. *The Quarterly Journal of Economics*, 108(3), 577–98.
Laursen, K., and Salter, A.J. (2014). The paradox of openness: Appropriability, external search and collaboration. *Research Policy*, 43(5), 867–78.
Link, A.N., and Siegel, D. S. (2007). *Innovation, Entrepreneurship, and Technological Change*. Oxford: Oxford University Press.
Marshall, A. (1890). *Principles of Economics*. London: Macmillan.
Mason, C.M., and Kwok, J. (2010). Investment readiness programmes and access to finance: A critical review of design issues. *Local Economy*, 25(4), 269–92.
Negassi S. (2004). R&D cooperation and innovation: A microeconometric study on French firms. *Research Policy*, 33, 365–84.
Storey, D.J. (2003). Entrepreneurship, small and medium sized enterprises and public policies. In *Handbook of Entrepreneurship Research* (pp. 473–511). Boston, MA: Springer.
van Beers, C., and Zand F. (2014). R&D cooperation, partner diversity, and innovation performance: An empirical analysis. *Journal of Product Innovation Management*, 31(2), 292–312.
West, J., Salter, A.J., Vanhaverbeke, W., and Chesbrough, H. (2014). Open innovation: The next decade. *Research Policy*, 43(5), 805–11.
Wiklund, J., and Shepherd, D. (2005). Entrepreneurial orientation and small business performance: A configurational approach. *Journal of Business Venturing*, 20(1), 71–91.
Wooldridge, J.M. (2009). *Introductory Econometrics: A Modern Approach*. 4th edn. Mason, OH: South-Western.

APPENDIX 3A.1

Table 3A.1 Three samples sector divisions (by SIC 2007)

Sector divisions	Baseline	%	Innovative	%
1 – Mining & Quarrying	486	0.65	175	0.81
2 – Manufacturing basic	4 025	5.41	1 277	5.88
3 – High-tech manufacturing	11 682	15.70	4 218	19.44
4 – Utility	780	1.05	170	0.78
5 – Construction	7 370	9.90	2 229	10.27
6 – Wholesale, retail trade	12 530	16.84	3 481	16.04
7 – Transport, storage	4 792	6.44	1 195	5.51
8 – Hotels & restaurants	5 400	7.26	1 174	5.41
9 – ICT	4 441	5.97	1 434	6.61
10 – Financial intermediation	2 651	3.56	850	3.92
11 – Real estate & other business activities	10 728	14.41	2 682	12.36
12 – Public admin, defense	8 305	11.16	2 196	10.12
13 – Education	213	0.29	152	0.70
16 – Other community, social activity	1 024	1.38	469	2.16
Total observations	74 427	100.00	21 702	100.00

Note: Due to missing values on firm's sector the total amount of observations (once controlled for sectors) in the baseline sample is 74,427 observations.
Source: UK Innovation Survey, 1994–2016; Business Structure Database, 1997–2017.

APPENDIX 3A.2

Table 3A.2 *Three samples regional distribution (by ten UK regions, Scotland and Northern Ireland) and distribution over survey waves*

Regions	Baseline	%	Innovative	%
North East	4 731	5.28	1 171	5.40
North West	8 506	9.50	1 997	9.20
Yorkshire and Humber	7 142	7.98	1 758	8.10
East Midlands	6 708	7.49	1 749	8.06
West Midlands	7 562	8.45	1 890	8.71
Eastern England	7 776	8.69	1 946	8.97
London	11 369	12.70	2 064	9.51
South East	10 353	11.57	2 367	10.91
South West	7 229	8.08	1 813	8.35
Wales	5 203	5.81	1 432	6.60
Scotland	7 487	8.36	1 700	7.83
Northern Ireland	5 452	6.09	1 815	8.36
Total	89 518	100.00	21 702	100.00
Years				
UKIS4 (2005)	16 445	18.37	12 557	57.86
UKIS5 (2007)	14 872	16.61	2 425	11.17
UKIS6 (2009)	14 281	15.95	1 454	6.70
UKIS7 (2011)	14 342	16.02	2 773	12.78
UKIS8 (2013)	14 487	16.18	1 174	5.41
UKIS9 (2015)	15 091	16.86	1 319	6.08
Total observations	89 518	100.00	21 702	100.00

Source: UK Innovation Survey, 1994–2016; Business Structure Database, 1997–2017.

PART II

INDIVIDUALS

4. Principal investigators and knowledge management: a micro-foundational conceptual framework

James A. Cunningham, Manlio Del Giudice, Melita Nicotra, Conor O'Kane and Marco Romano

4.1 INTRODUCTION

Science-based knowledge discovery and exploitation is a fundamental aspect of societal and industrial progress. Over the last number of decades, governments have increasingly tried to ensure they are securing multiple impacts on their investment in public science and research and development (R&D). In the last decade in particular, there has been a shift towards mission-orientated research where policy makers direct public R&D investment at major societal issues such as climate change (Mazzucato, 2018; Sampat, 2012). Policy makers are also involved in designing public sector entrepreneurship programmes designed to bring together diverse groups of scientific and industrial actors to create knowledge and to support them with respect to technology transfer and knowledge exploitation that has economic, technological and societal impacts within the innovation and entrepreneurship ecosystem (Audretsch et al., 2019). One of the beneficial aspects of public sector entrepreneurship programmes is that they can mobilize different actors and their knowledge and know-how to further advance science-based knowledge discovery and exploitation. Increasingly, public funded research requires collaboration between university or public research organization researchers and industry partners (Dooley and Kirk, 2007; Perkmann and Walsh, 2009; Welsh et al., 2008). Furthermore, companies are increasingly showing interest in research & development (R&D) outsourcing and external cooperation. Therefore, external knowledge exploitation, as well as knowledge transfer, have become some of the most critical knowledge management processes inside the contemporary worldwide business environment (Bollinger and Smith, 2001; Campanella et al., 2020; Krylova et al., 2016; Zhang et al., 2020).

Consequently, public R&D programmes vary in size and scale of partnerships and at the individual project level, this adds a lot of project complexity for the principal investigator (PI). Governments and policy makers continue to develop public R&D programmes that are designed to develop and enhance the human capital that is necessary for science-based knowledge discovery. In the European Union, the European Research Council and Marie Sklodowska-Curie Actions programmes are examples of such public investment in human capital. Public R&D programmes can contribute as potential drivers for change in different ways: the creation of new knowledge, its transmission through education, its dissemination through information and communication technologies, and its adoption in new industrial processes or services (Carayannis et al., 2017; Etzkowitz and Leydesdorff, 2000; Orlando et al., 2020). For this reason, several authors have analysed how scientific and industrial actors leverage knowledge management (Natalicchio et al., 2019).

Knowledge management is a multidimensional construct. Several researchers have attempted to define its dimensions in many ways through an integrative perspective of organizational boundaries (Ardito et al., 2019; Chesbrough, 2006; West and Bogers, 2017). They considered exploration, retention and exploitation of knowledge inside and outside the organization's boundaries.

Exploitation and exploration are two fundamentally different learning activities between which firms divide their attention and resources (Carayannis et al., 2017; March, 1991) While exploration involves "experimentation with new alternatives" with returns that are "uncertain, and distant", exploitation is the "refinement and extension of existing competencies, technologies and paradigms" with returns that are "proximate and predictable" (March, 1991, p. 85). The general agreement in this literature is that the capability of both exploiting existing competencies as well as exploring new opportunities with equal dexterity – *ambidexterity* – is an antecedent of organizational performances (Campanella et al., 2020; Carayannis et al., 2017; Soto-Acosta et al., 2018; Zhang et al., 2020).

In this vein, to support knowledge exploitation, national governments use R&D tax credit systems to support firm-level exploitation of knowledge created in-house or in collaboration with other firms, higher educational institutions or public research organizations (see Becker, 2015; EU Commission, 2017; OECD, 2018).

Against this backdrop, HEIs have responded to these new challenges. This is reflected in the creation of new organizational structures, programmes and initiatives that are designed to support increasing collaborations with industry partners as well as ensuring that higher education institutions (HEIs) contribute effectively as a core institutional actor within entrepreneurial and innovation ecosystems (Kruss, 2008; Perkmann et al., 2013). The entrepreneurial university paradigm attempts to capture and reflect such institutional changes with HEIs (Clark, 1998; Cunningham et al., 2017; Guerrero et al., 2014; Kirby, 2006). In practical terms, from an organizational structural perspective this has meant the universities have created new organizational functions centred on technology and knowledge transfer, researcher development and support and entrepreneurship. Technology Transfer Offices (TTOs) that have emanated from the US HEIs as the result of the Bayh-Dole Act (Audretsch et al., 2014; Link and Siegel, 2005; Link and van Hasselt, 2019; Siegel et al., 2004) have become commonplace within HEIs across Europe and Asia (Cunningham et al., 2020a). The primary mission of TTOs is to protect and exploit intellectual property created by their university community, through a variety of technology transfer mechanisms such as spin-out firms, material transfer agreements, patents and licences. This means that HEIs need to have in place effective policies and procedures that capture and protect effectively knowledge that is created through science and research discovery. Depending on the HEI institutional context, the policy environment and the local, regional and national economic and societal contexts, this shapes the mission of TTOs and ultimately the services they provide to the university communities they serve and how they structure and sustain their relations with industry (see Calcagnini and Favaretto, 2016; Fitzgerald and Cunningham, 2016). It also shapes how HEIs exploit the knowledge that is created (Del Giudice et al., 2014). An ongoing challenge for TTOs is knowledge leakage and therefore the HEI is not able to protect and exploit such knowledge (Goel and Göktepe-Hultén, 2018).

Aside from TTOs, HEIs have initiated the creation of other organizational structures to support knowledge creation and exploitation. This has led to the creation of cooperative research centres, entrepreneurship research centres, dedicated entrepreneurship centres (see Boardman and Gray, 2010; Dolan et al., 2019; Maas and Jones, 2017). The mission of these has been either focused on the exploitation of knowledge through entrepreneurship or supporting collaborations with industry partners focused on knowledge creation and exploitation such as cooperative research centres. Moreover, some HEIs have gone further. They have created dedicated incubation programmes, incubation spaces and are co-founders or owners of science parks (Albahari et al., 2019; Calza et al., 2014; Lecluyse et al., 2019; Romano et al., 2014; Schillaci et al., 2011). Other initiatives that can be found in HEIs to support knowledge creation and exploitation are visiting positions, entrepreneurship in residence programmes, venture funds (Croce et al., 2014; George et al., 2010). Within HEIs there has been a growth of innovation and entrepreneurship degree programmes and modules across faculties with a centred focus of activity within business schools. Some of these offerings are core, and some are extracurricular such as business plan competitions (Watson and McGowan, 2017). Concurrently, this has seen an increase in focus with respect to entrepreneurial education and a variety of pedagogical approaches designed to equip student cohorts with entrepreneurial skills and know-how to further enhance their own personal development and growth (Hasan et al., 2017; Turner and Gianiodis, 2018).

These changes have happened in tandem with HEIs expanding their human capital through faculty appointments and doctoral programmes. This has led to HEIs setting up dedicated centralized research career centres and formal researcher development programmes targeted at the doctoral and early research careers (see Jacob et al., 2015). Such researcher development programmes are designed to complement traditional doctoral training and to support individual researchers in developing their own skills and competencies beyond scientific excellence. The demand for these additional institutional-based supports is a result of HEIs increasing their research funding base which brings an expansion of researchers.

Against these macro and meso developments, at the micro level the scientist is at the core of leading and driving knowledge creation and exploitation. In order to undertake these core activities, scientists apply for funding through publicly funded programmes and when successful they take on the role of being a PI and the associated responsibilities. However, scientists do not receive any formal training for this strategic role and in essence they learn on the job (Cunningham et al., 2014; Modic and Yoshioka-Kobayashi, 2020). On paper, the role can appear to be an administrative one but, in practice, the role is demanding and challenging due to project complexities derived from factors such as number of project partners – academic, industry (multinational enterprises, small and medium-sized enterprises) – disciplines, cultures and so on. The PI role requires and demands research leadership and management. The various role responsibilities mean that PIs have to have a threshold knowledge management and know-how capacity that contributes to role effectiveness and successful project completion as well as realizing multiple impacts. The four roles of PIs identified by O'Kane et al. (2020) – research manager, contractor, networker and entrepreneur – means the PI not only initiates projects and collaborations but generates new knowledge, manages research teams and transfers the knowledge.

Due to this level of complexity, the effectiveness of the interaction process between scientific and industrial actors depends on their "skills of knowledge" as the abilities to manage different knowledge processes. As a matter of fact, to capture internal and external knowledge exploration, retention and exploitation, several knowledge capacities are needed. Inventive capacity allows for internal exploring of new knowledge (Argote et al., 2003; Dell'Anno and Del Giudice, 2015; Lane et al., 2006). Absorptive capacity enables people and organizations to explore external knowledge. On the other hand, connective capacity comprises the process stages of maintaining knowledge in interorganizational relationships and subsequently reactivating this knowledge (Xia and Roper, 2008). However, to maintain internally developed technology over time, transformative capacity is essential (Garud and Nayyar, 1994).

With our focus on the micro level, our next section reviews some growing literature on PIs and examines the PI and knowledge brokering. We follow this by outlining our micro-foundational conceptual model of PI knowledge management. We conclude the chapter with some future avenues for research.

4.2 PRINCIPAL INVESTIGATORS: STRATEGIC ACTORS IN KNOWLEDGE CREATION

Within the research communities across all disciplines, there is a common understanding of the PI role. For the individual scientists, to become a PI can add to their scientific reputation (Cunningham et al., 2014; Romano et al., 2017). According to a large-scale study of Slovenian PIs, being a PI results in a "longer and more fruitful career" (see Kastrin et al., 2018). More importantly, it provides them with resources that they need to pursue and advance a line of research based on originality and novelty. In order to secure the role through a competitive peer review selection process, the scientist has to demonstrate scientific excellence and articulation of a research programme that addresses funders' criteria with respect to impact – scientific, technological, human, economic, societal and so on (Cunningham and O'Reilly,

2019). Carl (2020) argues that PIs within entrepreneurial ecosystems can shape innovation processes from technological breakthroughs to social process and thereby the role is one of a transformative actor (Carl, 2020). The scientist has to illicit the support of their own institution and engage with other academics within and outside their institutional environment and also with other stakeholders, the predominant one being industry. The expression of an original idea and the groupage of a variety of partners is through a project plan and a research programme that details the resources required – financial, infrastructure and human – and control measures with respect to finance, intellectual property, data management and so on. In reaching this career point, the career path requires experiences such as being an independent researcher and being a co-PI on another project with some scientists having some industry experience (see Cunningham et al., 2020a; Niebur, 2009).

In defining the PI role, Cunningham et al. (2016c) state: "A PI is the person charged with the direct responsibility for completion of a funded project, directing the research and reporting directly to the funding agency" and involves a variety of role activities include knowledge brokerage. Del Guidice et al. (2017, p. 321) succinctly define PIs: "as the individual responsible for the scientific or technical direction of research projects". Studies on PIs have found that they have a clear long-term vision as to the scientific contribution they wish to make and in doing so they take a proactive strategic approach. PIs are open to all forms of collaboration to achieve their vision, and some are highly selective in relation to pursuing research funding (O'Kane et al., 2020a). In addition, PIs are motivated by new knowledge creation, consistently strategizing and accumulating lots of practices in the role (Casati and Genet, 2014; Cunningham et al., 2016c; Menter, 2016; O'Kane et al., 2015). Scientists in the PI role need to be adept at opportunity recognition to support their knowledge creation and exploitation (see Cunningham, 2019) and need to adopt entrepreneurial behaviours to deal with the challenges of the role (O'Kane et al., 2020a). In the role, PIs experience barriers at the institutional and project levels particularly with respect to technology and knowledge transfer (see Cunningham et al., 2014; O'Kane, 2018; O'Kane et al., 2017; O'Reilly and Cunningham, 2017). Furthermore, the country culture where the PI resides exerts some influence on their "academic and business orientation" (Del Guidice et al., 2017, p. 334). PIs also allocate their time on different activities within funded projects and rank the importance to technology and market impact more when they spend more time on research (see Cunningham et al., 2016b). In managing publicly funded projects, PIs experience managerial challenges in the form of project management and project adaptability (Cunningham et al., 2015).

In responding to public funding calls, PIs need to adapt by assembling the most effective research consortium in order to maximize their chances of securing funding. They have to be adept at identifying research and market opportunities. This means that PIs have to be able to demonstrate and then create value for all stakeholders – other academics, industry partners and so on. In essence, they have to identify the value motives and value orientation of each stakeholder (Cunningham et al., 2019). PIs use two value capture mechanisms in the role of boundary spanning and brokerage (O'Kane et al., 2020). In becoming a PI, a scientist experiences a change in role identity and takes on new role identities. These include being a research contractor, project manager, science networker and an entrepreneur (O'Kane et al., 2020b). Cunningham and O'Reilly (2019) in a contingency model of PIs, outline threshold responsibilities of a PI as research leader, resource allocator, innovation enabler, project coordinator and management, and a boundary spanner.

Some gender differences are evident in the small number of studies of PIs that focus on this issue (see Cunningham et al., 2020c; McCarren and Goldman, 2012). For example, gender differences with respect to securing grants based on a study of the US National Institutes of Health programme in the US (see Oliveira et al., 2019). In the fields of ecology and zoology, Salerno et al. (2019) found significant gender disparities in relation to co-authorship and authorship. Feeney and Welch's (2014) study of PIs and co-PIs who received funding from the National Science Foundation found gender and productivity differences and concluded that women PIs: "are able to convert the resources into productive outputs at even higher rates then men" (p. 131). Funding, commercial experience and numbers of projects were some of the gender differences found in an Irish study of PIs (Cunningham et al., 2016c).

4.3 PI AND KNOWLEDGE BROKERING

In the growing studies of scientists in the PI role, it is evident that they are involved in knowledge brokerage and the associated governance. Some of the definitions of PIs such as O'Kane et al. (2020b, p. 403) – "PI are distinct members of the research community who excel in knowledge generation and are well positioned for knowledge sharing. Such characteristics enable PIs to pursue both exploration and exploitation" – highlight their knowledge generation, management and brokering role. In one of the earliest studies that was based on a small sample of PIs, Kidwell (2013, p. 218) identified their knowledge brokering role and describes this as:

> These PI are not only investigators and inventors, but they are innovators. They are creating value by bridging structural holes and building trust through specific brokering activities. They emerged as knowledge brokers identifying structural holes and creating trust between the lab and industry through activities we categorized as extrapolation, seeking, aligning and anticipating.

A foundation stone of knowledge brokering is effective knowledge management.

In this context, the ability to exploit technological opportunities emerging from the environment depends on two main factors: the firm's knowledge base and its learning process (Carayannis et al., 2017; Scuotto et al., 2017a; Scuotto et al., 2017b). Knowledge sharing occurs as the "activity of transferring or disseminating knowledge from one person, group or organization to another" (Lin and Lee, 2004, p. 324). When designing the process of knowledge sharing, a firm ought to consider those constraints emerging from the characteristics of the knowledge itself. In fact, knowledge is deemed to occur in two different forms, tacit or explicit, distinguished by whether it is possible or not to code the knowledge and to what extent. The tacit dimension of knowledge refers to the interlace between cognitive and technical elements such as experience, opinions and know-how, which are hard to isolate and capture in a monistic and codified representation. By contrast, the explicit dimension of knowledge is codifiable and suitable to be articulated and communicated in a symbolic form and/or a natural language (Della Peruta et al., 2014; Nonaka and Takeuchi, 1995). An effective knowledge-sharing strategy might positively impact innovative performance (Della Peruta et al., 2014; Soliman et al., 2016).

For example, in a lab context PIs need to instil a lab culture and management practices to ensure research rigour and compliance associated with the norms of their discipline (see Antes et al., 2019). In pursuing knowledge brokering, prior commercial experience can influence PIs in how they transfer technology and commercialize research (see Cunningham et al., 2020b). As a matter of fact, accumulated prior knowledge could increase the ability to accrue more knowledge (Bhatt, 2001).

Set in a Spanish university context, Foncubierta-Rodríguez et al. (2020) identified research knowledge, innovation skills, critical skills, research performance ability and open-minded research ability as the key human factor skills for PIs. To support scientific knowledge commercialization, PIs are better placed to act as boundary spanners than TTOs (see Boehm and Hogan, 2014). PIs, given their ability to create knowledge, have the potential ability to influence and shape the markets and science through their boundary-spanning activities (Managematin et al., 2014). Casati and Genet (2014), in their study of PIs, emphasize how PIs not only shape science with knowledge generation but also by reshaping knowledge, market and individual organizations. Given the array of project partners for an entrepreneurial ecosystem that PIs typically are drawn together in responding to public funding calls, this requires excellent networks and boundary-spanning capabilities along with an ability to create and manage complex governance arrangements (see Cunningham et al., 2019).

The effectiveness of this approach depends on the willingness of actors to collaborate. There is a general agreement that collaboration in entrepreneurial ecosystems could foster new technologies exploration (Bresciani et al., 2018). Coherently, Gkypali et al. (2018) highlighted that absorptive capacity could improve R&D collaborations, innovation and performance. However, the role of individuals and how their actions can impact the effectiveness of collaborative efforts remained underexposed (Scuotto et al., 2020, p. 8).

4.4 A MICRO-FOUNDATIONAL CONCEPTUAL MODEL OF PI KNOWLEDGE MANAGEMENT

The study of micro-foundations of knowledge management might help to clearly identify organizational processes and competencies which support the development of innovation capabilities. However, studies examining the exploration and exploitation activity at a micro level are exiguous (Scuotto et al., 2020, p. 2).

The growing body of empirical studies of PIs has unearthed many different aspects of the PI role and several of these acknowledge how knowledge creation and exploitation is a core part of the PI role. Moreover, there have been calls in the literature for the need to take a more micro level approach in studies (see Cunningham and O'Reilly, 2018; Cunningham and Menter, 2020). Drawing on the micro-foundational research within the strategy field (see Barney and Felin, 2013; Foss and Lindenberg, 2013), the conceptual model of PI knowledge management seeks to configure core aspects of knowledge management that are essential for this core role within entrepreneurial and innovation ecosystems. We use March's (1991) categorization of knowledge – *exploration and exploitation* (Figure 4.1).

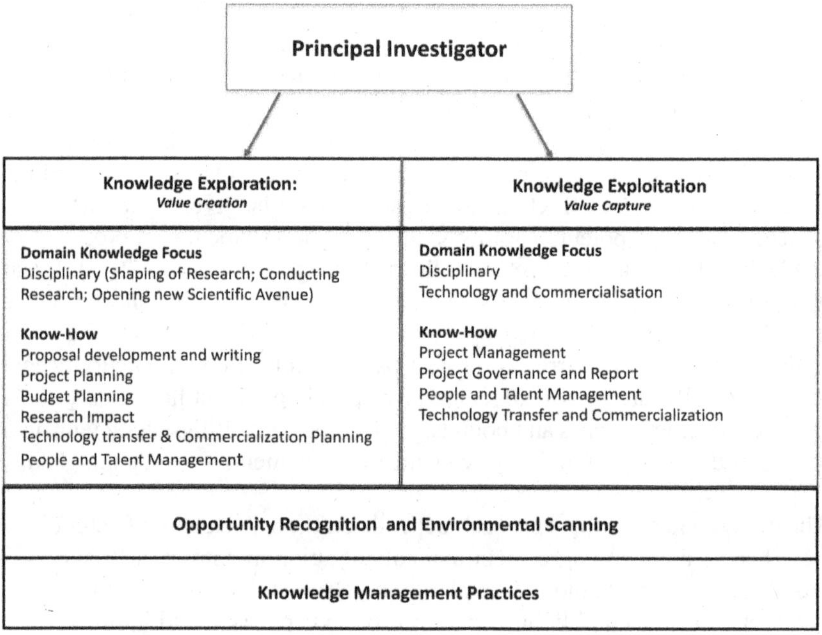

Figure 4.1 *A micro-foundational conceptual framework of principal investigator and knowledge management*

4.4.1 Exploration – Knowledge Domain Focus and Know-how

When PIs are successfully funded, value is captured through the exchange of resources – expertise for funding. The value captured through this exchange can be conceptualized as knowledge exploration (O'Kane et al., 2020b), of which the essence is "experimentation with new alternatives" (March, 1991, p. 85). In comparison to privately funded research which often results in less novelty, shorter time frames and more incremental innovation, publicly funded research typically mobilizes science teams around longer-term activities with high outcome uncertainty. With the expectation from public funders that new knowledge will be generated, original knowledge trajectories and innovative ideas are proposed, explored and discovered through wide search, experimentation and risk taking (Levinthal and March, 1993; March, 1991).

The exploration process for PIs can often begin long before any formal public funding call announcement where their focus is on building their knowledge creation and generation through focusing on pursuing science knowledge discovery within the domain/discipline experience. The knowledge domain focus centres on their discipline. The proprietary knowledge that they create is codified through peer review and published in the form of scholarly outputs such as journal articles, monographs, book chapters and so on, normally within their domain discipline. Part of this knowledge creation and generation process can involve presenting some of this knowledge at their disciplinary academic conferences to ellicit feedback. Also, they use these events to develop and build academic networks within their own or complementary domains. In this vein, several authors have analyzed the role of psychological precursors of knowledge exploration (Sarala et al., 2016; Scuotto et al., 2020).

In responding to a funding call, the scientist begins the process of becoming a PI by utilizing their discipline-based knowledge domain expertise, drawing on their codified knowledge and that of others to begin the process of creating a project proposal that is original and novel in nature so as to secure funding through a competitive peer review process. In order to enhance their chances and also to meet funder requirements they begin to select academic and industry partners and appropriate and assemble their domain expertise into a formal research programme and project plan. When a scientist is successful – they are awarded the public grant – they then become a PI. As they initiate the project plan and work with partners, the predominant knowledge domain focus is disciplinary and also putting in place project structures that enable different disciplinary expertise to be brought together in order to collaborate effectively to realize the overall project objectives.

Utilizing their disciplinary expertise to undertake all the main activities requires a lot of know-how at the project planning and implementation stages. According to the micro-foundational perspective, the ability to rapidly access new knowledge and integrate it into current processes assumes a strategic role (Buganza and Verganti, 2009).

For the exploration stage, PIs need know-how with respect to proposal development and writing, project planning, finance, impact and technology transfer and commercialization. Some of this know-how depends on how experienced scientists are – they can possess some or all of this know-how capacity. However, those scientists that are less experienced rely on colleagues and centralized institutional supports provided by research and technology transfer offices to support them in preparing a competitive proposal. When a scientist is awarded a competitive research grant, they need to draw on more know-how. This centres on know-how with respect to recruiting, developing, managing and leading a research team – people talent management. Project management know-how is needed to ensure the project deliverables are delivered against the funded research programme. A further focus of know-how is on university and funders' policies and procedures with respect to a range of issues such as financial, data management, research ethics, research integrity and so on. Depending on the project team structure and the institutional resources available to the PI, they rely on other colleagues and stakeholders to support them in mitigating against any of these know-how deficits.

4.4.2 Exploitation – Knowledge Domain Focus and Know-how

A decisive issue for knowledge transfer is the ability to guarantee the effectiveness and the best exploitation of market opportunities (Miller et al., 2018; Natalicchio et al., 2019). PIs utilize two values captured in use mechanisms – boundary spanning and brokering – to exploit and widen the impact and reach of knowledge exploration. Through brokering, PIs integrate and balance various combinations of ecosystem actors' interests identified among policy makers, funding bodies, industry, community end-users and universities. This brokering has some benefits that are less accessible through boundary-crossing. Specifically, although more complex, when done effectively brokering, avoids excessive stakeholder salience in terms of the value captured and simultaneously provides the opportunity to significantly widen the overall reach of the value in use captured from publicly funded research. In essence, wide search and effective brokering presents PIs with the chance to identify a greater number of "interest" recombination opportunities across various ecosystem actors for exploitation through their research (O'Kane et al., 2020b).

For knowledge exploitation for the PI, there is a continued knowledge domain focus on their own disciplinary knowledge domain. The additional knowledge domain of consortia partners – academic and industry – has been integrated effectively to attain the overall project mission. At this stage, the knowledge domain focus also turns to effective knowledge exploitation and this is where partners within this knowledge domain and expertise come to the fore. PIs with previous commercial experience may have somewhat of a knowledge advantage over PIs that don't have such commercial experience. For exploitation to work effectively in the exploration stage, PIs before project initiation and during the project proposal development phase would have created an overarching knowledge exploitation plan that all partners had agreed to. The integration of knowledge transfer can be accrued via knowledge sharing between parties (Lin and Lee, 2004, p. XXX). Another strategic factor is the role of people in the process of knowledge creation (Chatenier et al., 2010). In fact, as knowledge creation is unpredictable, some of these knowledge exploitation overarching plans may change due to changes in knowledge creation and associated progress. Also, changes in relation to the external market opportunity, a change in the business needs and priorities of industry partners can change the overall project exploitation plans.

Commercialization and technology transfer can increasingly come to the fore with respect to know-how as PIs seek to exploit their knowledge creation, thereby meeting or exceeding the funders' expectations. If PIs have previous experience in this domain and have developed some know-how capacity they can be independent whereas other PIs will be reliant on their institutional TTO office and their partners to support them with respect to commercialization and knowledge transfer. Given the ongoing project demand, continued know-how is required in relation to project management, people and talent management, project governance and reporting. In addition, PIs may have to draw on a combination of project partner and their own institutional know-how with respect to soft knowledge transfer and dissemination through events, project website, targeted stakeholder workshops, social media campaigns and so on.

4.4.3 Opportunity and Environmental Scanning

In order to be effective in the role, from knowledge exploration through knowledge exploitation, PIs need to be constantly identifying opportunities and conducting environmental scanning in a formal, systematic manner or through informal means. This requires PIs to have multiple knowledge sources. Such trusted knowledge sources can come from codified knowledge scientific stakeholders, while industry needs can come from industry networks or end-users. From these multiple knowledge sources, PIs can recognize opportunities and also begin to anticipate the industry and end-user needs. It is through their boundary-spanning activities with ecosystem stakeholders that PIs and their research teams can undertake such environmental scanning. Some PIs can create structured mechanisms within their research teams where they can share and disseminate knowledge that they have gathered from environmental scanning through attending conferences and engaging with different academic, industry, policy networks. Other PIs can take a more intuitive and informal approach to environmental scanning and opportunity recognition and this in turn influences their cognitive approach to knowledge exploitation and exploration.

4.4.4 Knowledge Management Practices

The norms of research and scientific practices provide the foundational elements of knowledge management practices among PIs. These practices pervade data collection, research integrity, management and sharing. Depending on the disciplinary focus of PIs, then further scientific norms apply such as in labs that are a fundamental element of effective knowledge management practices. PIs can adopt other knowledge management practices that they have adopted from other environments or that have been demanded by funders and other partners.

4.5 CONCLUDING REMARKS AND FUTURE AVENUES FOR RESEARCH

PIs are invisible but strategic actors in knowledge creation and exploitation within innovation and entrepreneurial ecosystems. There is a need from empirical, policy and practice perspectives to gain real insights into how PIs across different disciplines, public funding instruments and institutional and context settings – HEIs, public research organizations – manage knowledge and how they create value from the knowledge they create. The unique and proprietary knowledge they create is a unique and valuable resource that has the potential to achieve wider benefits for economies and societies.

To this end, future research on PIs and knowledge management should examine the approaches that PIs take in exploring and exploiting knowledge in the role. For example, how does know-how influence their choices with respect to knowledge transfer and commercialization? What approaches do PIs use to create effective knowledge management practices within their research teams and among their project partners? How do PIs conduct opportunity recognition activities and how does environmental scanning influence knowledge exploration and exploitation? Are there gender and diversity differences evident among PIs with respect to knowledge management? What managerial approaches do PIs use in synthesizing and integrating knowledge domain expertise from project partners to focus on project objectives? What strategies do PIs adopt to prevent knowledge leakages? These are some of the many research questions that will advance our understanding of PIs and knowledge management. Our micro-foundational conceptual model of PI knowledge management provides an initial frame and point of reference for future fruitful research that has the potential to advance knowledge and also to inform the contextual practices of scientists in the PI role.

ACKNOWLEDGEMENTS

James A. Cunningham wishes to acknowledge the funding support of Science Foundation Ireland (SFI) and co-funding under the European Regional Development Fund under Grant Number 13/RC/2073.

REFERENCES

Albahari, A., Klofsten, M., and Rubio-Romero, J.C. (2019). Science and technology parks: A study of value creation for park tenants. *The Journal of Technology Transfer*, 44(4), 1256–72.

Antes, A.L., Kuykendall, A., and DuBois, J.M. (2019). The lab management practices of "Research Exemplars" that foster research rigor and regulatory compliance: A qualitative study of successful principal investigators. PLoS One, 14(4), e0214595.

Ardito, L., Ferraris, A., Messeni Petruzzelli, A., Bresciani, S., and Del Giudice, M. (2019). The role of universities in the knowledge management of smart city projects. *Technological Forecasting and Social Change*, 142, 312–21.

Argote, L., McEvily, B., and Reagans, R. (2003). Managing knowledge in organizations: An integrative framework and review of emerging themes. *Management Science*, 49(4), 571–82

Audretsch, D.B., Lehmann, E.E., and Wright, M. (2014). Technology transfer in a global economy. *The Journal of Technology Transfer*, 39(3), 301–12.

Audretsch, D.B., Cunningham, J.A., Kuratko, D.F., Lehmann, E E., and Menter, M. (2019). Entrepreneurial ecosystems: Economic, technological, and societal impacts. *The Journal of Technology Transfer*, 44(2), 313–25.

Barney, J.B., and Felin, T. (2013). What are microfoundations? *Academy of Management Perspectives*, 27(2), 138–55.

Becker, B. (2015). Public R&D policies and private R&D investment: A survey of the empirical evidence. *Journal of Economic Surveys*, 29(5), 917–42.

Bhatt, G.D. (2001). Knowledge management in organizations: Examining the interaction between technologies, techniques, and people. *Journal of Knowledge Management*, 5(1), 68–75.

Boardman, C., and Gray, D. (2010). The new science and engineering management: Cooperative research centers as government policies, industry strategies, and organizations. *The Journal of Technology Transfer*, 35(5), 445–59.

Boehm, D.N., and Hogan, T. (2014). "A jack of all trades": The role of PIs in the establishment and management of collaborative networks in scientific knowledge commercialisation. *The Journal of Technology Transfer*, 39(1), 134–49.

Bollinger, A.S., and Smith, R.D. (2001) Managing organizational knowledge as a strategic asset. *Journal of Knowledge Management*, 5(1). 8–18.

Bresciani, S., Ferraris, A., and Del Giudice, M. (2018). The management of organizational ambidexterity through alliances in a new context of analysis: Internet of Things (IoT) smart city projects. *Technological Forecasting and Social Change*, 136, 331–8.

Buganza, T., and Verganti, R. (2009). Open innovation process to inbound knowledge: Collaboration with universities in four leading firms. *European Journal of Innovation Management*, 12(3), 306–25.

Calcagnini, G., and Favaretto, I. (2016). Models of university technology transfer: Analyses and policies. *The Journal of Technology Transfer*, 41(4), 655–60.

Calza, F., Dezi, L., Schiavone, F., and Simoni, M. (2014). The intellectual capital of business incubators. *Journal of Intellectual Capital*, 15(4), 597–610.

Campanella, F., Del Giudice, M., Thrassou, A., and Vrontis, D. (2020). Ambidextrous organizations in the banking sector: An empirical verification of banks' performance and conceptual development. *International Journal of Human Resource Management*, 31(2), 272–302.

Carayannis, E.G., Grigoroudis, E., Del Giudice, M., Della Peruta, M.R., and Sindakis, S. (2017). An exploration of contemporary organizational artifacts and routines in a sustainable excellence context. *Journal of Knowledge Management*, 21(1), 35–56.

Carl, J. (2020). From technological to social innovation – the changing role of principal investigators within entrepreneurial ecosystems. *Journal of Management Development*, 39(5), 739–52.

Casati, A., and Genet, C. (2014). Principal investigators as scientific entrepreneurs. T*he Journal of Technology Transfer*, 39(1), 11–32.

Chatenier, E.D., Verstegen, J.A., Biemans, H.J., Mulder, M., and Omta, O. S. F. (2010). Identification of competencies for professionals in open innovation teams. *R&D Management*, *40*(3), 271–80.

Chesbrough, H. (2006). *Open Business Models: How to Thrive in the New Innovation Landscape*. Cambridge, MA: Harvard Business Press.

Clark, B.R. (1998). *Creating Entrepreneurial Universities: Organizational Pathways of Transformation*. Issues in Higher Education. Bingley, UK: Elsevier Science.

Croce, A., Grilli, L., and Murtinu, S. (2014). Venture capital enters academia: An analysis of university-managed funds. *The Journal of Technology Transfer*, 39(5), 688–715.

Cunningham, J.A. (2019). Principal investigators and boundary spanning entrepreneurial opportunity recognition: A conceptual framework. In Audretsch, D.B., Lehmann, E.E., and Link, A.N. (eds), *A Research Agenda for Entrepreneurship and Innovation* (pp. 55–73). Cheltenham, UK and Northampton, MA, USA: Edward Elgar Publishing.

Cunningham, J.A., and O'Reilly, P. (2018). Macro, meso and micro perspectives of technology transfer. *The Journal of Technology Transfer*, 43(3), 545–57.

Cunningham, J.A., and O'Reilly, P. (2019). Roles and Responsibilities of Project Coordinators: A Contingency Model for Project Coordinator Effectiveness (No. JRC117576). Joint Research Centre.

Cunningham, J.A., O'Reilly, P., O'Kane, C., and Mangematin, V. (2014). The inhibiting factors that principal investigators experience in leading publicly funded research. *The Journal of Technology Transfer*, 39(1), 93–110.

Cunningham, J.A., O'Reilly, P., O'Kane, C., and Mangematin, V. (2015). Managerial challenges of publicly funded principal investigators. *International Journal of Technology Management*, 68(3–4), 176–202.

Cunningham, J.A., Mangematin, V., O'Kane, C., and O'Reilly, P. (2016a). At the frontiers of scientific advancement: The factors that influence scientists to become or choose to become publicly funded principal investigators. *The Journal of Technology Transfer*, 41(4), 778–97.

Cunningham, J.A., O'Reilly, P., Dolan, B., O'Kane, C., and Mangematin, V. (2016b). Publicly funded principal investigators' allocation of time for public sector entrepreneurship activities. *Economia e Politica Industriale*, 43(4), 383–408.

Cunningham, J.A., O'Reilly, P., O'Kane, C., and Mangematin, V. (2016c). Publicly funded principal investigators as transformative agents of public sector entrepreneurship. In Audretsch, D.B., and Link, A.N. (eds), *Essays in Public Sector Entrepreneurship* (pp. 67–94). Cham: Springer.

Cunningham, J. A., Guerrero, M., and Urbano, D. (2017). Entrepreneurial universities – overview, reflections, and future research agendas. In *The World Scientific Reference on Entrepreneurship: Volume 1: Entrepreneurial Universities Technology and Knowledge Transfer* (pp. 3–19). Singapore: World Scientific Publishing.

Cunningham, J.A., Menter, M., and O'Kane, C. (2018). Value creation in the quadruple helix: A micro level conceptual model of principal investigators as value creators. *R&D Management*, 48(1), 136–47.

Cunningham, J.A., Menter, M., and Wirsching, K. (2019). Entrepreneurial ecosystem governance: A principal investigator-centered governance framework. *Small Business Economics*, 52(2), 545–62.

Cunningham, J.A., Harney, B., and Fitzgerald, C. (2020a). *Effective Technology Transfer Offices: A Business Model Framework*. Springer.

Cunningham, J.A., Dolan, B., Menter, M., O'Kane, C., and O'Reilly, P. (2020b). How Principal Investigators' commercial experience influences technology transfer and market impacts. *Research-Technology Management*, 63(5), 49–58.

Cunningham, J.A., O'Reilly, P., Hooper, D., Nepelski, D., and Van Roy, V. (2020c). The Role of Project Coordinators in European Commission Framework Projects. Results of the Innovation Radar PC Survey in FP R&I Projects, EU 30131. EN Publications Office of the European Union, Luxembourg, ISBN 978–92–76–17304–5.

Del Giudice, M., Maggioni, V., Romano, M., and Nicotra, M. (2014). Knowledge creation and exploitation in Italian universities: The role of internal policies for patent activity. *Journal of Knowledge Management*, 18(5), 952–70.

Del Giudice, M., Nicotra, M., Romano, M., and Schillaci, C.E. (2017). Entrepreneurial performance of principal investigators and country culture: Relations and influences. *The Journal of Technology Transfer*, 42(2), 320–37.

Dell'Anno, D., and Del Giudice, M. (2015). Absorptive and desorptive capacity of actors within university-industry relations: Does technology transfer matter? *Journal of Innovation and Entrepreneurship*, 4(1), 13.

Della Peruta, M.R., Campanella, F., and Del Giudice, M (2014). Knowledge sharing and exchange of information within bank and firm networks: The role of the intangibles on the access to credit. *Journal of Knowledge Management*, 18(5), 1036–105.

Dolan, B., Cunningham, J.A., Menter, M., and McGregor, C. (2019). The role and function of cooperative research centers in entrepreneurial universities. *Management Decision*, 57(12), 3406–25.

Dooley, L., and Kirk, D. (2007). University-industry collaboration: Grafting the entrepreneurial paradigm onto academic structures. *European Journal of Innovation Management*, 10(3), 316–32.

Etzkowitz, H., and Leydesdorff, L. (2000). The dynamics of innovation: From national systems and "Mode 2" to a triple helix of university-industry-government relations. *Research Policy*, 29(2), 109–23.

EU Commission. (2017). R&D tax incentives how to make them most effective? Publication Office of the European Union, Luxembourg. doi: 10.2777/976217.

Feeney, M.K., and Welch, E.W. (2014). Academic outcomes among principal investigators, co-principal investigators, and non-PI researchers. *The Journal of Technology Transfer*, 39(1), 111–33.

Fitzgerald, C., and Cunningham, J.A. (2016). Inside the university technology transfer office: Mission statement analysis. *The Journal of Technology Transfer*, 41(5), 1235–46.

Foncubierta-Rodríguez, M.J., Martín-Alcázar, F., and Perea-Vicente, J.L. (2020). Measuring the human capital of scientists in the principal investigator role. *Journal of Management Development*, 39(5), 777–90.

Foss, N.J., and Lindenberg, S. (2013). Microfoundations for strategy: A goal-framing perspective on the drivers of value creation. *Academy of Management Perspectives*, 27(2), 85–102.

Garud, R., and Nayyar, P.R. (1994). Transformative capacity: Continual structuring by intertemporal technology transfer. *Strategic Management Journal*, 15(5), 365–85.

George, M., Gordon, I., and Hamilton, E. (2010). What is (the point of) an entrepreneur in residence? The Lancaster University experience, with some worldwide comparisons. *Industry and Higher Education*, 24(6), 495–503.

Gkypali, A., Arvanitis, S., and Tsekouras, K. (2018). Absorptive capacity, exporting activities, innovation openness and innovation performance: A SEM approach towards a unifying framework. *Technological Forecasting and Social Change*, 132(C), 143–55.

Goel, R.K., and Göktepe-Hultén, D. (2018). What drives academic patentees to bypass TTOs? Evidence from a large public research organisation. *The Journal of Technology Transfer*, 43(1), 240–58.

Guerrero, M., Urbano, D., Cunningham, J., and Organ, D. (2014). Entrepreneurial universities in two European regions: A case study comparison. *The Journal of Technology Transfer*, 39(3), 415–34.

Hasan, S.M., Khan, E.A., and Nabi, M.N.U. (2017). Entrepreneurial education at university level and entrepreneurship development. *Education + Training*, 59(7/9), 888–906.

Jacob, W.J., Xiong, W., and Ye, H. (2015). Professional development programmes at world-class universities. *Palgrave Communications*, 1(1), 1–27.

Kastrin, A., Klisara, J., Lužar, B., and Povh, J. (2018). Is science driven by principal investigators?. *Scientometrics*, 117(2), 1157–82.

Kidwell, D.K. (2013). Principal investigators as knowledge brokers: A multiple case study of the creative actions of PIs in entrepreneurial science. *Technological Forecasting and Social Change*, 80(2), 212–20.
Kirby, D.A. (2006). Creating entrepreneurial universities in the UK: Applying entrepreneurship theory to practice. *The Journal of Technology Transfer*, 31(5), 599–603.
Kruss, G. (2008). Balancing old and new organisational forms: Changing dynamics of government, industry and university interaction in South Africa. *Technology Analysis & Strategic Management*, 20(6), 667–82.
Krylova, K.O., Vera, D., and Crossan, M. (2016). Knowledge transfer in knowledge-intensive organizations: The crucial role of improvisation in transferring and protecting knowledge. *Journal of Knowledge Management*, 20(5), 1045–64.
Lane, P.J., Koka, B.R., and Pathak, S. (2006). The reification of absorptive capacity: A critical review and rejuvenation of the construct. *Academy of Management Review*, 31(4), 833–63.
Lecluyse, L., Knockaert, M., and Spithoven, A. (2019). The contribution of science parks: A literature review and future research agenda. *The Journal of Technology Transfer*, 44(2), 559–95.
Lee, J., and Win, H.N. (2004), Technology transfer between university research centers and industry in Singapore. *Technovation*, 24(5), 433–42.
Levinthal, D.A., and March, J.G. (1993). The myopia of learning. *Strategic Management Journal*, 14(S2), 95–112.
Lin, H.F., and Lee, G.G. (2004). Perception of senior managers toward knowledge-sharing behavior. *Management Decision*, 42, 108–25.
Link, A.N., and Siegel, D.S. (2005). University-based technology initiatives: Quantitative and qualitative evidence. *Research Policy*, 34(3), 253–7.
Link, A.N., and van Hasselt, M. (2019). On the transfer of technology from universities: The impact of the Bayh–Dole Act of 1980 on the institutionalization of university research. *European Economic Review*, 119, 472–81.
Maas, G., and Jones, P. (2017), *Entrepreneurship Centres: Global Perspectives on Their Contributions to Higher Education Institutions*, Heidelberg: Springer.
Mangematin, V., O'Reilly, P., and Cunningham, J. (2014). PIs as boundary spanners, science and market shapers. *The Journal of Technology Transfer*, 39(1), 1–10.
March, J.G. (1991). Exploration and exploitation in organizational learning. *Organization Science*, 2(1), 71–87.
Mazzucato, M. (2018). Mission-oriented innovation policies: Challenges and opportunities. *Industrial and Corporate Change*, 27(5), 803–15.
McCarren, M., and Goldman, S. (2012). Research leadership and investigators: Gender distribution in the federal government. *The American Journal of Medicine*, 125(8), 811–16.
Menter, M. (2016). Principal investigators and the commercialization of knowledge. In *University Evolution, Entrepreneurial Activity and Regional Competitiveness* (pp. 193–203). Cham: Springer.
Miller, K., McAdam, R., and McAdam, M. (2018). A systematic literature review of university technology transfer from a quadruple helix perspective: Toward a research agenda. *R&D Management*, 48(1), 7–24.
Modic, D., and Yoshioka-Kobayashi, T. (2020). Individual-level determinants of academic patent licensing to start-ups: Impacts of principal investigators' embeddedness in the industry. In Novotny, A., Rasmussen, E., Clausen, T.H., and Wiklund J. (eds), *Research Handbook on Start-up Incubation Ecosystems* (pp. 349–69). Cheltenham, UK and Northampton, MA, USA: Edward Elgar Publishing.
Natalicchio, A., Ardito, L., Messeni Petruzzelli, A., and Del Giudice, M. (2019). The origins of external knowledge inflows and the impact of university technologies. *R&D Management*, 49(4), 639–51.

Niebur, S.M. (2009). Principal Investigators and mission leadership. *Space Policy*, 25(3), 181–6.

Nonaka, I., and Takeuchi, H. (1995). *The Knowledge-creating Company: How Japanese Companies Create the Dynamics of Innovation*. Oxford: Oxford University Press.

O'Kane, C. (2018). Technology transfer executives' backwards integration: An examination of interactions between university technology transfer executives and principal investigators. *Technovation*, 76, 64–77.

O'Kane, C., Cunningham, J., Mangematin, V., and O'Reilly, P. (2015). Underpinning strategic behaviours and posture of principal investigators in transition/uncertain environments. *Long Range Planning*, 48(3), 200–14.

O'Kane, C., Zhang, J.A., Cunningham, J.A., and O'Reilly, P. (2017). What factors inhibit publicly funded principal investigators' commercialization activities? *Small Enterprise Research*, 24(3), 215–32.

O'Kane, C., Mangematin, V., Zhang, J.A., and Cunningham, J.A. (2020a). How university-based principal investigators shape a hybrid role identity. *Technological Forecasting and Social Change*, 159, 120179.

O'Kane, C., Zhang, J.A., Cunningham, J.A., and Dooley, L. (2020b). Value capture mechanisms in publicly funded research. *Industrial Marketing Management*, 90, 400–16.

O'Reilly, P., and Cunningham, J.A. (2017). Enablers and barriers to university technology transfer engagements with small- and medium-sized enterprises: Perspectives of principal investigators. *Small Enterprise Research*, 24(3), 274–89.

OECD. (2018). OECD Review of National R&D Tax Incentives and Estimates of R&D Tax Subsidy Rates, 2017. http://www.oecd.org/sti/rd-tax-stats-design-subsidy.pdf (accessed 16 October 2020).

Oliveira, D.F., Ma, Y., Woodruff, T.K., and Uzzi, B. (2019). Comparison of National Institutes of Health grant amounts to first-time male and female principal investigators. *JAMA*, 321(9), 898–900.

Orlando, B., Ballestra, L.V., Scuotto, V., Pironti, M., and Del Giudice, M. (2020). The impact of R&D investments on eco-innovation: A cross-cultural perspective of green technology management. *IEEE Transactions on Engineering Management*, doi: 10.1109/TEM.2020.3005525.

Perkmann, M., and Walsh, K. (2009). The two faces of collaboration: Impacts of university-industry relations on public research, 18(6), 1033–1065.

Perkmann, M., Tartari, V., McKelvey, M. et al. (2013). Academic engagement and commercialisation: A review of the literature on university-industry relations. *Research Policy*, 42(2), 423–42.

Romano, M., Catalfo, P., and Nicotra, M. (2014). Science parks and intellectual capital. *Journal of Intellectual Capital*, 15(4), 537–53.

Romano, M., Elita Schillaci, C., and Nicotra, M. (2017). Principal Investigators in entrepreneurial universities: A research framework. In Cunningham, J.A., Guerrero, M., and Urbano, D. (eds), *The World Scientific Reference on Entrepreneurship: Volume 1: Entrepreneurial Universities Technology and Knowledge Transfer* (pp. 165–84). Singapore: World Scientific Publishing.

Salerno, P.E., Páez-Vacas, M., Guayasamin, J.M., and Stynoski, J.L. (2019). Male principal investigators (almost) don't publish with women in ecology and zoology. *PLoS One*, 14(6), e0218598.

Sampat, B.N. (2012). Mission-oriented biomedical research at the NIH. *Research Policy*, 41(10), 1729–41.

Sarala, R.M., Junni, P., Cooper, Cary L.,, and Tarba, S. (2016). A sociocultural perspective on knowledge transfer in mergers and acquisitions. *Journal of Management*, 42(5), 1230–49.

Schillaci, C.E., Romano, M., and Longo, M.C. (2011). Academic entrepreneurship, university incubator and corporate governance. *Sinergie Italian Journal of Management*, 75, 89–107.

Scuotto, V., Del Giudice, M., Della Peruta, M.R., and Tarba, S. (2017a). The performance implications of leveraging internal innovation through social media networks: An empirical verification of the smart fashion industry. *Technological Forecasting and Social Change*, 120, 184–94.

Scuotto, V., Del Giudice, M., Bresciani, S., and Meissner, D. (2017b). Knowledge-driven preferences in informal inbound open innovation modes. An explorative view on small to medium enterprises. *Journal of Knowledge Management*, 21 (3), 640–55.

Scuotto, V., Orlando, B., Cillo, V., Nicotra, M., Di Gioia, L., and Farina Briamonte, M. (2020). Uncovering the micro-foundations of knowledge sharing in open innovation partnerships: An intention-based perspective of technology transfer. *Technological Forecasting and Social Change*, 152, 119906.

Siegel, D.S., Waldman, D.A., Atwater, L.E., and Link, A.N. (2004). Toward a model of the effective transfer of scientific knowledge from academicians to practitioners: Qualitative evidence from the commercialization of university technologies. *Journal of Engineering and Technology Management*, 21(1–2), 115–42.

Solima, L., Della Peruta, M.R., and Del Giudice, M. (2016). Object-generated content and knowledge sharing: The forthcoming impact of the internet of things. *Journal of Knowledge Economy*, 7(3), 738–52.

Soto-Acosta, P., Popa, S., and Martinez-Conesa, I., 2018. Information technology, knowledge management and environmental dynamism as drivers of innovation ambidexterity: A study in SMEs. *Journal of Knowledge Management*, 22(4), 931–48.

Turner, T., and Gianiodis, P. (2018). Entrepreneurship unleashed: Understanding entrepreneurial education outside of the business school. *Journal of Small Business Management*, 56(1), 131–49.

Watson, K., and McGowan, P. (2017). Technology nascent entrepreneur experiences of start-up competition participation. In *Technology-based Nascent Entrepreneurship* (pp. 279–308). New York: Palgrave Macmillan.

Welsh, R., Glenna, L., Lacy, W., and Biscotti, D. (2008). Close enough but not too far: Assessing the effects of university-industry research relationships and the rise of academic capitalism. *Research Policy*, 37(10), 1854–64.

West, J., and Bogers, M. (2017). Open innovation: Current status and research opportunities. *Innovation*, 19(1),43–50.

Xia, T., and Roper, S. (2008). From capability to connectivity – absorptive capacity and exploratory alliances in biopharmaceutical firms: A US–Europe comparison. *Technovation*, 28(11), 776–85.

Zhang, X., Liu, Y., Tarba, S.Y., and Del Giudice, M. (2020). The micro-foundations of strategic ambidexterity: Chinese cross-border M&As, mid-view thinking and integration management. *International Business Review*, 29(6), 101710.

5. Factors facilitating the inventing academics' transition from nascent entrepreneurs to business owners[1]

João Ricardo Faria, Rajeev K. Goel and Devrim Göktepe-Hultén

5.1 INTRODUCTION

Innovation system approaches have been providing insights into the economic performance and competitiveness in the Organisation for Economic Co-operation and Development (OECD) countries. They identified that innovation and technical progress are the result of a complex set of relationships among actors producing, distributing and applying various kinds of knowledge. These actors are primarily private enterprises, universities and public research institutes and the people within them (Freeman 1987; Lundvall 1992). The innovative performance of a country depends to a large extent on how these actors relate to each other as elements of a collective system of knowledge creation and use-related technologies. Innovation policy and the related analysis have traditionally focused on inputs (such as R&D investments and research personnel) and outputs (such as scientific publications, patents and high-tech start-ups). The interactions among these actors are as important as investments in research and development, and they are key to translating the inputs into useful outputs. The flow of knowledge between enterprises, universities and research institutions is fundamental for enhancing innovative performance. Sharing of both tacit and/or codified knowledge (in publications, patents and other sources) can take different forms such as joint research, personnel mobility, co-patenting, and other less formal collaborations (Goel and Grimpe, 2012). Firms are often at the center of the innovation system and the innovative firm is seen as operating within a complex network of cooperating and competing firms and other institutions, building on a range of joint ventures and close linkages with suppliers and customers (Lundvall 1992, 2010). In some cases, specific knowledge (e.g., patents or scientific papers) transferred is not so important, but rather the general quality of researchers' knowledge and skills to solve problems. As knowledge embodied in scientists as "human capital" is central to knowledge production, the movement of people and the knowledge they carry with them (often termed "tacit knowledge") is a key flow in national innovation systems. Public policymakers are increasingly urging scientists to collaborate with industry to increase the innovativeness and competitiveness of their economies.

This discourse, especially in Europe, has paved the way for the promotion of the "third task activities," that is, the inclusion of a commercial and economic development mandate for universities in addition to their traditional tasks of education and research (Etzkowitz, 1997). In addition to teaching and research tasks, universities are expected to actively participate in the diffusion and application of knowledge to other sectors of the society. The starting point is the assimilation of basic research into codified knowledge and thus into usable information (Laredo, 2007). In recent times, there has been a movement toward an interpretation of third task activities as commercialization of academic research (Jacob et al., 2003). The argument for this specification of "third task" is that the propensity to commercial activities at universities is lower in Europe compared to the high levels of scientific performance and investments. This perception is partly exacerbated by the impression that universities in the USA have higher performance in commercializing their research results due to the enactment of the Bayh-Dole Act of 1980. Since its enactment, the Bayh-Dole Act has resulted in a rapid increase in university patenting activities. Despite some disagreements and skepticism, commercialization of scientific research results, in particular via patents and licensing and academic start-ups, is seen largely as a potential source of revenue, and an avenue to enhance economic growth and development (Etzkowitz, 2001; Mowery et al., 2004). It is, thus, not surprising that several countries, both developing and developed, have taken steps to initiate policy tools to promote academic entrepreneurship and to facilitate relations between universities, firms and public agencies.

Commercialization of scientific research results can be achieved in many different ways, but much of the literature has focused on the role of patents and licensing (Bercovitz and Feldman, 2008; Siegel et al., 2004) and academic spinoffs have also received a substantial amount of attention from researchers (Di Gregorio and Shane, 2003). Studies on academic entrepreneurship have focused on a few themes like the characteristics of the scientists (human capital aspects); the environment surrounding the scientists, the role of scientists (social or human capital) in the new ventures created and the process of new venture creation (Zucker et al., 1998).

The contribution of this chapter is twofold. First, it yields a theoretical model that highlights the integration among three modern sectors responsible for the production of science and technology and its commercialization, namely, universities, laboratories and private companies. The model yields a full characterization of the integration of commercial activities into a knowledge-producing environment. It provides optimal equilibrium levels for the number of scientists, patents, as well as the determination of scientists' pay in laboratories and for-profit firms, and the optimal market value of patents. In addition, the model also determines the optimal number of entrepreneurs, distinguishing the ones with scientific background from others. The second contribution of this chapter is the estimation of the optimal number of academic entrepreneurs derived from the model, using the dataset that comes from a large survey of researchers at the Max Planck Society (MPS) in Germany. The empirical estimation takes account of the sequential stages of academic entrepreneurship, with nascent entrepreneurship frequently preceding business ownership.

Regarding scientists' scholarly versus entrepreneurial activities, Jain et al. (2009) suggested that academic entrepreneurs have an entrenched and usually well-developed academic identity that is more valued than the newly adopted "entrepreneurial persona." They proposed two separation mechanisms, buffering and delegation, that help protect and defend the academic self from entrepreneurial "contamination." Karhunen et al. (2017) challenged this view by affirming that, in the Finnish context, academic entrepreneurs viewed their two roles as compatible and complementary and that science-based entrepreneurship is described as an integral part of the research process. Some scholars have tried to identify why (and how) scientists choose a particular mode for technology transfer and have aimed to distinguish academic entrepreneurs as "active versus passive commercializer." While the former type tries to implement their research results in the form of university spinoffs by starting their own business, the latter refers to scientists who adapt their basic research agendas to the new funding sources (industry) but without a financial growth motive, forming a start-up or any perspective of leaving academia (Göktepe-Hultén, 2008; Meyer, 2005). They have referred to financial incentives and support provided by institutions and organizations to explain the entrepreneurial activities of researchers. Göktepe-Hultén and Mahagaonkar (2010) and Lam (2010) investigated the motivations of scientists to patent and commercialize their scientific research results.

High level of mobility of qualified personnel contributes to the overall skill level of the labor force as well as to the innovative performance of the economy (Stenberg et al., 1996). Although the evolution of pure scientists to become academic entrepreneurs has been the subject of several studies (Etzkowitz, 2001; Meyer, 2005), science and technology policy analysts have paid relatively little attention to independent or lone inventors, in particular the transition from inventorship to entrepreneurship. Factors that may influence inventors and/or aspiring (nascent) academic entrepreneurs to become business owners are less well understood compared to the praise or criticism that academic entrepreneurship has received. That is the gap that this chapter tries to address.

The chapter is structured as follows. The next section provides the background and develops a theoretical framework on the relationship between the different types of factors on the academic entrepreneurship, followed by the empirical model and estimation results. The chapter ends by discussing some policy implications.

5.2 LITERATURE REVIEW

Universities and Public Research Organizations (PROs hereafter) have been traditionally tasked with the key roles of education and producing basic research. This understanding changed when several policy initiatives began pushing in the direction of adding a "third mission" of technology transfer (Etzkowitz, 2001). This is supported by reference to a number of important developments of successful cases such as Stanford and MIT as sources of university income maximization (Etzkowitz ,2001) and to compensate the need for new sources of funding for academic research (Geuna, 2001), as well as the need to close the gap between discovery and utilization of university research by industry (Mowery et al., 2004) and a tool for making a social and economic impact (Grimaldi et al., 2011). The process of innovation, through which knowledge is created, is increasingly a collective endeavor, shaped by institutional and knowledge-sharing systems. The movement of scientists carrying the "tacit knowledge" from one epistemic sphere to another epistemic sphere where knowledge is being applied and commodified is key in national innovation systems. Personal interactions, whether on a formal or informal basis, are an important channel of knowledge transfer within industry and between the public and private sectors. Despite some policy measures, the mobility rate of university researchers is low and research institute personnel's movement to industry is low. A majority of scientists entering the business sector quit research and switch to other activities within industry (Stenberg et al., 1996).

Studies of academic entrepreneurship have focused on a few themes like the scientists' values and attitudes toward commercialization, traits of the scientists (demographics, age, experience, position, citizenship and other related human capital aspects); the environment surrounding the scientists (peers and leadership toward commercialization), as well as pecuniary and non-pecuniary rewards of commercialization and the support provided by institutions and organizations to explain the entrepreneurial activities of researchers in the new ventures created and the process of new venture creation (Zucker et al., 1998). In light of these studies, we aim to investigate the factors behind the transition of individual scientists' engagement in commercial acts to entrepreneurial and commercial acts. Which factors significantly impact entrepreneurship propensities of academics, and especially their transition from nascent entrepreneurs to business owners?

5.2.1 Impacts of Values and Attitudes

While considerable attention has been devoted to institutional and contextual factors when investigating the concept of academic entrepreneurship, individual-level studies remain relatively few (Rothaermel et al., 2007). Most policy suggestions often inaccurately assume that scientists have not developed their own values, implying that once laws, institutions and organizations are established, scientists will be satisfied and thus follow the rules and regulations automatically. Although most academics aren't religiously adhering to Merton's norms in practice, they nevertheless have a major normative significance for the community and influence scientists' identity (Lam, 2010). Merton (1957) defined cultural expectations related to academic work through four different norms of behavior: universalism (or the dissociation between the scientific statements and the personal characteristics of the scientist), communality (or the duty to share one's research results with all other researchers, thus allowing for its wide dissemination and replication), disinterestedness (which implies that research is conducted only for the sake of pursuing the truth and intellectual progress and not for financial or personal motives), and organized skepticism (or the need for critical review of scientific statements to ensure the correctness of deduction and conclusions).

Latour (1987) argued scientists acquire a range of investments – skills, beliefs, past achievements, objectives – throughout their career trajectory, which they intrinsically try to preserve and enhance. Such investments could be offered as explanations of scientists' decisions and judgments in their endeavors. Scientists may not be influenced directly by policy tools as their own values may prevail over the alleged impacts of institutions and organizations (Shapin, 1988). Scientists may want others to follow their values and objectives and impose their behaviors on others.

The commercialization of scientific research results is risky and uncertain (Audretsch and Stephan, 1999). Attitudes toward risk are shaped partly by a person's psychological bent and partly by past experiences (Goel and Göktepe-Hultén, 2013). Scientists' attitudes toward risk and uncertainty have an impact on their transition from inventing (invention disclosure to start-up formation and business ownership). The scientific reputation and skills of scientists provide credibility and capability to any anticipated commercial project (Audretsch et al., 2006). Due to the low risks of losing their image and credibility, scientists with strong scientific reputations have higher incentives to patent. The propensity of a scientist to engage in patenting is positively related to the amount of the expected rewards. Negative tendencies would be more pronounced, *ceteris paribus*, for more risk-averse inventors. Due to the uncertainties regarding commercialization of science, one would expect greater uncertainty to lower tendencies toward invention disclosure and patenting (Goel and Göktepe-Hultén, 2013).

5.2.2 Impacts of Internal Experience

Institutional experience can lower transaction costs of entrepreneurship via familiarity with rules and norms, yet it can also lead to lethargy. Stephan and Levin (2005) investigated whether scientists' personal characteristics, age (life cycle), citizenship status, gender and receipt of federal funding were related to patenting activities. They found little evidence of age effects, yet they found that tenured scientists are more likely to patent than non-tenured ones (Levin and Stephan, 1991). Goel and Göktepe-Hultén (2018) consider the behavior of academic leaders.

Bercovitz and Feldman (2008) found that certain high-opportunity departments, such as genetics and pharmacology, showed high levels of disclosure events per researcher. Faculty who are involved in multidisciplinary fields or who collaborate with industrial partners are more likely to patent. In emerging fields, scientists are more open and motivated to patent in order to establish their emerging fields or ideas or to start a spinoff company when there is no industrial interest in order to launch an emerging field or embryonic idea.

While scientific discipline plays an important role in the commercialization of research results, local group norms, cultural social norms and the university's traditional encouragement of entrepreneurship were important determinants of successful and widespread entrepreneurship at MIT (Louis et al., 1989; Roberts, 1991). In addition, professional training does more than simply transfer technical knowledge; it actively socializes people to value certain things above others (Bercovitz and Feldman, 2008).

From the early 1980s, government agencies have encouraged collaboration through requiring such training as a condition for funding, especially between universities and industries (Geuna, 2001) and technology transfer policies have facilitated collaboration among researchers and industrial R&D organizations (Goel et al., 2022).

University-industry collaboration may play a unique role in each country, based on the specific conditions of firms, universities and the socioeconomic context (Azagra-Caro, 2007; Bergman, 2010). For instance, Freitas et al. (2013) discuss the role of university-industry collaboration for the development of innovation in mature and emergent industries in newly industrialized countries. They show that university-industry collaboration in Brazilian industries is more likely to be initiated by current and former students, and to be funded from several sources, including postgraduate grants. Industrial interactions with universities may be facilitated either by sending employees to a university for training or by firms investing in networking with a university. Faria et al. (2018), however, show that innovation in Brazilian universities is negligible and that universities' market evaluation is negatively related to innovation, suggesting that the focus on patenting and on the transfer of knowledge/technology to industrial processes is still precarious in Brazil. A study by Cowan and Zinovyeva (2013) investigates whether the development of a university system affected local industry innovation in Italy between 1985 and 2000. They find that opening of new schools increased regional innovation activity within five years, thus suggesting that local industry innovation is mostly caused by the high-quality scientific research brought to the region with new schools. A similar study by Kafouros et al. (2015) examines how collaborations with universities influence the ability of Chinese emerging-market enterprises to develop innovations. These authors show that subnational institutional disparities have a deep impact on the relationship between academic collaborations and firms' innovation performance.

Scientists who have stronger social networks may have easier access to complementary skills and resources that may be necessary for patenting. They may, therefore, patent more than scientists who do not enjoy the same social networks. Researchers who change jobs between academia, industry and government, sometimes changing sectors, or are working in multiple settings simultaneously, should have developed more diverse skills and networks that may motivate them to patent (Dietz and Bozeman, 2005). Such meaningful relations and linkages that scientists have with others are important. Goel and Göktepe-Hultén (2013) find that both industrial cooperation and consulting boost patenting by academic researchers, with the influence of cooperation being more statistically robust.

5.2.3 Other Factors: Citizenship, Gender, Leadership

We consider other factors, based on the personal attributes of researchers. These include their citizenship, gender, internal leadership.

5.2.3.1 Citizenship

Alongside scientific skills, demographic factors also influence the transition of scientists from one type of commercialization to another. Innovation productivity of foreign researchers and the human capital that empowers them is an important issue for policymakers due to implications for technology transfer and knowledge dissemination. Previous studies report that foreign-born and foreign-educated scientists are likely to be more productive for several reasons (Levin and Stephan, 1999; Teichler, 2015; see also Goel and Göktepe-Hultén 2021 for more information about discussions on brain-drain, brain-grain, brain-circulation). However, despite new geographical location, certain important cultural, social, institutional and organizational barriers remain for foreign (non-citizen) scientists. Success for foreign scientists in innovation activities depends heavily on becoming integrated in the host country's knowledge and commercial flows and networks. In order to address such challenges, an increasing number of countries are implementing measures to empower the foreign researchers' role in the overall national innovation system of the country, especially in nations with a significant presence of expatriate researchers. Understanding "hybridization of foreign scientists" (i.e., the transition from one role to another) is therefore important for science and innovation policy.

5.2.3.2 Gender

As the persisting gender gap to women's disadvantage in scientific productivity is well documented, it is further hypothesized that commercialization of scientific outcomes (research results, publications, knowledge) in terms of industrial and commercial uses will also be disadvantageous to women (Goel et al., 2015). Despite the political and institutional pressures on women scientists to be competitive in the commercial and innovation front, it is still hard for women scientists to combine different tasks. Well-known human and social factors that are often utilized to explain the likelihood of becoming an entrepreneur indicate advantages for male faculty. Studies have shown that more established faculty (e.g., having a tenured position) with higher scientific productivity (publications) or with industrial work or collaboration/consultancy experience) and who have more resources are found to be mostly men and they are generally found to be more successful in commercializing research results (Bozeman and Gaughan, 2011; Bozeman et al., 2013; Link, 2017). Understanding "hybridization of women scientists" (i.e., transition from one role to another) is therefore important for science and innovation policy.

5.2.3.3 Leadership

Academic leaders often understand the organizational complexity of PROs and universities and the wisdom in shared decision-making. Effective leaders avoid the exercise of unilateral power; they have a clear understanding of who has the authority and/or responsibility for decision-making, and they respect the roles of various groups and constituents (faculty, staff, students, alumni, legislators, etc.) in the process (Morris, 2016). Academic leaders not only shape the norms and culture but will follow the institutional rules and regulations and expectations. They know whom to consult – formally and informally – and when to bring about change.

In comparison to researchers, academic leaders (Principal Investigators (PIs), research group leaders, directors) may have stronger social networks and may have easier access to complementary skills, financial support and resources that may be necessary for commercial engagements. They may leverage their social capital in a way that helps them to commercialize their research results more than scientists who do not enjoy the similar social networks (Goel and Göktepe-Hultén, 2018). Likewise, understanding the relevance of leadership on the "hybridization of scientists" (i.e., transition from one role to another) is important for science and innovation policy.

To sum up, we have reviewed the literature on university technology transfer, and there are a number of studies that examine *ex post* entrepreneurial behavior based on data from actual entrepreneurs (Goel and Göktepe-Hultén, 2013; Goel and Grimpe, 2012; Krabel and Mueller, 2009). Accordingly, we aim to control for impacts of internal (institute) experience, perceived commercialization costs, spillovers from institute spinoffs, effects of risk aversion. If factors facilitating or inhibiting the tendencies toward the transition from one type of commercial endeavor to entrepreneurship can be identified, then institutional support and technology policies can be appropriately (tailored) adopted. This should result in a greater number of academics who do eventually become entrepreneurs.

5.3 MAX PLANCK SOCIETY: BACKGROUND AND RESEARCH CONTEXT

This study is based on a European Public Research Organization, the Max Planck Society (MPS hereafter). MPS was founded in the late 1940s in Germany, and it is funded to a large extent by both the federal and state governments. Although the aim is to conduct basic research in the interest of the general public in natural sciences, life sciences, social sciences and the humanities, the institutes associated with MPS scientists pursue new and innovative ideas that the teaching-intensive German universities are not in a position to conduct adequately. In order to accelerate the transfer of research results to potential applications, MPS has recently launched several initiatives to facilitate collaboration between MPS scientists and industrial firms. Research contracts and cooperation with commercial enterprises and consultancy agreements in connection with research contracts are encouraged, yet the institutes are encouraged to keep a degree of scientific freedom in these cooperative endeavors. Moreover, cooperative projects and joint activities with other PROs (e.g., Fraunhofer Society) in certain fields have been initiated.

Patenting and start-up companies have become important channels for transferring technology and intellectual knowledge originating from the MPS to industry. MPS established one of the first technology transfer offices in Germany; formerly named Garching Instrumente, this company is now known as Max Planck Innovation GmbH. MP-Innovation (hereafter) provides legal advice and support to those interested in patenting or establishing a start-up company. Over the last 50 years, MP-Innovation has supervised more than 4,500 inventions and concluded 2,500 license agreements. Almost 80 percent of the roughly 160 spinoffs supervised by MP-Innovation are still in business today; seven of them have even made it to the stock exchange. Measured by its income of approximately EUR 500 million, MP-Innovation is one of Germany's leading technology transfer organizations, the other being Fraunhofer (another PRO in Germany). Additional details can be found at https://www.max-planck-innovation.com/files/Downloads/MPR_2020_Sp_EN.pdf.

5.4 DATA: MPS SCIENTISTS' SURVEY

This study is part of a larger project on the identification of commercial activities among scientists employed at MPS. The data collection was accomplished through a screening survey of all scientists associated with different institutes of MPS and follow-up surveys with them. Specifically, the survey first identified more than 2,500 scientists' engagement in different forms of academic entrepreneurship and collected in-depth data about their work environments; perceptions on the commercialization of science. The survey was conducted in the last part of 2007 and during 2008 at around 80 institutes specialized in different scientific disciplines and located in different cities in Germany. This rich dataset has been comprehensively used in previous studies, for example, Krabel and Mueller (2009), Göktepe-Hultén and Mahagaonkar (2010), Krabel et al. (2012), Goel and Göktepe-Hultén (2013).

Before performing the survey, the executive directors of each institute were asked for permission to interview the scientists; 67 out of 80 institutes agreed to participate in the study and provided us with the necessary contact information to scientists. The population for the survey consisted of 7,808 scientists of 67 Max Planck Institutes, and 2,604 responded to the survey denoting a response rate of 33.35 percent (see Krabel and Mueller, 2009 for details).

The survey questions were particularly designed to analyze the entrepreneurial activities of scientists. Accordingly, besides the questions explicitly focusing on entrepreneurship, a broad range of individual characteristics (work experience, skills and competences; risk aversion and psychological traits; social, family and demographic factors etc.) and professional characteristics (position; type and innovativeness of research; scientific excellence; industry contacts; institutional characteristics etc.) were also collected which might be related to both the propensity to start a business and to the development and success of the ventures (see Max Planck Innovation GmbH, 2020 and Max Planck Research, 2020).

5.5 THE THEORETICAL MODEL

The model shows the integration among three modern sectors responsible for the production of science and technology and its commercialization, namely, universities, laboratories and private companies. Universities, by offering a number of scientific graduate programs, are responsible for the formation of PhDs. Laboratories hire PhDs who are responsible for patenting efforts, and these efforts face the trade-off between public and private science. Sequentially, firms in the process after universities and laboratories. They use scientists and patents to create spinoffs that depend on entrepreneurial vision and organization. This formulation, however, ignores the possibility of where firms and laboratories may be integrated (Goel, 1992). The model allows us to distinguish between academic entrepreneurs (the ones holding a PhD degree) from others.

The model provides a full characterization of the integration of commercial activities into a knowledge-producing environment. It determines the equilibrium levels for the number of scientists, scientists' pay in laboratories and for-profit firms, and the market value of patents. In addition, the model also determines the optimal number of entrepreneurs, allowing for the distinction between academic entrepreneurs and others. Next, we model the representative university's problem.

5.5.1 The University's Problem

The representative university aims at maximizing the formation of human capital, that is, the number of newly minted PhDs, N. The faculty of graduate programs at time t is composed of people with PhDs in time $t-j$, where $j = 1,2,\ldots$ We assume quadratic costs to form (produce) a PhD, so that the university maximizes:

$$U=\left(aN_{t-j}\right)N_t-0.5\vartheta N_t^2 \tag{5.1}$$

Where parameter a captures the marginal productivity of PhDs, and the parameter ϑ its marginal cost. Without loss of generality, let us assume $j = 1$, so that we can rewrite Equation (5.1) as:

$$U=\left(aN_{t-1}\right)N_t-0.5\vartheta N_t^2 \tag{5.1'}$$

The first-order condition for a maximum yields the optimum number of PhDs supplied by the university:

$$N_t=\frac{aN_{t-1}}{\vartheta} \tag{5.2}$$

Let us assume that the variation of newly minted PhDs is constant, equal to *g*:

$$g = N_t - N_{t-1} \tag{5.3}$$[2]

Advancing Equation (5.2) in one period and using Equation (5.3) yields the optimal number of PhDs:

$$N_t = \frac{\vartheta g}{a - \vartheta} \tag{5.4}$$

Notice that the right-hand side of Equation (5.4) is constant, therefore it yields the optimal number of PhDs, N^*:

$$N^* = \frac{\vartheta g}{a - \vartheta} \tag{5.4'}$$

Note that $(\partial N^*/\partial a) < 0$, implying that with a higher marginal product, fewer PhDs are needed,[3] and $(\partial N^*/\partial \vartheta) > 0$, which follows from the assumption of a constant growth in PhDs in Equation (5.3). In the empirical analysis below, we show whether a scientist holds a PhD (variable PHD in Table 5.1a).

5.5.2 The Laboratory's Problem

The representative laboratory may have two different objectives: (1) to produce basic science that is a public good; and/or (2) to produce applied science that may yield patents that are a private good. The laboratory uses qualified personnel, mostly PhD holders,[4] to achieve either or both objectives:

$$L = \tau D N_t S_t - 0.5 \beta S_t^2 - w N_t \tag{5.5}$$

Where S is science, D is the scientific area (medicine, engineering, etc.), w is the salary paid to PhD personnel, τ is patenting effort, β is academic effort to produce pure science. In our empirical section, we have information on the academic discipline, with 44 percent of the respondents being in biology or medicine, 47 percent in physics, chemistry or technics, and the rest in humanities (Table 5.1a). The laboratory faces convex costs, given by the term $0.5\beta S_t^2$ associated with the production of basic or pure science.

The optimal choice of S yields:

$$S_t = \frac{\tau D N_t}{\beta} \tag{5.6}$$

Inserting Equation (5.4) into (5.6) yields the optimal amount of science produced by the laboratory:

$$S^* = \frac{\tau D}{\beta}\left(\frac{\vartheta g}{a-\vartheta}\right) \tag{5.7}$$

Note that $(\partial S^*/\partial g) > 0$, implying that the production of science increases as the growth of PhDs increases.
Maximizing Equation (5.5) with respect to N and using Equations (5.4), (5.6) and (5.7), one can determine the optimal wage of scientists working for laboratories as:

$$w^* = \tau D S_t = \tau^2 \frac{D^2 N_t}{\beta} = \tau^2 \frac{D^2}{\beta}\left(\frac{\vartheta g}{a-\vartheta}\right) \tag{5.8}$$[5]

The representative firm's (i.e., business owner's) problem in modeled next.

5.5.3 The Representative Firm's Problem

The representative firm also has two options at hand: (1) it may continue operating as usual, using entrepreneurial effort E and scientists to produce good Q; and/or (2) it may form a spinoff by producing innovations with the science created by the laboratory. The firm aims at maximizing profits:

$$\pi = E_t^b N_t^{1-b} + \varepsilon E_t N_t S_t - cE_t - r(S_t) - w' N_t \tag{5.9}$$

Where $E_t^b N_t^{1-b}$ is the production function of good Q (whose price is normalized equal to 1) and b and $(1-b)$ are the output elasticities of entrepreneurial effort and scientists, respectively. The term $\varepsilon E_t S_t$ is the introduction of innovation, where εE_t is the entrepreneurial effort to make the innovation happen, c is the unit cost of entrepreneurial effort, $r(.)$ is the patent pay and w' is the salary of a scientist employed by the firm.
The firms choose the optimal level of entrepreneurial effort that maximizes profit:

$$E_t = \left(\frac{bN_t^{1-b}}{c-\varepsilon S_t N_t}\right)^{1/(1-b)} \tag{5.10}$$

Plugging Equations (5.4) and (5.7) into Equation (5.10) yields the optimal level of entrepreneurial effort:

$$E^* = \left(\frac{b\left(\frac{\vartheta g}{a-\vartheta}\right)^{1-b}}{c - \varepsilon \frac{\tau D}{\beta}\left(\frac{\vartheta g}{a-\vartheta}\right)^2} \right)^{1/(1-b)} \tag{5.11}$$

One can determine the market value of the patent by maximizing Equation (5.9) with respect to *S*:

$$r'(S_t) = \varepsilon E_t N_t \tag{5.12}$$

Where $r'(S_t)$ is the market value of the patent (see Kamien and Tauman, 1986).

Introducing Equations (5.4) and (5.7) into (5.12) yields:

$$r'\left[\frac{\tau D}{\beta}\left(\frac{\vartheta g}{a-\vartheta}\right)\right] = \varepsilon E_t \left(\frac{\vartheta g}{a-\vartheta}\right) \tag{5.13}$$

In the same vein, one can find the optimal pay of scientists in the market, *w*', by maximizing Equation (5. 9) with respect to *N*:

$$w' = (1-b) E_t^b N_t^{-b} + \varepsilon E_t\, S_t \tag{5.14}$$

Substituting the optimal levels of *S*, *N* and *E* into Equation (5.14) yields:

$$w' = (1-b)\left(\frac{b\left(\frac{\vartheta g}{a-\vartheta}\right)^{1-b}}{c - \varepsilon \frac{\tau D}{\beta}\left(\frac{\vartheta g}{a-\vartheta}\right)^2} \right)^{b/(1-b)} \left(\frac{\vartheta g}{a-\vartheta}\right)^{-b} + \varepsilon \frac{\tau D}{\beta}\left(\frac{\vartheta g}{a-\vartheta}\right) \left(\frac{b\left(\frac{\vartheta g}{a-\vartheta}\right)^{1-b}}{c - \varepsilon \frac{\tau D}{\beta}\left(\frac{\vartheta g}{a-\vartheta}\right)^2} \right)^{1/(1-b)} \tag{5.15}$$

If scientists working for laboratories are perfect substitutes for scientists working for private firms, then $w^* = w'$, and we have from Equations (5.8) and (5.15)[6]

$$\frac{\tau D}{\beta}\left[\tau-\varepsilon\left(\frac{b\left(\frac{\vartheta g}{a-\vartheta}\right)^{1-b}}{c-\varepsilon\frac{\tau D}{\beta}\left(\frac{\vartheta g}{a-\vartheta}\right)^{2}}\right)^{1/(1-b)}\right]=(1-b)\left(\frac{b\left(\frac{\vartheta g}{a-\vartheta}\right)^{1-b}}{c-\varepsilon\frac{\tau D}{\beta}\left(\frac{\vartheta g}{a-\vartheta}\right)^{2}}\right)^{b/(1-b)}\left(\frac{\vartheta g}{a-\vartheta}\right)^{-b-1} \tag{5.16}$$

This model provides a full characterization of the integration of commercial activities into a knowledge-producing environment. It provides the optimal equilibrium levels for the number of scientists, N^*, patents, s^* (assuming a fixed proportion y between science production and the number of patents e, e.g., $s^*=yS^*$), as well as the determination of scientists' pay in laboratories and for-profit firms, and the optimal market value of patents. In addition, the model also determines the optimal number of entrepreneurs e^*, which is the object of the next section.

5.5.4 Entrepreneurs

The model allows us to distinguish between academic entrepreneurs (the ones holding a PhD degree) and others. First, we assume a proportion x between entrepreneurial effort and the number of entrepreneurs e:

$$e^* = xE^* \tag{5.17}$$

The term x can be decomposed into two terms, the first relates to the number of academic entrepreneurs, scientists that become entrepreneurs, and the second term depends on the other types of entrepreneurs (denoted by M), without a PhD degree and deep scientific knowledge:

$$x=\alpha N^*+(1-\alpha)M \tag{5.18}$$

Where the parameter α captures risk and/or commercialization attitudes of the academic entrepreneur.
Plugging Equation (5.18) into Equation (5.17) and using Equations (5.4) and (5.11) yields the optimal number of entrepreneurs:

$$e^*=xE^*=\left[\alpha N^*+(1-\alpha)M\right]E^*=\left[\alpha\frac{\vartheta g}{a-\vartheta}+(1-\alpha)M\right]\left(\frac{b\left(\frac{\vartheta g}{a-\vartheta}\right)^{1-b}}{c-\varepsilon\frac{\tau D}{\beta}\left(\frac{\vartheta g}{a-\vartheta}\right)^{2}}\right)^{1/(1-b)} \tag{5.19}$$

From Equations (5.18) and (5.19) one can have the determination of the optimal number of academic entrepreneurs:

$$\alpha N^{*}E^{*}=\left[\alpha\frac{\vartheta g}{a-\vartheta}\right]\left(\frac{b\left(\dfrac{\vartheta g}{a-\vartheta}\right)^{1-b}}{c-\varepsilon\dfrac{\tau D}{\beta}\left(\dfrac{\vartheta g}{a-\vartheta}\right)^{2}}\right)^{1/(1-b)} \qquad (5.20)$$

Equation (5.20) is the focus of the empirical section of this chapter. Therefore, a detailed analysis of it is necessary. According to Equation (5.20), the optimal number of academic entrepreneurs depends positively on the following parameters: the marginal cost of forming newly minted PhDs, ϑ; the growth rate of newly minted PhDs, g; the scientific area in which the academic entrepreneur holds a PhD, D; patenting effort, τ; entrepreneurial effort to make the innovation happen, ε; the risk and/ or commercialization attitudes of the academic entrepreneur, α.

The optimal number of academic entrepreneurs is negatively impacted by the unit cost of entrepreneurial effort, c; the universities' marginal productivity of forming PhDs, a; academic effort to produce pure science, β; and by the output elasticity of scientists, $1-b$. The formal empirical model follows.

5.6 THE EMPIRICAL MODEL

Our empirical model accounts for the sequential stages in entrepreneurship, with nascent entrepreneurship preceding business ownership. While the empirical section will address the issues raised above, the correspondence with the theoretical model is imperfect due to the detail in the survey data (the survey was not conducted with this study in mind). About 6 percent of the survey respondents in our sample were business owners (BUSown), which was double the percentage for nascent entrepreneurs (NASCENTent) (Table 5.1a). This makes sense since not all business owners are necessarily nascent entrepreneurs earlier. Based on the data described in Section 5.4 and Table 5.1a, the estimated model takes the following form.

Probability of business ownership: $p_1(\text{BUSown} = 1|X_1) = f_1(X_1'\psi_1)$ (5.21)

and

Probability of nascent entrepreneurship: $p_2(\text{NASCENTent} = 1|X_2) = f_2(X_2'\,\psi_2)$ (5.22)

Where

X_1 = NASCENTent. INDwork, ATTRACTent, GROUPlead, PHD, FEMALE, GERMAN, AGE, Discipline1, Discipline2, *Zm*;

and

X_2 = INVENTION, ATTRACTent, GROUPlead, PHD, FEMALE, GERMAN, AGE, Discipline1, Discipline2, Zm

m = YEARSMPS, COMMcost, SPINOFFS, RISKaverse

Complete details about these variables, including definitions and summary statistics, are in Table 5.1a.

Table 5.1a Variable definitions, summary statistics and data sources

Variable	Definition (mean; std. dev.)
BUSown	Respondent was a business owner or started a business in the past, (= 1; 0 otherwise), (0.055; 0.227)
NASCENTent	Respondent is a nascent entrepreneur, (= 1; 0 otherwise), (0.032; 0.177)
INVENTION	Respondent "disclosed invention to MPS", (= 1; 0 otherwise), (0.111; 0.314)
ATTRACTent	Respondent's perceptions about "Attractiveness of starting a business", (1–5 scale, with 1 = Not attractive at all; 5 = Highly attractive), (2.725, 1.193)
GROUPlead	Respondent is group leader at MPS, (= 1; 0 otherwise), (0.131; 0.337)
COMMcost	Respondent's perceptions about commercialization costs, "Cost of commercialization high", (1–5 scale, with 1 = strongly disagree; 5 = strongly agree), (3.809, 0.778)
RISKaverse	"Researcher does not invest any money", 1 = yes; 0 = no, (0.249, 0.432)
SPINOFFS	Number of spinoffs at the respondent's institute, (1.855; 3.122)
YEARSMPS	Experience working at Max Planck Society, years, (5.918; 7.070)
INDwork	Respondent has industry work experience = 1; 0 otherwise, (0.250; 0.586)
PHD	Respondent has a PhD degree (= 1; 0 otherwise), (0.509; 0.500)
FEMALE	Respondent is a female = 1; 0 otherwise, (0.321; 0.467)
AGE	Respondent's age, years, (35.458; 9.573)
GERMAN	Respondent is a German citizen = 1; 0 otherwise, (0.609; 0.488)
Discipline1	Respondent's discipline is biology or medicine, (= 1; 0 otherwise), (0.442; 0.497)
Discipline2	Respondent's discipline is chemistry, physics or technics, (= 1; 0 otherwise), (0.474; 0.499)

Note: The data come from a large survey of researchers at the Max Planck Society (MPS) in Germany. Humanities is the default academic discipline. For details, see *Max Planck Society: Annual Report 2008*. http://www.mpg.de/7313642/Annual_Report_2008.pdf. Accessed June 2019.

Given the sequential and interdependent nature of business ownership and nascent entrepreneurship, with nascent entrepreneurship frequently preceding business ownership, we use bivariate probit regressions. INVENTION is the offset variable in the NASCENTent in Equation (5.22), while industry work experience (INDwork) is the offset variable in the business ownership (Equation (5.21)). Tying to the theoretical model above, our dependent variables can be seen as following from Equation (5.2) above.

Nascent entrepreneurs generally do not have production facilities set up and are generally testing the waters to see if their venture is viable, before becoming full-fledged business owners. Thus, prior industry experience is less likely to be relevant in someone's propensity to become a nascent entrepreneur (see Dietz and Bozeman, 2005; Goel and Göktepe-Hultén, 2013).

The main explanatory variable of interest is respondents' attitudes toward entrepreneurship (ATTRACTent). Other things being the same, one would expect greater attractiveness of entrepreneurship to lead to greater propensities to be business owners and nascent entrepreneurs. This variable varied between 1 and 5, with higher values signifying greater attractiveness. In our sample, the average value of ATTRACTent was 2.7.

The other control variables, related to scientists' personal attributes and training, include scientists' internal leadership position (i.e., whether the scientist is a group leader (GROUPlead), education (PHD),[7] gender (FEMALE), citizenship (GERMAN) and experience (AGE or YEARSMPS). These have bearings on entrepreneurship costs and networking (Bozeman et al., 2013).

The consideration of age is tied to the life cycle productivity of scientists (Levin and Stephan, 1991), whereas education and leadership are tied to human capital (Zucker et al., 1998). Furthermore, German citizens likely have lower transaction costs due to greater familiarity with the language and institutions (Goel and Göktepe-Hultén, 2021). Many of these aspects have been reviewed in reviews of the literature – see, for example, Bozeman et al. (2013) and Rothermael et al. (2007).

Discipline1 and Discipline2 account for academic disciplines, with humanities being the default group. Academic disciplines are denoted by the parameter "*D*" in the theoretical model (Equation (5.5)). Research output from certain disciplines (e.g., the applied sciences) is more readily commercialized than others.

Later variations of the estimated models also account for perceived commercialization costs (COMMcost), risk aversion (RISKaverse) and spillovers from spinoffs (SPINOFFS). Higher commercialization costs can be seen as a barrier to entrepreneurship, while greater risk aversion will make entrepreneurship less likely (Goel and Göktepe-Hultén, 2019).[8] The consideration of spinoffs is tied to spillovers (see Audretsch and Stephan, 1999).

The correlation between BUSown and NASCENTent in our sample was about 0.2 (Table 5.1b). In our sample, about 13 percent of the respondents were group leaders, 61 percent were German citizens, and a third were females (Table 5.1a). The estimation results are discussed next.

Table 5.1b Correlation matrix of key variables

	NASCENTent	BUSown	INVENTION	SPINOFFS
NASCENTent	1.000			
BUSown	0.167	1.000		
INVENTION	0.180	0.187	1.000	
SPINOFFS	–0.015	–0.002	0.072	1.000

Note: See Table 5.1a for variable definitions. Observations: 2579.

5.7 RESULTS

5.7.1 Baseline Models

Table 5.2 presents the results for the baseline models. The cross-equation correlation (rho) is significant and different from zero.

Table 5.2 Inventing academics' transition from nascent entrepreneurs to business owners: baseline models

Dependent variable→	BUSown	NASCENTent	Marginal effects
ATTRACTent	0.34**	0.40**	0.002**
	(0.04)	(0.06)	(0.001)
GROUPlead	0.02	0.10	0.0003
	(0.13)	(0.18)	(0.001)
PHD	0.23**	–0.26*	–0.0001
	(0.11)	(0.14)	(0.0004)
FEMALE	–0.30**	0.14	–0.0003
	(0.13)	(0.13)	(0.0004)
GERMAN	0.03	–0.41**	–0.001#
	(0.10)	(0.12)	
AGE	0.03**	0.001	0.0001*
	(0.005)	(0.01)	(0.00003)
Discipline1	–0.06	–0.29	–0.0008
	(0.18)	(0.21)	(0.001)
Discipline2	–0.11	–0.18	–0.0007
	(0.18)	(0.22)	(0.001)
N	2549		
Wald chi²	202.41**		
rho	0.26**		
	(0.086)		
Wald test of rho = 0 (chi. sq.)	8.11**		

Note: See Table 5.1a for variable details. Constant included but not reported in these bivariate probit regressions. INVENTION is the offset variable in the NASCENTent equation; INDwork is the offset variable in the BUSown equation. Discipline3 (Humanities) is the default group. # not reported by STATA. The numbers in parentheses are robust standard errors. * and **, respectively, denote statistical significance at the 10 percent and 5 percent (or better) levels.

We see that attractiveness of entrepreneurship, doctoral degrees and age make business ownership more likely while being a female has the opposite effect. The challenges facing female academics in becoming entrepreneurs have been well considered in the literature (Bozeman and Gaughan, 2011; Goel et al., 2015; Link, 2017; Stephan and Levin, 2005). Doctoral degrees add to human capital while also sending positive signals for raising finance. Another aspect of human capital may be tied to age via experience. The effects of academic discipline, German citizenship and being a group leader are statistically insignificant.

However, doctoral education makes nascent entrepreneurship less likely, as does German citizenship, with the impact of entrepreneurship attractiveness remaining positive. These findings may be reflective of different opportunity costs for scientists with doctoral degrees and for German citizens.

In terms of the marginal effects, only ATTRACTent and AGE have positive and significant marginal effects (with BUSown = 1; NASCENTent = 1).

5.7.2 Impacts of Internal (Institute) Experience

Table 5.3 considers the effect of years of internal experience. Greater internal experience, besides adding human capital, would make researchers familiar with the incentives for entrepreneurship. The average work experience among the sample respondents was about six years (Table 5.1a).

Table 5.3 *Inventing academics' transition from nascent entrepreneurs to business owners: impacts of internal experience*

Dependent variable→	BUSown	NASCENTent	Marginal effects
ATTRACTent	0.33**	0.39**	0.002**
	(0.04)	(0.06)	(0.001)
GROUPlead	0.05	0.14	0.001
	(0.13)	(0.18)	(0.001)
PHD	0.39**	-0.24*	0.0003
	(0.10)	(0.13)	(0.0004)
FEMALE	−0.33**	0.13	−0.0004
	(0.12)	(0.13)	(0.0004)
GERMAN	0.09	−0.38**	−0.001
	(0.10)	(0.12)	(0.001)
YEARSMPS	0.02**	−0.01	0.00002
	(0.01)	(0.01)	(0.00004)
Discipline1	−0.12	−0.27	−0.001
	(0.18)	(0.22)	(0.001)
Discipline2	−0.17	−0.17	−0.001
	(0.18)	(0.22)	(0.001)
N	2558		
Wald chi^2	189.49**		
rho	0.26**		
	(0.09)		
Wald test of rho = 0 (chi. sq.)	8.49**		

Note: See Table 5.2.

The results show that greater internal experience (YEARSMPS) makes business ownership more likely, but does not have a significant effect on nascent entrepreneurship. The effects of internal experience are similar to those of age in Table 5.2. Both age and experience can impact the costs of entrepreneurship, as denoted by c in the theoretical model. The findings with respect to the other control variables were similar to that presented in Table 5.2.

5.7.3 Impacts of Perceived Commercialization Costs

As another dimension, we consider the effects of perceived commercialization costs on the incentives of academics to become business owners and nascent entrepreneurs. In Table 5.4, while the coefficient on COMMcost is, as expected, negative in both cases, it is statistically significant (at the 10 percent level) in the NASCENTent equation. These findings corroborate the theoretical predictions. Specifically, from Equation (5.20), one can see that the optimal number of academic entrepreneurs (αN^*E^*) would go down as the costs of commercialization increase, that is, $(\partial \alpha N^*E^*/\partial c) < 0$.

Table 5.4 *Inventing academics' transition from nascent entrepreneurs to business owners: impact of perceived commercialization costs*

Dependent variable→	BUSown	NASCENTent	Marginal effects
ATTRACTent	0.34**	0.43**	0.002**
	(0.05)	(0.06)	(0.001)
GROUPlead	0.02	0.11	0.001
	(0.14)	(0.18)	(0.001)
PHD	0.24**	−0.22	−0.00001
	(0.11)	(0.15)	(0.0004)
FEMALE	−0.32**	0.18	−0.0002
	(0.13)	(0.13)	(0.0004)
GERMAN	0.05	−0.36**	−0.001
	(0.11)	(0.12)	(0.0005)
AGE	0.03**	0.002	0.0001*
	(0.005)	(0.01)	(0.00003)
COMMcost	−0.08	−0.13*	−0.0004*
	(0.06)	(0.07)	(0.0003)
Discipline1	−0.07	−0.29	−0.001
	(0.18)	(0.23)	(0.001)
Discipline2	−0.15	−0.22	−0.001
	(0.18)	(0.23)	(0.001)
N	2366		
Wald chi^2	209.24**		
rho	0.22**		
	(0.09)		
Wald test of rho = 0 (chi. sq.)	5.56**		

Note: See Table 5.2.

Higher perceived commercialization costs dissuade nascent entrepreneurship, but not necessarily business ownership. This may be due to the fact that entrepreneurs expect to make profits under business ownership, but not necessarily under nascent entrepreneurship.

5.7.4 Spillovers from Institute Spinoffs

The total spinoffs at an institute might generate positive or negative spillovers for academic entrepreneurs. Such spillovers would be positive when spinoffs add to the researcher's reputation and make securing external credit easier while at the same time having strengthened internal support such as technology transfer offices. On the other hand, greater spinoffs might crowd out funding for new ventures or the administration might not feel the need to support new budding ventures when spinoffs are taking place on their own. The corresponding findings in Table 5.5 are qualitatively similar to those in Table 5.4 – the coefficient on SPINOFFS is negative in both cases, but (marginally) statistically significant in the case of nascent entrepreneurship. This finding is consistent with the crowding out story.

Table 5.5 *Inventing academics' transition from nascent entrepreneurs to business owners: spillovers from institute spinoffs*

Dependent variable→	BUSown	NASCENTent	Marginal effects
ATTRACTent	0.35**	0.40**	0.002**
	(0.04)	(0.06)	(0.001)
GROUPlead	0.01	0.10	0.0003
	(0.13)	(0.18)	(0.001)
PHD	0.22**	−0.27*	−0.0002
	(0.11)	(0.14)	(0.0004)
FEMALE	−0.30**	0.14	−0.0003
	(0.13)	(0.13)	(0.0004)
GERMAN	0.04	−0.40**	−0.001
	(0.10)	(0.12)	(0.001)
AGE	0.03**	0.002	0.0001**
	(0.005)	(0.01)	(0.00003)
SPINOFFS	−0.02	−0.04*	−0.0001*
	(0.02)	(0.02)	(0.0001)
Discipline1	0.002	−0.18	−0.0004
	(0.19)	(0.22)	(0.001)
Discipline2	−0.09	−0.14	−0.0005
	(0.18)	(0.22)	(0.001)
N	2549		
Wald chi^2	207.62**		
rho	0.26**		
	(0.08)		
Wald test of rho = 0 (chi. sq.)	8.52**		

Note: See Table 5.2.

5.7.5 Effects of Risk Aversion

Finally, in Table 5.6, we examine if greater risk aversion made entrepreneurship less likely by entrepreneurs. We incorporate risk in the theoretical model above via the parameter α. The results show the impact of risk aversion to be insignificant, with the other findings remaining similar to the baseline model in Table 5.2. Goel and Göktepe-Hultén (2019) use different measures of risk aversion and find that risk aversion has a negative effect on invention disclosures, but not on patenting.

Table 5.6 *Inventing academics' transition from nascent entrepreneurs to business owners: impact of risk aversion*

Dependent variable→	BUSown	NASCENTent	Marginal effects
ATTRACTent	0.34**	0.40**	0.002**
	(0.05)	(0.06)	(0.001)
GROUPlead	0.02	0.10	0.0003
	(0.13)	(0.18)	(0.001)
PHD	0.23**	−0.26*	−0.0001
	(0.11)	(0.14)	(0.0004)
FEMALE	−0.30**	0.14	−0.0002
	(0.13)	(0.13)	(0.0004)
GERMAN	0.03	−0.41**	−0.001#
	(0.10)	(0.12)	
AGE	0.03**	0.001	0.0001*
	(0.005)	(0.01)	(0.00003)
RISKaverse	0.06	0.05	0.0002
	(0.11)	(0.13)	(0.0004)
Discipline1	−0.06	−0.28	−0.001
	(0.18)	(0.22)	(0.001)
Discipline2	−0.10	−0.18	−0.001
	(0.18)	(0.22)	(0.001)
N	2549		
Wald chi²	203.01**		
rho	0.26**		
	(0.09)		
Wald test of rho = 0 (chi. sq.)	8.05**		

Notes See Table 5.2. # not reported by STATA.

In sum, we see that attractiveness of entrepreneurship consistently increases the propensity to become a nascent entrepreneur and business owner in all cases, with age/internal experience having a positive effect on business ownership but not on nascent entrepreneurship. On the other hand, greater perceived commercializing costs made nascent entrepreneurship less likely, but not necessarily business ownership. Interestingly, doctoral degrees had opposite effects on the two forms of entrepreneurship, and being a German citizen actually turned out to be a handicap with regard to nascent entrepreneurship.

5.8 CONCLUSIONS

The concept of the innovation system argues for the importance of the flow of technology and information among people, enterprises and institutions as key to an innovative process. Innovation does not occur in a perfectly linear sequence, but through feedback loops and interaction within this system. It contains the interactions between the actors needed in order to turn an idea into a process, product or service on the market (Lundvall, 1992). Firms, universities, laboratories (PROs) and other public organizations are the key stakeholders within the innovation systems. As innovation and entrepreneurship are widely seen as remedies for increasing unemployment and economic problems, public policymakers are increasingly urging scientists to actively participate in the diffusion and application of knowledge to other sectors of society and commercialize their research results in addition to teaching and research tasks. As a result, scientists (universities) are expected to offer a number of scientific graduate programs which are responsible for the formation of PhDs; as well as new knowledge, laboratories are responsible for patenting efforts, and these efforts face the trade-off between public and private science. Investors and/or firms often use scientists, researchers and patents to create new products or spinoffs that depend on entrepreneurial vision and organization. The mechanisms for knowledge flows include joint industry research, public/private sector partnerships, technology diffusion and movement of personnel. Yet, each actor aims to pursue their own *raison d'être* which may eventually lead to conflicts of interest and may hinder the alleged flow and interaction between the actors of innovation systems. While the creation and retention of scientific staff (e.g., PhDs and scientific inventors) are fundamental roles of universities and PROs, transition of skilled scientists as business owners is an aspect of the mobility of human resources; it is a subset of the overall mobility experience in the innovation system. Loss of valuable human resources (scientists) is a challenge for universities and PROs.

Considering the sequential nature of nascent entrepreneurship and business ownership, this chapter examines the propensities of academic entrepreneurs to be business owners. A theoretical model sets up the empirical analysis based on survey data from a large German public research institute. Traditionally, scientists and entrepreneurs have been seen to occupy opposite ends of a spectrum in terms of their role in the innovation.

By bringing the analysis from the level of social behavior and roles to the level of knowledge production, we can better address questions such as: How is knowledge in the interfaces of epistemic communities produced? How can such knowledge be organized and sustained? How do individuals reconcile ostensibly incommensurable epistemic norms? And how can relations between individuals on "opposing sides" be constructively managed?

The theoretical model we develop provides a full characterization of the integration of commercial activities into a knowledge-producing environment. It determines the equilibrium levels for the number of scientists, scientists' pay in laboratories and for-profit firms, and the market value of patents. In addition, the model also determines the optimal number of entrepreneurs, allowing for the distinction between academic entrepreneurs from others.

Using survey data on scientists from the MPS,[9] in the empirical model, we take account of the stages of academic entrepreneurship, with nascent entrepreneurship preceding business ownership. The empirical results show that scientists' positive attitudes toward commercialization of results consistently contribute to tendencies toward academic entrepreneurship; however, the academic discipline and risk aversion did not have a statistically significant impact. Having a doctoral degree lowered the propensities toward nascent entrepreneurship, but had the opposite effect on business ownership. This result might signify different opportunity costs of the two entrepreneurship types. The result regarding the negative impacts of commercialization costs suggests the need for government subsidies to boost academic entrepreneurship. Finally, age and experience made business ownership more likely. These findings are consistent with experience lowering the transaction costs of business ownership. While the survey was not conducted with this study in mind, some of the predictions of the theoretical model hold up to empirical verification (e.g., the negative effects of commercialization costs on entrepreneurship).

According to the innovation systems approach, innovation and technology development are the result of a complex set of relationships among the key stakeholders, enterprises, universities and government research institutes (Lundvall, 2010). Knowledge flows primarily through joint research activities and technical collaborations, co-patenting, co-publications as well as through more informal linkages and personnel mobility. The lack of interaction between the actors in the system, mismatches between basic research in the public sector and more applied research in industry, malfunctioning of technology transfer organizations, and information and absorptive deficiencies on the part of enterprises may all contribute to poor innovative performance in a country. However, with such broad definition, it is nebulous for policymakers to develop an understanding of how knowledge is actually produced and shared among different epistemic spheres. These findings underscore the specific avenues that policymakers can target in promoting transition from basic science to the market. This would increase the social returns from investment in R&D and basic science.

NOTES

1. JEL classification: O33; O52; L26.
2. This constant growth of PhDs is assumed for analytical tractability.
3. Also, $(\partial^2 N^*/\partial a^2) > 0$, signifying a convex relation.
4. More than half of the respondents in our sample held a PhD degree (Table 5.1a).
5. Unfortunately, our data sample does not have information on scientists' salaries.
6. This aspect is somewhat accounted for in our empirical analysis by including the variable INDwork that identifies scientists with industry work experience.
7. The variable PHD corresponds to N in Equations (5.4) and (5.4') of the theoretical model.
8. Incidentally, the parameter α in the theoretical section captures the risk and/or commercialization attitudes of academic entrepreneurs.
9. Additional details about MPS can be found at https://www.max-planck-innovation.com/files/Downloads/MPR_2020_Sp_EN.pdf. Accessed 30 July 2020.

REFERENCES

Audretsch, D.B., and Stephan, P.E. (1999). Knowledge spillovers in biotechnology: Sources and incentives. *Journal of Evolutionary Economics*, 19(1), 97–107.

Audretsch, D., Aldridge, T., and Oettl, A. (2006). The knowledge filter and economic growth: The role of scientist entrepreneurship. Working Paper, Max Planck Institute of Economics, Entrepreneurship, Growth and Public Policy Group, Jena.

Azagra-Caro, J.M. (2007). What type of faculty member interacts with what type of firm? Some reasons for the delocalization of university–industry interaction. *Technovation*, 27(11), 704–15.

Bercovitz, J., and Feldman, M. (2008). Academic entrepreneurs: Organizational change at the individual level. *Organization Science*, 19(1), 69–89.

Bergman, E.M. (2010). Knowledge links between European universities and firms: A review. *Papers in Regional Science*, 89(2), 311–33.

Bozeman, B., and Gaughan, M. (2011). How do men and women differ in research collaborations? An analysis of the collaboration motives and strategies of academic researchers. *Research Policy*, 40(10), 1393–402.

Bozeman, B., Fay, D., and Slade, C.P. (2013). Research collaboration in universities and academic entrepreneurship: The-state-of-the-art. *Journal of Technology Transfer*, 38(1), 1–67.

Cowan, R., and Zinovyeva, N. (2013). University effects on regional innovation. *Research Policy*, 42(3), 788–800.

Di Gregorio, D., and Shane, S. (2003). Why do some universities generate more start-ups than others? *Research Policy*, 32(2), 209–27.

Dietz, J.S., and Bozeman, B. (2005). Academic careers, patents, and productivity: industry experience as scientific and technical human capital. *Research Policy*, 34(3), 349–67.

Etzkowitz, H. (1997). The entrepreneurial university and the emergence of demoratic corporatism. In Leydesdorff, L., and Etzkowitz, H. (eds), *Universities and the Global Knowledge Economy: A Triple Helix of University-Industry-Government Relations* (pp. 141–52). London: Cassell.

Etzkowitz, H. (2001). The second academic revolution and the rise of entrepreneurial science. *IEEE Technology and Society*, 22(2), 18–29.

Faria, J.R., Wanke, P., Ferreira, J., and Mixon, F. (2018). Research and innovation in higher education: Empirical evidence from research and patenting in Brazil. *Scientometrics*, 116(1), 487–504.
Freeman, C. (1987). *Technology and Economic Performance: Lessons from Japan*. London: Pinter.
Freitas, I., Marques, R., and Silva, E. (2013). University–industry collaboration and innovation in emergent and mature industries in new industrialized countries. *Research Policy*, 42(2), 443–53.
Geuna, A. (2001). The changing rationale for European university research funding: Are there negative unintended consequences? *Journal of Economic Issues*, 35(3), 607–32.
Goel, R.K. (1992). On vertical integration into R&D. *The Quarterly Review of Economics and Finance*, 32(3), 54–9.
Goel, R.K., and Göktepe-Hultén, D. (2013). Industrial interactions and academic patenting: Evidence from German scientists. *Economics of Innovation and New Technology*, 22(6), 551–65.
Goel, R.K., and Göktepe-Hultén, D. (2018). Academic leadership and commercial activities at research institutes: German evidence. *Managerial and Decision Economics*, 39(5), 601–9.
Goel, R.K., and Göktepe-Hultén, D. (2019). Risk attitudes, patenting and invention disclosures by academic researchers. *Journal of Technology Transfer*, 44(1), 155–66.
Goel, R.K., and Göktepe-Hultén, D. (2021). Innovation by foreign researchers: Relative influences of internal versus external human capital. *Journal of Technology Transfer*, 46(1), 258–76.
Goel, R.K., Göktepe-Hultén, D. and Nelson, M.A. (2022). Drivers of international research spending. *Annals of Science and Technology Policy*, forthcoming.
Goel, R.K., Göktepe-Hultén, D., and Ram, R. (2015). Academics' entrepreneurship propensities and gender differences. *Journal of Technology Transfer*, 40(1), 161–77.
Goel, R.K., and Grimpe, C. (2012). Are all academic entrepreneurs created alike? Evidence from Germany. *Economics of Innovation and New Technology*, 21(3), 247–66.
Göktepe-Hultén, D. (2008). Academic inventors and research groups: Entrepreneurial cultures at universities. *Science and Public Policy*, 35(9), 657–67.
Göktepe-Hultén, D., and Mahagaonkar, P. (2010). Inventing and patenting activities of scientists: In the expectation of money or reputation? *Journal of Technology Transfer*, 35(4), 401–23.
Grimaldi, R., Kenney, M., Siegel, D.S., and Wright, M. (2011). 30 years after Bayh–Dole: Reassessing academic entrepreneurship. *Research Policy*, 40(8), 1045–57.
Jacob, M., Lundqvist, M., and Hellsmark, H. (2003). Entrepreneurial transformations in the Swedish university system: The case of Chalmers University of Technology. *Research Policy*, 32(9), 1555–68.
Jain, S., George, G., and Maltarich, M. (2009). Academics or entrepreneurs? Investigating role identity modification of university scientists involved in commercialization activity. *Research Policy*, 38(6), 922–35.
Kafouros, M., Wang, C., Piperopoulos, P., and Zhang, M. (2015). Academic collaborations and firm innovation performance in China: The role of region-specific institutions. *Research Policy*, 44(3), 803–17.
Kamien, M.I., and Tauman, Y. (1986). Fees versus royalties and the private value of a patent. *The Quarterly Journal of Economics*, 101(3), 471–91,
Karhunen, P., Olimpieva, I., and Hytti, U. (2017). Identity work of science-based entrepreneurs in Finland and in Russia. *Entrepreneurship & Regional Development*, 29(5–6), 544–66.

Krabel, S., and Mueller, P. (2009). What drives scientists to start their own company? An empirical investigation of Max Planck Society scientists. *Research Policy*, 38(6), 947–56.
Krabel, S., Siegel, D.S., and Slavtchev, V. (2012). The internationalization of science and its influence on academic entrepreneurship. *Journal of Technology Transfer*, 37(2), 192–212.
Lam, A. (2010). From "ivory tower traditionalists" to "entrepreneurial scientists"? Academic scientists in fuzzy university-industry boundaries. *Social Studies of Science*, 40(2), 307–40.
Laredo, P. (2007). Revisiting the third mission of universities: toward a renewed categorization of university activities. *Higher Education Policy*, 20(4), 441–56.
Latour, B. (1987). *Science in Action: How to Follow Scientists and Engineers through Society*. Cambridge, MA: Harvard University Press.
Levin, S.G., and Stephan, P.E. (1991). Research productivity over the life cycle: Evidence for academic scientists. *American Economic Review*, 81(1), 114–32.
Levin, S.G., and Stephan, P.E. (1999). Are the foreign born a source of strength for U.S. science? *Science*, 285(5431), 1213–14.
Link, A.N. (ed.). (2017). *Gender and Entrepreneurial Activity*. Cheltenham, UK and Northampton, MA, USA: Edward Elgar Publishing.
Louis, K.S., Blumenthal, D., Gluck, M.E., and Stoto, M.A. (1989). Entrepreneurs in academe: An exploration of behaviours among life scientists. *Administrative Science Quarterly*, 34(1), 110–31.
Lundvall, B-Å. (ed.) (1992). *National Systems of Innovation: Towards a Theory of Innovation and Interactive Learning*. London: Pinter
Lundvall, B-Å. (ed.) (2010). *National Systems of Innovation. Toward a Theory of Innovation and Interactive Learning*. London: Anthem Press.
Max Planck Innovation GmbH. (2020). Technology transfer since 1970. Retrieved from https://www.max-planck-innovation.com/max-planck-innovation.html. Accessed 30 July 2020.
Max Planck Research. (2020). Max Planck Innovation, the Special Issue, The Science Magazine of the Max Planck Society. Retrieved from https://www.max-planck-innovation.com/files/Downloads/MPR_2020_Sp_EN.pdf. Accessed 30 July 2020.
Merton, R.K. (1957). Priorities in scientific discovery: A chapter in the sociology of science. *American Sociological Review*, 22(6), 635–59.
Meyer, M. (2005). Independent inventors and public support measures: Insights from 33 case studies in Finland. *World Patent Information*, 27(2), 113–23.
Morris, L.V. (2016). Management and leadership in colleges and universities. *Innovative Higher Education*, 41(1), 1–3
Mowery, D.C., Nelson, R.R., Sampat, B.N., and Ziedonis, A.A. (eds). (2004). *Ivory Tower and Industrial innovation: University-Industry Technology Transfer before and after the Bayh-Dole Act*. Stanford, CA: Stanford Business Books.
Roberts, E.B. (1991). *Entrepreneurs in High Technology – Lessons from MIT and Beyond*. New York: Oxford University Press.
Rothermael, F.T., Agung, S.D., and Jian, L. (2007). University entrepreneurship: A taxonomy of the literature. *Industrial and Corporate Change*, 16(4), 691–792.
Shapin, S. (1988). Following scientists around. *Social Studies of Science*, 18(3), 533–50.
Siegel, D.S., Waldman, D., Atwater, L., and Link, A.N. (2004). Toward a model of the effective transfer of scientific knowledge from academicians to practitioners: Qualitative evidence from the commercialization of university technologies, *Journal of Engineering and Technology Management*, 21(1/2), 115–42.

Stenberg, L., Gustafsson, E., and Marklund, G. (1996). Use of human resource data for analysis of the structure and dynamics of the Swedish innovation system. *Research Evaluation*, 6(2), 121–32.

Stephan, P.E., and Levin, S.G. (2005). Leaving careers in IT: Gender differences in retention. *Journal of Technology Transfer*, 30(4), 383–96.

Teichler, U. (2015). Academic mobility and migration: What we know and what we do not know. *European Review*, 23(S1), S6–S37.

Zucker, L., Darby, M.R., and Brewer, M.B. (1998). Intellectual human capital and the birth of U.S. biotechnology enterprises. *American Economic Review*, 88(1), 290–336.

6. The role of work-family initiatives in fostering technology transfer: research opportunities on family and non-family SMEs

Katerina Vasilevska, Mara Brumana and Tommaso Minola

6.1 INTRODUCTION

Innovation, in its broader conceptualization, is one of the most important dimensions for companies to achieve a competitive advantage in front of their competitors. In addition, the creation of new knowledge and efficient technology transfer is considered as a crucial element for economic growth (Bengoa et al., 2021). A company's ability to innovate and reinvent itself is crucial for its survival and for sustaining a competitive advantage (Lengnick-Hall, 1992; Keupp et al., 2012). There are many factors that influence a company's innovation and technology transfer goals. Among them, recent research has emphasized the role of the workforce and employees (Pfeffer, 1995; Laursen and Foss, 2003; Shipton et al., 2005; Jiménez-Jiménez and Sanz-Valle, 2008; Chen et al., 2016; Seeck and Diehl, 2017; Wei et al., 2020).

Management, psychology and economics literature have been dedicating growing attention to the topic of "work-family initiatives" (Lobel et al., 1999; Lambert, 2000; Eaton, 2003; Wang and Walumbwa, 2007; Maniam et al., 2016; Perrigino et al., 2018). The work-family initiatives (WFIs thereafter) are "deliberate organizational changes in policies, practices, or the target culture that intend to reduce work–family conflict and/or support employees' lives outside of work" (Kelly et al., 2008, p. 310). Organizations of all sizes have been adopting WFIs for several decades (Kossek et al., 2010). The increasing trend in the adoption of such initiatives has been mainly explained by demographic and social changes. For instance, the needs and expectations arising from the family life, such as time and space constraints or care for dependents of employees are nowadays a real issue with many families struggling to balance their family and professional life (European Survey of Quality of Life, 2016;[1] Managing work-life balance during the Covid-19 crisis, 2020[2]). Constraints arising from the conflict between work and family can produce negative effects on companies such as lack of motivation, decreased levels of job satisfaction, lack of commitment or even lack of creative and devoted work. To avoid such drawbacks, companies are responding with different solutions. The WFIs are, indeed, a voluntary organizational response to address the above-stated issues.

These initiatives have been associated with a wide range of effects on employees. By offering WFIs, companies show care toward their employees and their non-work concerns and in that way are motivating them to participate more actively and more creatively in the work activities. These initiatives also influence organizations as such. For example, they can lead to improved share prices, improved overall performance and even improved innovation (Arthur, 2003; Chen et al., 2016). In their literature review from 2017, Seeck and Diehl point out that practices that foster employee commitment, loyalty, learning and motivation may have an impact on innovation. However, even though WFIs have been researched broadly and with growing emphasis by academics (Goodstein, 1994; Osterman, 1995; Milliken et al., 1998; Arthur, 2003; Bloom et al., 2011; Bourdeau et al., 2019), their effect on innovation seems rather overlooked. That is surprising, given Seeck and Diehl's (2017) arguments and knowing how important it is nowadays for companies to be competitive and particularly to accumulate, process and transfer knowledge by means of their workforce (Jiménez-Jiménez and Sanz-Valle, 2008; Bengoa et al., 2021).

In particular, the literature on the topic reveals at least three gaps. First, it reveals that until now the focus has mainly been on individual-level studies (i.e., studying variables such as employee commitment, worker's productivity or employee's job satisfaction), while fewer studies focus on the firm-level of analysis (i.e., studying the effects of WFIs on variables such as company's performance, turnover rate or organization's attractiveness). In addition, there is only scarce information on how the individual-level mechanisms contribute to firm-level outcomes. Second, until now, the focus has mainly been on large companies, while the context of the small and medium-sized enterprises (SMEs) is slightly left behind. Yet, addressing the above-stated issues in the context of SMEs is especially relevant. The importance of innovation and technology transfer as a driver of performance and competitive advantage in this context is indisputable. Previous studies show that an organization's capacity to innovate does not only depend on technology-driven antecedents but also on its employees' capabilities and motivation (Jiménez-Jiménez and Sanz-Valle, 2008). While large organizations exploit employees' capabilities and motivation through complex knowledge management systems, SMEs need to compensate the absence of such knowledge systems with a focus on the individual employees (Camuffo and Comacchio, 2005). WFIs represent a mechanism that may support SMEs to focus on their employees and thus to improve their innovation capacity. Therefore, exploring the WFIs and their impact on innovation certainly adds value to the debate on SMEs' growth and survival. Third, most SMEs are family-controlled (European Commission, 2009). Even though the literature has identified that managing the tension between work and family systems is quite intense in the family business context (Tagiuri and Davis, 1996; Beehr et al., 1997; Foley and Powell, 1997; Chua et al., 1999; Caputo et al., 2018; Neckebrouck et al., 2018; Michael-Tsabari et al., 2020), it remains silent on how those initiatives are exploited in family SMEs. Instead, elaborating on family firm distinctiveness and the difference with non-family firms represents an important avenue to advance scholarly knowledge in management disciplines (Sharma, 2004; Zahra et al., 2004).

In line with the above reasoning, the objective of this chapter is twofold. First, it aims to shed light on the role of WFIs in the context of SMEs and on their effect on innovation as explored by the literature until now. Second, it aims to offer a research agenda on the effects of WFIs on innovation and technology transfer in SMEs, and particularly in family SMEs. The chapter is organized as follows. First, it provides an overview of the WFIs as a driver for innovation and technology transfer. Second, it offers a historical, terminological and literature review of the WFIs, focusing on the relevance of these initiatives and the various impacts they have on individuals and companies. Third, it discusses the contexts of SMEs and of family SMEs as particularly relevant and offers some guidance and questions for future research.

6.2 WFIS AS A DRIVER FOR INNOVATION AND TECHNOLOGY TRANSFER

Becoming competitive nowadays means going through a complex path, paved with many unknowns. One of the broadly accepted dimensions for being competitive is developing an ability to innovate, to adapt to changes in the business environment (Lengnick-Hall, 1992), as well as to create and apply new knowledge (Bengoa et al., 2021). While technology-driven antecedents do not seem to provide all answers on how companies achieve or improve innovation and technology transfer, previous studies show that an organization's capacity to innovate also largely resides in its employees' capabilities and motivation (Jiménez-Jiménez and Sanz-Valle, 2008). Achieving competitive advantage through people fundamentally changes the perception on the workforce as such. It means achieving success by investing in people, rather than replacing them or limiting their activities. In that way, "the workforce itself becomes a source of strategic advantage, rather than a cost to be minimized or avoided" (Pfeffer, 2005, p. 95). Indeed, firms that took such an approach seem to have managed to perform better than their competitors (Pfeffer, 1995). Therefore, actions that improve an employee's connection to the company and that are attracting their attention toward the company's goals and objectives start to become more important and to gain more attention. According to Pfeffer (1981), actions that symbolize organizational concern or special treatment can provide intangible benefits to organizations. By showing care toward their employees and their non-work concerns, companies are motivating them to participate more actively and more creatively in the work activities. This is indeed supported by various research findings, which show that the employee treatment is positively related to firm's performance (Ichniowski et al., 1994; MacDuffie, 1995; Arthur, 2003; Arthur and Cook, 2004; Kelly et al., 2008; Bloom et al., 2011). In their literature review of 2017, Seeck and Diehl point out that practices that foster employee commitment, loyalty, learning, and intrinsic motivation are not only leading to general benefits related to a company's performance but may also be conducive to innovation and technology transfer.

Despite the academic debate on the relationship between human resource (HR) practices and innovation (Chowhan, 2016; Shipton et al., 2017; Seeck and Diehl, 2017), this research body seems to have neglected an important subgroup of practices that is particularly on the rise in the last years – that of the WFIs. Even though WFIs have been broadly researched by academics (Goodstein, 1994; Osterman, 1995; Milliken et al., 1998; Arthur, 2003; Bloom et al., 2011; Bourdeau et al., 2019), their effect on innovation is still unknown. This is surprising if one accepts as plausible the idea that an organization's capacity to innovate resides in its employees' capabilities and motivation (Jiménez-Jiménez and Sanz-Valle, 2008), while those are strongly related to the work-life balance of the employees (Bui et al., 2016; Aruldoss et al., 2020).

Being socially responsible toward employees is a strategic option that merits more attention because efforts that companies make in this direction seem to provide them with a sustainable development, better management and, thereby, competitive success. The interest in the development of socially responsible initiatives has increased significantly in the twenty-first century. Companies have been increasingly adopting WFIs to meet the needs of their employees (Clifton and Shepard, 2004). Following this trend, WFIs are attracting the attention of national-level policies as well. It is over a decade since the European Union promoted work-life balance by supporting parental leave, childcare subsidies, flexible working hours and in such way encourages changes in the environment, structure, and organization of work (Riedmann et al., 2006). Also, a growing public pressure has led leading companies to take actions that will lead to improvement of the work-life balance of their employees to such an extent that many companies now highlight such concerns as a priority social issue (Riedmann, 2006). However, to better understand the interest that companies, national policies and academic circles have today, it is important to take a look backwards and understand in more detail exactly how the evolution of WFIs happened, from a historical, terminological and literature point of view.

6.3 A REVIEW OF WFIS

6.3.1 Historical Review of WFIs

Organizational involvement in work-family issues has evolved as a function of significant social, economic and political changes (Goodstein, 1994). The evolution of the WFIs went through a few historically important stages, and among other factors, it is also tightly related to the notion of national state welfare. Until the 1970s, the welfare model remained concentrated in the state, both in the US and in Europe. However, after the US faced a certain decline in productivity, in 1975, Congress adopted the National Productivity and Quality of Working Life (QWL) Act. However, as a response to the general decline in productivity, in 1975 Congress adopted the National Productivity and Quality of Working Life Act. Indeed, to foster companies' productivity, novel working models had to be introduced. A model of such nature is the enhancement of the work environment in which the employee functions gain importance and attention. The changes introduced by the US government led to a situation where already in the 1980s, the WFIs represented a common feature in many US organizations (Fulmer and Coleman, 1984). Driven by the need to enrich the jobs of the workers, many organizations have developed programs that aim to improve to quality of work life. More than a decade later, after the year 2000, as the employment relationship started to take a different turn, and the standard model of a 40-hour working week started to change in favor of more complex and more flexible working conditions, the welfare model dominated by the state started to fade, while the state started to transfer its responsibility to the private sector. Today, the rise in the adoption of WFIs is almost undisputable.

As the trend of the WFIs was increasing, the types of WFIs adopted were changing from year to year, depending on the current societal needs. For example, as reported by Michelle Arthur in 2003, throughout the 1970s and 1980s, firms were adopting more lower-cost alternatives to on-site childcare such as emergency and sick-childcare programs, shared childcare centers and childcare referral services, in the late 1980s, elder care services and family-counseling services emerged as concerns, while from the 1990s onwards, firms started addressing flexible work arrangements, to meet perhaps the most serious conflict created by work-family tensions. Programs such as flextime, job sharing, compressed work weeks, and most recently, telecommuting were adopted.

However, the continuously increasing trend of adoption of WFIs is not a simple reaction of the organizations to follow up on new regulations imposed by local governments, but also an effect of recent social and business trends that affect the work-life balance. In their study from 1999, Lobel et al. claim that in the future, the environmental trends having a major impact on the interference between work and family will revolve around two main business-related trends: globalization and increasing needs for organizational flexibility; and two external trends: changing family structure and technological change.

6.3.2 Terminological Review of WFIs

The WFIs have been present in practice, but also in the literature, since the 1980s and even before. However, a literature review on the topic of WFIs has revealed a significant lack of terminological clarity. What we call here "work-family initiatives" exists under different terms in different fields, different countries or geographical contexts. For example, in Germany they are related with the term "*betriebliche Sozialpolitik*" (occupational social policy); in the UK a frequently used term is "*occupational welfare*," whereas in Italy the most broadly accepted term is "*welfare aziendale*" (company welfare). The common denominator for all is that they are well distinguished from the welfare state and the social welfare given that their purpose is to support exclusively employees, unlike that of the social rights for citizens in the welfare state or that of the fiscal welfare granted by states to companies (Marshall, 1950 [1963, 1965]; Titmuss, 1976).

Table 6.1 *Terms and definitions*

Term	Definition	Reference
Work-life programs	Initiatives adopted by organizations to help employees manage the interface between their paid work and other important life activities, including family.	Lobel et al., 1999
Work-family bundle	A group of complementary, highly related and, in some cases, overlapping human resource policies that may help employees manage non-work roles.	Perry-Smith and Blum, 2000
Family-friendly work practices	Work arrangements that facilitate the reconciliation of work and family life.	Gray and Tudball, 2003
Family-friendly policies	Business strategies intended to respond to the concerns of employees with family responsibilities.	Albrecht, 2003
Work-life initiatives	Control mechanisms that help organizations attain their goals by facilitating employees' fulfillment of their professional responsibilities.	Bourdeau et al., 2019
Family-friendly benefits	For the organization, a means for maintaining a competitive advantage, raising morale, and attracting and retaining a dedicated workforce within today's turbulent work environment. For the employee, these benefits are designed to alleviate the difficulty inherent in coordinating and managing multiple life roles.	Allen et al., 2001
Family-friendly working policies	Policies intended to ease tensions that emerge from work–family conflict (i.e., "a form of interrole conflict in which role pressures from the work and family domains are mutually incompatible in some respect).	Ko et al., 2013

To provide a more accurate overview of the different definitions used in the literature, but also to illustrate the terminological diversity through the years, Table 6.1 reports some of the most relevant definitions used in the literature to date.

As seen from the non-exhaustive list of definitions reported above, the WFIs not only differ in the definition itself, but also in the label, which includes variations between the words work-family, work-life, family-friendly from one side and policy, practice, program, benefit, initiative, bundle from the other side. All those words have different meanings: *policy* – a set of ideas or a plan of what to do in particular situations that has been agreed officially by a group of people, a business organization, a government or a political party. Policies are covenants we collectively choose to live by, as articulated in legislation and regulation; *practice* – something that is usually or regularly done, often as a habit, tradition or custom; *program* – a group of activities or things to be achieved. Program is a short-term intervention that creates temporary improvements in the wake of challenges; *benefit* – an advantage, such as medical insurance, life insurance and sick pay, that employees receive from their employer in addition to money; *initiative* – a new plan or process to achieve something or solve a problem. Furthermore, in the literature, we observe a certain prevalence for the use of *work-family* label in favor of *work-life* or *family-friendly*, and on the other side the prevalence of the use of the term *policy* in favor of the rest of the alternatives.

For this chapter and to avoid any further misunderstanding or overlaps in the meaning of the label, we adopt the term and the definition applied by Kelly et al. (2008, p. 310), according to which:

> Work-family initiatives are deliberate organizational changes – in policies, practices, or the target culture – to reduce work–family conflict and/or support employees' lives outside of work.

Similar to the heterogeneity of the term itself, the list of WFIs remains broad and non-definitive, based on the preference and the resources of each company. Given their variety, in the literature, there are few efforts to classify the WFIs. For example, in their study from 2001, Allen et al. divide the WFIs into two main categories: flexible work arrangements, in which they include initiatives such as flextime, compressed work week, telecommuting, and part-time work, and dependent care supports, in which they include initiatives such as on-site childcare center, subsidized local childcare, childcare information/referral services, paid maternity leave, paid paternity leave, and elder care. In addition, Bourdeau et al. (2019) offer a second classification of WFIs, which is based on the way they operate, that is, whether they operate in an enabling way, giving employees some latitude over when, where and how much they work, or in an enclosing way, promoting longer hours on work premises. In that respect, the enabling work-life policies give employees latitude over when, where and how much they work, which enables them to spend time and energy outside of work to take care of themselves and their family members whereas the enclosing work-life policies maximize employees' availability for work, as well as the time they spend on work premises. However, the classification of WFIs remains an open topic that is to be adopted or decided depending on the purpose of the research.

Apart from the classifications present in the literature, the WFIs may also differ with respect to the purpose for which they are intended to serve. To illustrate this, in Table 6.2, we provide a non-exhaustive list of WFIs that appear as most frequently used in the literature, provisionally clustered in two categories: generic and family-driven initiatives. Generic initiatives are those that serve a general purpose of employees, such as wellbeing or facilitating the work modes in a general way, while "family-driven" initiatives are those initiatives that aim to alleviate the overlaps between work and family life of the employees.

Table 6.2 List of WFIs

Generic WFIs	Initiatives that enhance workplace flexibility (e.g., flexible hours, work-at-home programs, option to sometimes work remotely or at home)
	Entitlement to switch from full-time to part-time work
	Entitlement to job-sharing scheme
	Career breaks
	Compressed work weeks (total work hours in 4 days, not 5)
	Variable starting and quitting times
	Fitness center
	Psychological counseling
	Resource referral programs
	Shopping vouchers
	Medical check-up visits
Family-driven WFIs	Providing fully or partially subsidized on-site childcare
	Employee childcare expenses, e.g., by contracting with external organizations or making benefit contributions
	Providing childcare information and referral services that link employees to childcare providers
	Workplace nursery
	Extra-statutory maternity leave
	Extra-statutory paternity leave
	Flexibility in case of unexpected childcare emergency
	Assistance to employees with aging parents
	Tuition reimbursement
	Family insurance/savings plan

6.3.3 Literature Review of WFIs

To closely understand the debate around the WFIs, we performed a literature review on the topic. The literature review included a total number of 369 journal articles available on the Scopus database, with a detailed analysis of 70 most relevant articles. The following elements have been considered when analyzing the articles: geographical region, publication year, sample size and type (individual or firm-level), terminology used, research question/hypothesis, theoretical perspective/conceptual contributions, relationship explored (for empirical articles), results obtained as well as the measures/statistical methods used for the purpose of the research. More details on the literature review strategy are provided in the Appendix. The final sample consisted of 49 empirical articles, 16 conceptual and five literature reviews. Among the concerned geographical regions, the following countries have been detected: Kenya, Australia, China, Thailand, Hong Kong, India, New Zealand, Finland, Italy, UK, France, Germany and US. As for the analysis levels, 23 articles integrated an individual-level of analysis, 12 a firm-level of analysis, while in the rest of the articles the analysis level was either mixed or unclear. Overall, the findings of the literature review show that the WFIs have been shown to have both positive and negative effects, both on an individual-level (e.g., affective commitment and turnover intention (Grover and Crooker, 1995); employee loyalty (Roehling et al., 2001), and on an organizational-level (shareholder return (Arthur, 2003) or firm's rate of personnel turnover (Lee and Hong, 2011)).

Regarding the studies that have been exploring the *WFIs as an outcome*, in an article from 1995, Paul Osterman states that firms that value employee commitment and have high commitment systems are more likely to adopt work-family programs; Milliken et al. (1998) show that companies that have human resource executives for whom work and family issues are salient will tend to have higher levels of work-family responsiveness, while Thompson et al. (1999) show that employees who are female, married or have children living with them will be more likely to utilize work-family benefits than employees who are male, unmarried or childless. Konrad and Mangel (2000) state that firms employing a larger percentage of professionals develop more extensive work-life programs and when a firm employs a higher percentage of professionals, the extensiveness of its work-life programs is more strongly positively related to productivity, while Bloom et al. (2011) show that firms with good management practices offer more WFIs.

With respect to the *WFIs as an antecedent to individual-level outcomes*, Grover and Crooker (1995) conclude that family-responsive benefits will be positively related to affective commitment and turnover intention and that people who benefit from such initiatives will have greater affective commitment to the organization. Lambert (2000) shows that the more useful workers find work-life benefits, the more likely they are to engage in citizenship behavior at both the organizational-level and the individual-level, while Roehling et al. (2001) show that the presence of flexible-time policies is positively related to employee loyalty. Wang and Walumbwa (2007) state that employees' perception of availability of childcare-related family-friendly programs will be positively related to employee organizational commitment. Lee and Hong (2011) indicate that an employee's average level of satisfaction with a childcare subsidy policy is negatively associated with the turnover rate of the firm. Ko et al. (2013) find out that satisfaction with flexible work scheduling and dependent care programs is positively related to the job satisfaction of the employees and perceived organizational performance, while a year later, in 2014, James finds out that 61 percent of managers indicated an improved worker productivity because of the adoption of WFIs, while 54 percent indicated an improved corporate environment for learning and creativity. The effect of WFIs on *organizational level* has also been largely explored. For example, in their article from 1996, Delaney and Huselid find that the types of HR practices that are affecting employee skills, motivation and the structure of work are positively related to organizational performance. Perry-Smith and Blum in 2000 conclude that firms with more comprehensive bundles of work-family policies will have higher organizational performance than firms with less comprehensive bundles, while Arthur (2003) finds that firm announcements of WFIs positively affect the shareholder return.

The WFIs have been explored through various *theoretical perspectives*, depending on the focus of the study and level of analysis. For example, the social exchange theory has been used to support the view that employees feel obligated to reciprocate by adopting a more positive attitude toward the workplace when they are exposed to a favorable work environment and desirable benefits and has been mainly associated with an individual-level analysis (Wang and Walumbwa, 2007; Lee and Hong, 2011; Ko et al., 2013). Moshavi and Koch (2005) referred to the resource dependence theory to point out that work-family demands are particularly strong in organizations that employ relatively large numbers of women. Institutional theory has been used to study how and why employers decide to provide work-family benefits to their employees (Goodstein, 1994; Ingram and Simons, 1995) or to point out the isomorphic reasons, that is, the practices of "similar firms" (e.g., those in the same industry, of equal size, etc.) to reduce uncertainty and increase their legitimacy and survival capabilities (DiMaggio and Powell, 1983). Apart from these, other theories such as equity theory, border theory, conflict theory, resource-based view or signaling theory have also been used while exploring the topic of WFIs.

6.4 OPPORTUNITIES FOR FUTURE RESEARCH ON WFIS: INNOVATION AND TECHNOLOGY TRANSFER IN FAMILY AND NON-FAMILY SMES

6.4.1 WFIs, Innovation and Technology Transfer

As shown above, the literature on the effects of WFIs is quite broad; however, their effects on innovation remain underexplored. Yet, exploring this relationship is of great importance, given that in firms' struggle to innovate, all factors but especially those that are related to their employees count and should be further explored. Until now, there are some indicative findings, even though they could be found in a more generic debate related to the effect of various generic human resources management (HRM) practices or employee treatment on innovation. Even though HRM practices and "employee treatment" are both more generic constructs that are also terminologically different from WFIs, the findings in this field are relevant because they indicate that providing proper conditions to the personnel indeed leads the company toward various positive outcomes. For example, Laursen and Foss (2003, p. 258) conclude that the implementation of HR practices has a positive effect on the probability of a company to innovate. Shipton et al. (2005) show that the HRM systems have a positive effect on product innovation, while Jiménez-Jiménez and Sanz-Valle (2008) find that an HRM system influences every type of innovation. Further to this discussion, in 2016, Chen et al. find that firms with better employee treatment schemes produce patents that improve market valuation and facilitate better future operating performance, while Wei et al. (2020) find that companies with more intense employee welfare perform better in terms of innovation. Certainly, this debate already provides fruitful indications that treatment of the employees in a broad sense also contributes to innovation.

However, future research questions particularly focused on the WFIs could open possibilities for fruitful contributions and research perspectives on the topic. For example, it would be interesting to investigate what is the role of the WFIs on innovation and technology transfer, how particular types of WFIs affect innovation or what are the individual-level mechanisms that "mediate" the relationship between WFIs on the one hand and innovation and technology transfer on the other. First, as lenses to understand the WFI-innovation/technology transfer relationship, we suggest elaborating on micro-level dimensions. While the *innovation* literature has emphasized the role of psychological ownership (i.e., the state in which individuals feel as though the target of ownership or a piece of it is "theirs" (Pierce et al., 2001), and job satisfaction (a feeling that appears as a result of the perception that the job enables the material and psychological needs (Aziri, 2011)), the literature on *knowledge and technology transfer* has focused on creativity (the production of novel and useful ideas) by an individual or small groups of individuals working together (Amabile, 1988) and knowledge transfer/sharing (the process through which one unit (e.g., group, department or division) is affected by the experience of another (Argote and Ingram, 2000)). Second, we highlight two important contexts that, albeit somehow neglected by the literature, might be fruitful and promising because in these contexts the WFI-innovation and technology transfer relationship might take a different turn or lead to interesting, unexpected outcomes. As anticipated, the first one is the context of the SMEs, and the second one is the specific case of the family SMEs.

6.4.2 Focus on SMEs

In the literature, it is broadly accepted that large organizations are more likely to have formal HR mechanisms for dealing with their HR challenges (Konrad and Mangel, 2000), in comparison to smaller size companies. Therefore, the literature on WFIs within SMEs has been historically underdeveloped, as studies have tended to focus upon larger organizations. This is possibly the case because the larger companies also have access to more resources and to accompanying specialization in management functions, including the design and implementation of HR strategies (Deshpande and Golhar, 1994; Chandler and McEvoy, 2000; Cassell et al., 2002).

However, the significant numbers of SMEs worldwide have a great strategic value for all countries and regions since these companies play a major role in the overall economic growth (González-Loureiro and Pita-Castelo, 2012; Klewitz and Hansen, 2014). Therefore, they deserve the needed support and guidance to continue operating at an acceptable level of performance (Hodorogel, 2009; Soininen et al., 2012). The few studies that explore WFIs in SMEs show intriguing results. For example, in their study from 2019, Moneva-Abadía and colleagues found that in times of economic crisis, socially responsible strategies may be an important factor in SMEs' competitiveness. Moreover, their results indicate that WFIs have the potential to assist SMEs in responding to periods of constrained demand.

Like other firms, conditions such as weaker market demand typically cause SMEs to consider pursuing cost reductions. Yet, in contrast to larger firms, SMEs are expected to find a response without the benefit provided by the degree of resources and management specialization traditionally available to larger firms (Deshpande and Golhar, 1994). For example, the use of WFIs is different in SMEs than in larger firms, because of the former having generally leaner and more unified organizational structures, differential patterns of ownership, and fewer management changes as well as a more informal and more reactive view in their business strategies (Matlay, 2002; Tocher and Rutherford, 2009). Furthermore, they are also less likely to have the expertise and knowledge that enable larger companies to introduce WFIs (Dex and Scheibl, 2001). It is a common understanding not just that large companies have easier access to and implement more WFIs, but also that large companies perform better in terms of innovation (Wei et al., 2020). Yet, as the "engine of driving revenue growth" (Patterson and Kerrin, 2014) and "the cornerstone of organizational survival" (Hurley and Hult, 1998), innovation appears to be even more crucial when it comes to SMEs because they need to confront the bigger companies that have a bigger market share, more resources and easier access to them.

Given this quite apparent gap, the literature has for some time suggested that a focus on SMEs is needed (Chandler and McEvoy, 2000; Tocher and Rutherford, 2009). Addressing the WFIs and their relation to innovation becomes particularly relevant in the context of SMEs because the success of the SMEs largely depends on their human capital, but also on their flexibility to adapt to the new demographic and social context. This certainly opens a space for development of many research questions focused on exploring the relationship between innovation and WFIs in SMEs. The literature has explored at least three dimensions that make the context of SMEs particularly intriguing: resources, governance and goals (Brunninge et al., 2007; Sui and Baum, 2014). In order to explore how those dimensions resonate with the above-explained dimensions of innovation (psychological ownership and job satisfaction) and technology transfer (creativity and knowledge transfer/sharing), we suggest a few examples of research questions: Does the adoption of job-sharing initiatives discriminate radical versus incremental innovation-oriented SMEs under resource constraints? Can psychological ownership explain these differences? How do flexible working hours affect the relationship between creativity and technology transfer in SMEs? Do organizational configurations (e.g., horizontal versus vertical organization) affect this relationship? Does the smart working initiative affect the knowledge-transfer-intense SMEs more than other types of SMEs?

6.4.3 Focus on Family SMEs

Most SMEs are family-owned (European Commission, 2009). Therefore, the previously exposed arguments become even more important and distinctive in the context of the family SMEs. According to the definition by Chua et al. (1999, p. 25), a family firm is "a business governed and/or managed with the intention to shape and pursue the vision of the business held by a dominant coalition controlled by members of the same family or a small number of families in a manner that is potentially sustainable across generations of the family or families." As such, family firms differ from non-family firms along the dimensions of ownership, governance and management;

and these dimensions, in turn, affect the degree and the forms of innovation (De Massis et al., 2013; Duran et al., 2016).

While the relationship between family firms and innovation is quite well explored in the literature (De Massis et al., 2013; Duran et al., 2016), that of their relation to the WFIs appears as significantly less explored. The literature has identified that interdependences between work and family are quite intense in the family business context (Beehr et al., 1997; Foley and Powell, 1997; Lansberg, 1999). Previous work-family research (Milliken et al., 1998) indicates that even when firms face similar institutional and resource-dependence contingencies, they may still vary in their degree of work-family responsiveness. One possible explanation for this difference in firm responsiveness, not currently addressed in the extant literature, is type of ownership. There are indications from the family business literature that studying family ownership may be an avenue for understanding different nuances in the adoption of WFIs (Michael-Tsabari et al., 2020). While it is intuitively suggested that family-owned firms may be supportive of employees' work-family conflicts, more recent research seems to suggest that family firms may not be as "family-friendly" as thought (Moshavi and Koch, 2005; Neckebrouck et al., 2018). Several family business researchers have observed, for instance, that family business owners and managers are more used to the overlap between work and family systems than non-family-owned business managers. As a result, they may undertake less formally recognized actions to manage the relationship between the two systems than their non-family business counterparts. Exactly for that reason, the research suggests that family-owned businesses tend to address employee issues with a high degree of informality (Kets de Vries, 1996) or to manage work-family conflict issues on a case-by-case basis rather than to formally implement policies (Lansberg, 1999), for example, through informal meetings rather than formal business plans to set policy (Astrachan and Kolenko, 1994). Hence, these findings suggest that more formalized WFIs are less likely to appear in family-owned businesses than in non-family-owned businesses.

However, the low number of studies on this topic indicates that the topic is insufficiently developed. Therefore, further exploration of the relationship between WFIs and innovation in the context of family SMEs certainly deserves attention. Similar to the context of SMEs, the literature has proven that, on the one hand, the typical dimensions of SMEs introduced in the previous section are also relevant for the family SMEs, but with their own particularities (Kotey, 2005; Basly and Saunier, 2020; Randolph et al., 2019), while on the other hand, another characteristic that makes family SMEs a particular context for studying is their "family dynamics" (Aldrich and Cliff, 2003). To open a debate on how those characteristics of the family SMEs affect the individual-level dimensions introduced above, we suggest a few examples of research questions for future scholars, such as: Do work-family initiatives increase psychological ownership more in firms whose founder is childless than when the founder has children? Does the gender composition of the board of directors play a role in the relationship between work-family initiatives and innovation? Does the network of multigenerational families affect the relationship between job sharing and knowledge transfer in family SMEs?

6.5 CONCLUSION

This chapter provides an overview of the relation between WFIs and innovation/technology transfer, presenting the WFIs as a driver for companies to improve their innovation performance and thus to improve their competitive advantage. We support the idea that drivers for innovation and technology transfer that focus on employees deserve proper attention in the literature and should be further explored. We have shown that WFIs started gaining shape and attention in the 1980s while showing their historical and terminological evolution from then onwards. Furthermore, to provide a clear image on the debate of WFIs, we have shown what the literature has considered so far as antecedents and effects of the WFIs to various individual-level and firm-level outcomes. We have then introduced that the relationship between WFIs and innovation is an important missing gap in the literature while pointing out that the debate on the general HR and innovation has already provided some fruitful indications that nurture and strongly motivate our discussion. Lastly, we have introduced a novel research agenda with respect to the contexts of SMEs and family SMEs and by exploring certain individual-level dimensions of innovation and technology transfer. We believe that such a research agenda shall open a promising path for further exploration of the relationship between innovation/technology transfer and WFIs and shall inspire future scholars to engage even more in this topic.

ACKNOWLEDGEMENTS

We would like to express our deepest gratitude to the Italian insurance company Assimoco for supporting this research project.

NOTES

1. https://www.eurofound.europa.eu/surveys/european-quality-of-life-surveys/european-quality-of-life-survey-2016 (accessed 31 May 2022).
2. https://www.eurofound.europa.eu/publications/report/2020/living-working-and-covid-19 (accessed 31 May 2022).

REFERENCES

Albrecht, G.H. (2003). How friendly are family friendly policies?. *Business Ethics Quarterly*, 13(2), 177–92.

Aldrich, H.E., and Cliff, J.E. (2003). The pervasive effects of family on entrepreneurship: Toward a family embeddedness perspective. *Journal of Business Venturing*, 18(5), 573–96.

Allen, T.D. (2001). Family-supportive work environments: The role of organizational perceptions. *Journal of Vocational Behavior*, 58(3), 414–35.

Amabile, T. M. (1988). From individual creativity to organizational innovation. In Gronhaug, K & Kaufmann, G (Eds.), Innovation: A cross-disciplinary perspective (pp. 139–166). Norwegian University Press.

Argote, L., and Ingram, P. (2000). Knowledge transfer: A basis for competitive advantage in firms. *Organizational Behavior and Human Decision Processes*, 82(1), 150–69.

Arthur, M.M. (2003). Share price reactions to work-family initiatives: An institutional perspective. *Academy of Management Journal*, 46(4), 497–505.

Arthur, M.M., and Cook, A. (2004). Taking stock of work-family initiatives: How announcements of "family-friendly" human resource decisions affect shareholder value. *ILR Review*, 57(4), 599–613.

Aruldoss, A., Kowalski, K.B., and Parayitam, S. (2020). The relationship between quality of work life and work-life-balance mediating role of job stress, job satisfaction and job commitment: Evidence from India. *Journal of Advances in Management Research*, 18(1), 36–62.

Astrachan, J.H., and Kolenko, T.A. (1994). A neglected factor explaining family business success: Human resource practices. *Family Business Review*, 7(3), 251–62.

Aziri, B. (2011). Job satisfaction: A literature review. *Management Research & Practice*, 3(4), 77–86

Basly, S., and Saunier, P.L. (2020). Familiness, socio-emotional goals and the internationalization of French family SMEs. *Journal of International Entrepreneurship*, 18(3), 270–311.

Beehr, T.A., Drexler Jr, J.A., and Faulkner, S. (1997). Working in small family businesses: Empirical comparisons to non-family businesses. *Journal of Organizational Behavior: The International Journal of Industrial, Occupational and Organizational Psychology and Behavior*, 18(3), 297–312.

Bengoa, A., Maseda, A., Iturralde, T., and Aparicio, G. (2021). A bibliometric review of the technology transfer literature. *The Journal of Technology Transfer*, 46(5), 1514–50.

Bloom, N., Kretschmer, T., and Van Reenen, J. (2011). Are family-friendly workplace practices a valuable firm resource? *Strategic Management Journal*, 32(4), 343–67.

Bourdeau, S., Ollier-Malaterre, A., and Houlfort, N. (2019). Not all work-life policies are created equal: Career consequences of using enabling versus enclosing work-life policies. *Academy of Management Review*, 44(1), 172–93.

Brunninge, O., Nordqvist, M., and Wiklund, J. (2007). Corporate governance and strategic change in SMEs: The effects of ownership, board composition and top management teams. *Small Business Economics*, 29(3), 295–308.

Bui, H.T., Liu, G., and Footner, S. (2016). Perceptions of HR practices on job motivation and work-life balance: Mixed drives and outcomes in a labor-intensive sector. *International Journal of Manpower*, 37(6), 1004–23.

Camuffo, A., and Comacchio, A. (2005). Linking intellectual capital and competitive advantage: A cross-firm competence model for north-east Italian SMEs in the manufacturing industry. *Human Resource Development International*, 8(3), 361–77.

Caputo, A., Marzi, G., Pellegrini, M.M., and Rialti, R. (2018). Conflict management in family businesses: a bibliometric analysis and systematic literature review. *International Journal of Conflict Management*, 29(4), 519–42

Cassell, C., Nadin, S., Gray, M., and Clegg, C. (2002). Exploring human resource management practices in small and medium sized enterprises. *Personnel Review*, 31(6), 671–92

Chandler, G.N., and McEvoy, G.M. (2000). Human resource management, TQM, and firm performance in small and medium-size enterprises. *Entrepreneurship Theory and Practice*, 25(1), 43–58.

Chen, C., Chen, Y., Hsu, P.-H., and Podolski, E.J. (2016). Be nice to your innovators: Employee treatment and corporate innovation performance. *Journal of Corporate Finance*, 39, 78–98.

Chowhan, J. (2016). Unpacking the black box: Understanding the relationship between strategy, HRM practices, innovation and organizational performance. *Human Resource Management Journal*, 26(2), 112–33.

Chua, J.H., Chrisman, J.J., and Sharma, P. (1999). Defining the family business by behavior. *Entrepreneurship Theory and Practice*, 23(4), 19–39.

Clifton, T.J., and Shepard, E. (2004). Work and family programs and productivity: Estimates applying a production function model. *International Journal of Manpower*, 25(8), 714–28.

De Massis, A., Frattini, F., and Lichtenthaler, U. (2013). Research on technological innovation in family firms: Present debates and future directions. *Family Business Review*, 26(1), 10–31.

Delaney, J.T., and Huselid, M.A. (1996). The impact of human resource management practices on perceptions of organizational performance. *Academy of Management Journal*, 39(4), 949–69.

Deshpande, S.P., and Golhar, D.Y. (1994). HRM practices in large and small manaufacturing firms: A comparative study. *Journal of Small Business Management*, 32(2), 42–54.

Dex, S., and Scheibl, F. (2001). Flexible and family-friendly working arrangements in UK-based SMEs: Business cases. *British Journal of Industrial Relations*, 39(3), 411–31.

DiMaggio, P.J., and Powell, W.W. (1983). The iron cage revisited: Institutional isomorphism and collective rationality in organizational fields. *American Sociological Review*, 48(2), 147–60.

Duran, P., Kammerlander, N., Van Essen, M., and Zellweger, T. (2016). Doing more with less: Innovation input and output in family firms. *Academy of Management Journal*, 59(4), 1224–64.

Eaton, S. C.(2003). If you can use them: Flexibility policies, organizational commitment, and perceived performance. *Industrial Relations: A Journal of Economy and Society*, 42(2), 145–67.

European Commission, Directorate-General for Enterprise and Industry, European SMEs under pressure: Annual report on EU small and medium-sized enterprises 2009, *Publications Office*, 2010.

Foley, S., and Powell, G N. (1997). Reconceptualizing work-family conflict for business/ marriage partners. *Journal of Small Business Management*, 35, 36–47.

Fulmer, W.E., and Coleman, J.J. (1984). Do quality-of-work-life programs violate Section 8 (a)(2)? *Labor Law Journal*, 35(11), 675.

González-Loureiro, M., and Pita-Castelo, J. (2012). A model for assessing the contribution of innovative SMEs to economic growth: The intangible approach. *Economics Letters*, 116(3), 312–15.

Goodstein, J.D. (1994). Institutional pressures and strategic responsiveness: Employer involvement in work-family issues. *Academy of Management Journal*, 37(2), 350–82.

Gray, M., and Tudball, J. (2003). Family-friendly work practices: Differences within and between workplaces. *Journal of Industrial Relations*, 45(3), 269–91.

Grover, S.L., and Crooker, K.J. (1995). Who appreciates family-responsive human resource policies: The impact of family-friendly policies on the organizational attachment of parents and non-parents. *Personnel Psychology*, 48(2), 271–88.

Hodorogel, R.G. (2009). The economic crisis and its effects on SMEs. *Theoretical & Applied Economics*, 16(5), 79–88.

Hurley, R.F., and Hult, G.T.M. (1998). Innovation, market orientation, and organizational learning: An integration and empirical examination. *Journal of Marketing*, 62(3), 42–54.

Ichniowski, C., Shaw, K., and Prennushi, G. (1994). The impact of human resource management practices on productivity. Columbia Business School.

Ingram, P., and Simons, T. (1995). Institutional and resource dependence determinants of responsiveness to work-family issues. *Academy of Management Journal*, 38(5), 1466–82.

James, A. (2014). Work–life "balance", recession and the gendered limits to learning and innovation (or, why it pays employers to care). *Gender, Work & Organization*, 21(3), 273–94.

Jiménez-Jiménez, D., and Sanz-Valle, R. (2008). Could HRM support organizational innovation? *The International Journal of Human Resource Management*, 19(7), 1208–21.

Kelly, E.L., Kossek, E.E., Hammer et al. (2008). Getting there from here: Research on the effects of work–family initiatives on work–family conflict and business outcomes. *Academy of Management Annals*, 2(1), 305–49.

Ketz de Vries, Manfred F.R. (1996). *Family Business: Human Dilemmas in the Family Firm*. International Thomson Business Press, UK.

Keupp, M.M., Palmié, M., and Gassmann, O. (2012). The strategic management of innovation: A systematic review and paths for future research. *International Journal of Management Reviews*, 14(4), 367–90.

Klewitz, J., and Hansen, E.G. (2014). Sustainability-oriented innovation of SMEs: A systematic review. *Journal of Cleaner Production*, 65, 57–75.

Ko, J., Hur, S., and Smith-Walter, A. (2013). Family-friendly work practices and job satisfaction and organizational performance: Moderating effects of managerial support and performance-oriented management. *Public Personnel Management*, 42(4), 545–65.

Konrad, A.M., and Mangel, R. (2000). The impact of work-life programs on firm productivity. *Strategic Management Journal*, 21(12), 1225–37.

Kossek, E., Lewis, S., and Hammer, L.B. (2010). Work—life initiatives and organizational change: Overcoming mixed messages to move from the margin to the mainstream. *Human Relations*, 63(1), 3–19.

Kotey, B. (2005). Goals, management practices, and performance of family SMEs. *International Journal of Entrepreneurial Behavior & Research*, 11(1), 3–24.

Lambert, S.J. (2000). Added benefits: The link between work-life benefits and organizational citizenship behavior. *Academy of Management Journal*, 43(5), 801–15.

Lansberg, I. (1999). *Succeeding Generations: Realizing the Dream of Families in Business*. Cambridge, MA: Harvard Business Review Press.

Laursen, K., and Foss, N.J. (2003). New human resource management practices, complementarities and the impact on innovation performance. *Cambridge Journal of Economics*, 27(2), 243–63.

Lee, S.Y., and Hong, J.H. (2011). Does family-friendly policy matter? Testing its impact on turnover and performance. *Public Administration Review*, 71(6), 870–9.

Lengnick-Hall, C.A. (1992). Innovation and competitive advantage: What we know and what we need to learn. *Journal of Management*, 18(2), 399–429.

Lobel, S.A., Googins, B.K., and Bankert, E. (1999). The future of work and family: Critical trends for policy, practice, and research. *Human Resource Management* (published in cooperation with the School of Business Administration, The University of Michigan and in alliance with the Society of Human Resources Management), 38(3), 243–54.

MacDuffie, J.P. (1995). Human resource bundles and manufacturing performance: Organizational logic and flexible production systems in the world auto industry. *ILR Review*, 48(2), 197–221.

Maniam, B., Leavell, H., and Pustejovsky, S. (2016). Family friendly work practices in the US revisited. *International Journal of Economics and Financial Issues*, 6(6S), 55–9.

Marshall, T.H. (1950). *Social Class and Citizenship*. Reprinted in *Sociology at the Crossroads and other Essays*. London: Heinemann, 1963 and in *Social Policy in the Twentieth Century*. London: Hutchinson, 1965.
Matlay, H. (2002). Training and HRD strategies in family and non-family owned small businesses: A comparative approach. *Education + Training*, 44(8/9), 357–69.
Michael-Tsabari, N., Houshmand, M., Strike, V.M., and Efrat Treister, D. (2020). Uncovering implicit assumptions: Reviewing the work–family interface in family business and offering opportunities for future research. *Family Business Review*, 33(1), 64–89.
Milliken, F.J., Martins, L.L., and Morgan, H. (1998). Explaining organizational responsiveness to work-family issues: The role of human resource executives as issue interpreters. *Academy of Management Journal*, 41(5), 580–92.
Moneva-Abadía, J.M., Gallardo-Vázquez, D., and Sánchez-Hernández, M.I. (2019). Corporate social responsibility as a strategic opportunity for small firms during economic crises. *Journal of Small Business Management*, 57, 172–99.
Moshavi, D., and Koch, M.J. (2005). The adoption of family-friendly practices in family-owned firms: Paragon or paradox? *Community, Work and Family*, 8(3), 237–49.
Neckebrouck, J., Schulze, W., and Zellweger, T. (2018). Are family firms good employers? *Academy of Management Journal*, 61(2), 553–85.
Osterman, P. (1995). Work/family programs and the employment relationship. *Administrative Science Quarterly*, 40(4), 681–700.
Patterson, F., and Kerrin, M. (2014). Characteristics and behaviours associated with innovative people in small-and medium-sized enterprises. In *Handbook of Research on Small Business and Entrepreneurship*. Cheltenham, UK and Northampton, MA, USA: Edward Elgar Publishing, 187–206.
Perrigino, M.B., Dunford, B.B., and Wilson, K.S. (2018). Work–family backlash: The "dark side" of work–life balance (WLB) policies. *Academy of Management Annals*, 12(2), 600–30.
Perry-Smith, J.E., and Blum, T.C. (2000). Work-family human resource bundles and perceived organizational performance. *Academy of Management Journal*, 43(6), 1107–17.
Pfeffer, J. (1981). Management as symbolic action: The creation and maintenance of organizational paradigms. *Research in Organizational Behavior*, 3, 1–52.
Pfeffer, J. (2005). Producing sustainable competitive advantage through the effective management of people. *Academy of Management Perspectives*, 19(4), 95–106.
Pierce, J.L., Kostova, T., and Dirks, K.T. (2001). Toward a theory of psychological ownership in organizations. *Academy of Management Review*, 26(2), 298–310.
Randolph, R.V., Alexander, B.N., Debicki, B.J., and Zajkowski, R. (2019). Untangling non-economic objectives in family & non-family SMEs: A goal systems approach. *Journal of Business Research*, 98, 317–27.
Riedmann, A. (2006). Working time and work-life balance in European companies: Establishment Survey on Working Time 2004–2005. *Office for Official Publications of the European Communities*.
Riedmann, A., Bielenski, H., Szczurowska, T., and Wagner, A. (2006). Working time and work-life balance in European companies, Eurofound. Retrieved from https://policycommons.net/artifacts/1843392/working-time-and-worklife-balance-in-european-companies/2587068/ (accessed 31 May 2022).
Roehling, P.V., Roehling, M.V., and Moen, P. (2001). The relationship between work-life policies and practices and employee loyalty: A life course perspective. *Journal of Family and Economic Issues*, 22(2), 141–70.
Seeck, H., and Diehl, M.R. (2017). A literature review on HRM and innovation – taking stock and future directions. *The International Journal of Human Resource Management*, 28(6), 913–44.

Sharma, P. (2004). An overview of the field of family business studies: Current status and directions for the future. *Family Business Review*, 17(1), 1–36.

Shipton, H., Fay, D., West, M., Patterson, M., and Birdi, K. (2005). Managing people to promote innovation. *Creativity and Innovation Management*, 14(2), 118–28.

Shipton, H., Sparrow, P., Budhwar, P., and Brown, A. (2017). HRM and innovation: Looking across levels. *Human Resource Management Journal*, 27(2), 246–63.

Soininen, J., Puumalainen, K., Sjögrén, H., and Syrjä, P. (2012). The impact of global economic crisis on SMEs. *Management Research Review*, 35(10), 927–44.

Sui, S., and Baum, M. (2014). Internationalization strategy, firm resources and the survival of SMEs in the export market. *Journal of International Business Studies*, 45(7), 821–41.

Tagiuri, R., and Davis, J. (1996). Bivalent attributes of the family firm. *Family Business Review*, 9(2), 199–208.

Thompson, C.A., Beauvais, L.L., and Lyness, K.S. (1999). When work–family benefits are not enough: The influence of work–family culture on benefit utilization, organizational attachment, and work–family conflict. *Journal of Vocational Behavior*, 54(3), 392–415.

Titmuss, R.M. (1976). *Commitment to Welfare*. London: Routledge.

Tocher, N., and Rutherford, M.W. (2009). Perceived acute human resource management problems in small and medium firms: An empirical examination. *Entrepreneurship Theory and Practice*, 33(2), 455–79.

Wang, P., and Walumbwa, F.O. (2007). Family-friendly programs, organizational commitment, and work withdrawal: The moderating role of transformational leadership. *Personnel Psychology*, 60(2), 397–427.

Wei, Y., Nan, H., and Wei, G. (2020). The impact of employee welfare on innovation performance: Evidence from China's manufacturing corporations. *International Journal of Production Economics*, 228, 107753.

Zahra, S.A., Hayton, J.C., and Salvato, C. (2004). Entrepreneurship in family vs. non–family firms: A resource–based analysis of the effect of organizational culture. *Entrepreneurship Theory and Practice*, 28(4), 363–81.

APPENDIX 6A.1

The literature review has been performed adopting the following strategy:

- Keyword and search strategy: Work-family/work-life/family-friendly AND practices, policies, programs, initiatives, benefits
- Search database: Scopus
- Literature review time span: 1961–2020
- Types of publications considered: Journal articles
- Number of journal articles retrieved: 369

After a qualitative analysis (journal, title, abstract and keywords) of the 369 journal articles, to be sure that they fall in the following categories: entrepreneurship/innovation/management/sociology/psychology/economics/gender, we selected a final sample of 70 journal articles.

On this final sample, we performed a descriptive analysis looking at:

- geographical region
- publication year
- sample size and type (individual-level or firm-level)
- terminology used
- research question/hypothesis
- theoretical perspective/conceptual contributions
- relationship explored (for empirical articles)
- results obtained
- measures/statistical methods used for the purpose of the research.

PART III

INSTITUTIONS

7. University-industry collaboration: drivers and barriers

Thomas Lauvås and Einar Rasmussen

7.1 INTRODUCTION

Firms have historically organized their research and development (R&D) internally (Mowery, 1983), but firms now increasingly follow an open innovation model by complementing their in-house knowledge through interorganizational collaborations (Hagedoorn, 2002; Chesbrough, 2003; Sampson, 2007). A prominent example is the growing number of university-industry collaborations (UICs)[1] (Caloghirou et al., 2001),[2] by which universities may contribute relevant expertise and knowledge to a firm's technological resource base and create new possibilities for innovation through research (Mansfield, 1991; Cohen et al., 2002; Dahlander and Gann, 2010).

Conducting research that yields both scientific progress and industrial innovations is challenging for both university and industry partners (Adler et al., 2009; Bruneel et al., 2010; Galán-Muros and Plewa, 2016). In general, firms desire more applied research, while universities strive for basic research (Lee, 2000; Nelson, 2004; D'Este and Perkmann, 2011). This divergence is related to the differing incentive structures and opposing logics of academic research and industrial commercialization (Bjerregaard, 2010; Steinmo, 2015; Galán-Muros and Plewa, 2016).

Although these collaborative challenges are widely acknowledged, prior studies have mainly explored the effects of university-industry links (Perkmann and Walsh, 2007; Wirsich et al., 2016), showing that UIC may lead to successful outcomes such as firm innovations (Mansfield, 1991; Cohen et al., 2002; Robin and Schubert, 2013), patents and licenses (Cohen et al., 2002; Perkmann and Walsh, 2007; Gulbrandsen et al., 2011), products (Kaiser and Kuhn, 2012), and academic publications (Gulbrandsen and Smeby, 2005; Banal-Estañol et al., 2015). Hence, the outcomes of UICs are well known, while the collaborative *dynamics*[3] and *processes* leading to these outcomes are less explored (Perkmann and Walsh, 2007; Thune and Gulbrandsen, 2014; Boardman and Bozeman, 2015). There is also limited knowledge of both success factors and how collaborations fail to achieve their targets (Giuliani and Arza, 2009; Bozeman et al., 2013; Steinmo and Rasmussen, 2016). Accordingly, this chapter is guided by the following research question: *What empirical evidence is available on the development of formal university-industry collaborations aiming at innovation?* By focusing on UICs, the emphasis is, therefore, on the processes that often precede or follow either the transfer of intellectual property (Perkmann and Walsh, 2007) or commercialization of research (Perkmann et al., 2013).

This chapter continues by defining UIC, before presenting different types of UIC; research partnerships and research services, and the partners' motives for establishing such UICs. Next, four drivers and barriers of UIC are described. Finally, we provide some reflections and implications for achieving successful UIC.

7.1.1 Defining University-Industry Collaboration

The term UIC may appear as self-explanatory, and most academic papers do not provide an explicit definition of UIC (Galán-Muros and Plewa, 2016). UIC is often considered as a homogeneous activity; however, in practice, there are different collaboration types (Perkmann and Walsh, 2007; Gulbrandsen et al., 2011; Ankrah and Al-Tabbaa, 2015). Hence, it is difficult to provide a definition that covers all types of UIC. The relatively broad definition by Ankrah and Al-Tabbaa (2015, p. 387) defines UIC as "the interaction between any parts of the higher educational system and industry aiming mainly to encourage knowledge and technology exchange." By focusing on the interaction between university and industry partners and their expected output, the definition covers most collaboration types between universities and industry partners.

7.2 DIFFERENT TYPES OF UIC: RESEARCH PARTNERSHIPS AND RESEARCH SERVICES

The literature provides several characterizations of UICs (Ankrah and Al-Tabbaa, 2015). Perkmann and Walsh (2007) distinguish between research partnerships and research services based on the degree of finalization of the research undertaken. Finalization refers to whether the research pursues a specific (technical, social or economic) purpose, or if new knowledge is a goal in itself (Perkmann and Walsh, 2007). As such, research services are more applied, while research partnerships entail more basic research.

Research partnerships are designed to generate outputs that are of high academic relevance that can be used in academic publications, and include collaborative research activities, also known as sponsored research and university-industry research centers. Such research partnerships can range from temporary small-scale projects to large-scale organizations (Perkmann and Walsh, 2007) with up to 50 members, such as in the Norwegian research centers of excellence, the US Engineering Research Centers and the "Cooperative Research Centers" program in Australia (Santoro and Gopalakrishnan, 2001; Boardman, 2012).

Research services are offered by university researchers for industrial clients and tend to be less relevant for academic publications. Contract research and most of academic consulting fall under this category, and involves research or consulting that are carried out by academics but paid for by industry (Perkmann and Walsh,

2007). In research services, there is often less room for academic freedom compared to research partnerships because research services target specific objectives and deliverables. Although the boundaries between contract research and consulting are blurred, contract research is generally commissioned to explore specific, previously unresearched challenges, while consulting mainly exploits existing knowledge (Perkmann and Walsh, 2007).

The next subsection explores who establish research partnerships and research services, and what motivates these establishments.

7.2.1 Who Establish Research Partnerships and Research Services, and What Are Their Motivations?

Perkmann and Walsh (2009) find that university partners often initiate and set the agenda for research partnerships that contain basic research (low degree of finalization) and that these projects grant more publications for the university partners than more applied collaborations. Accordingly, Lee (2000) and D'Este and Perkmann (2011) find that most academics collaborate with industry to further progress and finance their research and learn from them, rather than to commercialize their knowledge. Hence, university researchers are likely to choose research topics that are (1) perceived to be of interest by their peers, (2) advance the knowledge in their specific area through academic publications, and (3) help them rise up the academic ladder (Nelson, 2004). Hence, new university-industry research centres are often initiated by the university partners.

In contrast, research services that are targeted for problem-solving activities are mostly initiated by firms to obtain assistance from university researchers (Perkmann and Walsh, 2009). Calderini et al. (2007) find that the university researchers who conduct applied research are more likely to produce industrial applications than researchers who conduct basic research. However, applied research leads to fewer academic publications (Perkmann and Walsh, 2009) and therefore conflicts with the long-term goals of most university researchers (Lee, 2000; Nelson, 2004; D'Este and Perkmann, 2011).

These general findings differ between industries. Firms in science-intensive sectors such as biotechnology, pharmaceuticals, and chemicals have strong complementarities between their internal R&D and basic academic research, and tend to rely on research partnerships and, to some degree, on research services (Meyer-Krahmer and Schmoch, 1998; Cohen et al., 2002; Perkmann and Walsh, 2007). In contrast, sectors that rely less on scientific breakthroughs, such as software development, mechanical engineering or metals have a stronger preference for research services (Meyer-Krahmer and Schmoch, 1998; Cohen et al., 2002; Perkmann and Walsh, 2007).

The differences between research partnerships and research services may depend on the timing of these activities, where research services are especially required in the latter stages of the innovation process (Polt et al., 2001). Research partnerships, for example in research centers, often lead to research services that are conducted "outside" of the research center (Lauvås, 2017). A notable implication for firms is that research services have comparatively lower entry costs than research partnerships, and require lower levels of absorptive capacity (Perkmann and Walsh, 2007). Also, the fixed cost of specialized personnel and equipment makes research services a good starting point for industry (Perkmann and Walsh, 2008). Hence, firms seeking to engage in collaborations with university partners could start with research services before engaging in research partnerships (Lauvås, 2017). The remainder of this chapter focuses on research partnerships between university and industry partners, as these are the most complex types of UIC to handle for both university and firm partners (Perkmann and Walsh, 2007; Lauvås, 2017).

7.3 DRIVERS AND BARRIERS OF UIC – AND IMPLICATIONS FOR HOW TO SUCCEED

Both sides in a UIC are dependent on each other to obtain access to necessary resources and to generate synergies (Santoro and Chakrabarti, 1999; Carayol, 2003; Lind et al., 2013). Hence, the partners face an inherent challenge by relying on resources from other organizations that are known to have diverse perspectives, interests, and objects (Pfeffer and Salancik, 2003; Wry et al., 2013). There are a range of drivers of successful UIC, as well as barriers that make UIC challenging (Ankrah and Al-Tabbaa, 2015). The realization of the potential benefits of UIC depends on the partners' ability to overcome such barriers (Bruneel et al., 2010; Plewa et al., 2013b).

To examine the development of UICs, we examine the drivers and barriers along two main dimensions. First, drivers and barriers in UICs may be related to either structural characteristics of the collaboration or to relational issues. Second, some of the drivers and barriers are related to the more general particularities of universities and firms, and are found in UICs in general, while others are collaboration-specific issues that relate to the concrete dyadic relation between universities and firms in a particular UIC. Applying this in a two-by-two matrix, we can distinguish between four types of drivers and barriers labeled "connections," "organizational culture," "characteristics", and "relationships" as illustrated in Figure 7.1 and discussed in the following subsections.

7.3.1 Connections

Three connections between universities and firms have been identified as particularly important for initiating UIC: finding an appropriate partner and social and geographical proximity. The first challenging step in UIC is to find an appropriate partner (Muscio and Pozzali, 2013). The lack of awareness of, and connections to, potential university and industry partners is found to inhibit UIC (Galán-Muros and Plewa, 2016). In a large-scale survey of UK academics, D'Este and Patel (2007) find that the variety and frequency of interactions with industry are not explained by the rankings of the university departments but rather by the researchers' individual characteristics.

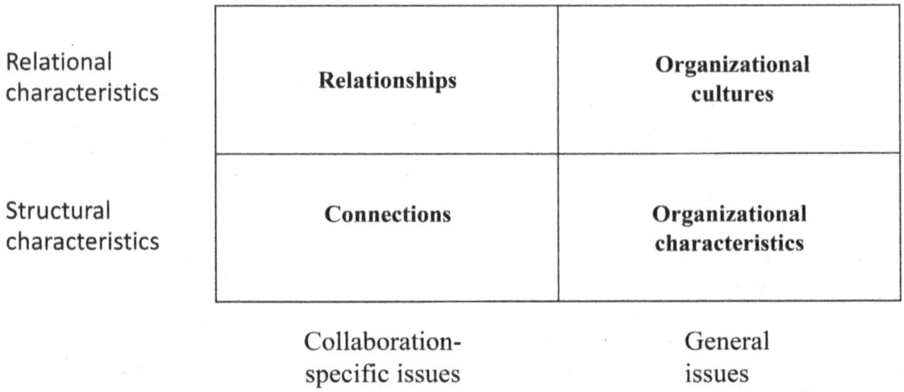

Figure 7.1 *Types of barriers and enablers affecting the collaboration process and outcomes*

Accordingly, firms are found to judge and choose their university partners at an individual, rather than an institutional level (Johnston and Huggins, 2018). These findings are in line with the literature on R&D alliances, where organizational-level collaborations often are found to be based on individual members' social relationships (Perkmann and Walsh, 2007).

Social proximity is therefore found to be central in UIC, referring to the actors' relations at the micro level, involving trust, friendship, and common experiences (Boschma, 2005). Social proximity is generally associated with past collaborations and repeated contacts between partners (Balland, 2011; Huber, 2012; Davids and Frenken, 2017), and is particularly central to the success of UIC innovation projects, and for the continuing of new UICs (Mora-Valentin et al., 2004; Belderbos et al., 2015; Steinmo and Rasmussen, 2016). Hence, UIC is to a large extent based on prior established relationships (Barnes et al., 2002; Mora-Valentin et al., 2004), and although there is limited evidence in the literature on how these relationships are created or maintained (Steinmo and Rasmussen, 2018), Lauvås and Steinmo (2019) recently illustrated that social proximity could be built over time by engaging in activities to build mutual understanding between the partners, such as ad hoc informal contact and industry visits.

Geographical proximity is also found to be of importance in the connection and establishment of UICs (Mora-Valentin et al., 2004), particularly for small firms (Slavtchev, 2013; Dornbusch and Neuhäusler, 2015). Steinmo and Rasmussen (2016), in their study of 15 successful innovation projects, found that engineering-based firms tend to rely on geographical and social proximity to universities, while science-based firms rely more heavily on cognitive proximity in establishing UICs. Further, Fitjar and Rodríguez-Pose (2013) find that firms with established collaborative linkages with geographically distant universities and research centers radically increased their innovation potential. These findings imply that science-based firms are more likely to find collaborative university partners across long geographical distances and generate innovations, while engineering-based firms in more peripheral regions with larger distances to universities would have greater difficulties engaging in UIC, as it is generally easier to establish UICs with partners in close geographical proximity (Nilsen and Lauvås, 2018).

7.3.2 Relationships

There has been extensive research on the specific relationship-related drivers and barriers in the collaboration process between university and industry partners in UIC, showing the importance of commitment, trust, and ongoing relations.

First, commitment is important in UIC (e.g., Santoro, 2000; Okamuro and Nishimura, 2018), while Knockaert et al. (2014) find that higher involvement in research centers affects the firm's network and competence additionality positively. Further, Okamuro and Nishimura (2018) find that the firm's commitment is important for commercialization in UICs, and Jarvenpaa and Valikangas (2016) found that continued participation over time is important if firms are to influence and reap the benefits from UICs. Lauvås and Steinmo (2019) extended these findings by illustrating that developing proximity along the dimensions of social and cognitive proximity and mutual commitment enabled UIC partners to develop both academic publications and innovations. Moreover, Thune and Gulbrandsen (2014) found that limited firm commitment could lead to "symbolic collaboration" and high firm exit rates.

Second, trust is critical for developing UICs (Barnes et al., 2002; Mora-Valentin et al., 2004), and is especially important for reducing tensions in UIC (Bruneel et al., 2010). Although university and industry partners may experience a "honeymoon period" in the early stages of the collaboration before their differences surface (Estrada et al., 2016), university and industry partners collaborate better over time in successful UICs (Lundberg and Andresen, 2012; Steinmo, 2015; Estrada et al., 2016). Hence, ongoing collaborations may reduce the tensions in UIC over time (Thune, 2011), and because previous collaboration promotes trust in UIC (Lhuillery and Pfister, 2009; Bruneel et al., 2010), building and maintaining trust is an important component for long-term UIC (Bruneel et al., 2010; Galán-Muros and Plewa, 2016). Hence, former collaboration is consistently found as one of the key factors for explaining successful UICs (Bruneel et al., 2010; Núñez-Sánchez et al., 2012; Bishop et al., 2011).

7.3.3 Organizational Characteristics

The organizational characteristics are found to matter for the initiation and continuation of UICs, which particularly affect a firm's ability to absorb research results, and relate to four issues: firm size, firms' absorptive capacity/cognitive proximity, firm representative(s), and openness of research results.

First, the size of the firms matters, where larger firms are most likely to collaborate with universities (Fontana et al., 2006; Roper and Hewitt-Dundas, 2013). In a study of Swedish university-industry research centers, McKelvey et al. (2015) found that the larger firms focused more on knowledge transfer from the center back to the firm, while the small firms used the research center to develop knowledge about customer needs in order to create market opportunities, especially through networking with large firms in the research center.

Second, Fontana et al. (2006) found that firms with high absorptive capacity more often collaborate with universities, while firms with low absorptive capacity have lower probability of collaborating with universities. When it comes to the similarities in the partners' knowledge bases, or cognitive proximity (Boschma, 2005), similar competencies and capabilities between firms and research partners are found to be important for successful UIC. However, too much similarity may lead to less valuable collaborations because complementary knowledge is required for innovation development (Petruzzelli, 2011). Accordingly, firms with high absorptive capacity and strong knowledge bases are more likely to diffuse the knowledge produced by the universities (Giuliani and Arza, 2009).

Third, firms often struggle to integrate the knowledge of research results because of the knowledge boundaries between firms and universities (Galán-Muros and Plewa, 2016; Alexander et al., 2018; Hayter et al., 2020). Prior research shows that firms are critically dependent on the skills and activities performed by their firm representatives to successfully manage the knowledge stemming from universities (Santoro and Chakrabarti, 2002; Knudsen et al., 2017; Takanashi and Lee, 2018). However, in their study of university-industry research centers, Santoro and Chakrabarti (2002) found that only five of 202 firms had more than one firm representative involved in the research centers' activities. That firms often engage one firm representative in UIC illustrates the importance of engaging knowledgeable firm representative(s). Hence, the knowledge integration activities performed by the firm representative are found to be important in UIC (Lauvås, 2017), as external knowledge and opportunities often require translation before people inside their organizations can understand them (Cyert and Goodman, 1997; Johansson et al., 2011; Monteiro and Birkinshaw, 2017).

Finally, universities' research results are often openly published, while firms' R&D strategies often involve secrecy and IP protection (Perkmann et al., 2011). Relatedly, firms often fear that confidential information will be disclosed in UICs (Hall et al., 2001), in particular when participating in consortia and research centers with competitors. Hence, to handle the issues of openness of R&D, firms often participate in UICs with competitors to develop knowledge on issues other than their core-technology, researching issues faced by the whole industry, such as regulations and environmental challenges (Jakobsen and Steinmo, 2016; Jakobsen et al., 2019). Such collaborations often happen in interdisciplinary research centers, seeking to overcome specific challenges in particular industries (Gulbrandsen et al., 2015; Villani et al., 2017) or to tackle some of the 'Grand Challenges' facing humanity (Hessels et al., 2014). However, in research centers, as it is for UIC in general, there is a difference in firms' motives for participating (Broström, 2012; Fitjar and Rodríguez-Pose, 2013), where large firms often participate in research centers to advance non-core technologies, while small firms want to advance core technologies (Santoro and Chakrabarti, 2002; McKelvey et al., 2015).

7.3.4 Organizational Cultures

The large cultural differences between universities and firms have been identified as a barrier for UIC, related to different time horizons, motivations, and communication modes (Bjerregaard, 2010; Steinmo, 2015; Galán-Muros and Plewa, 2016). First, universities often have a long-term orientation, whereas firms are more oriented toward short-term, applied research that can lead to solutions to current problems (Spithoven et al., 2011). This can lead to tensions between university and industry partners (Ambos et al., 2008), and firms' short-term perspectives are seen as the biggest disadvantage of UICs by university researchers (Meyer-Krahmer and Schmoch, 1998). In contrast, the long-term orientation of universities was seen as a significant barrier by more than two-thirds of firms in the UK (Bruneel et al., 2010).

Second, firms and university partners have different motivations and R&D requirements (Perkmann and Walsh, 2007; Bjerregaard, 2010; Steinmo, 2015), where firms generally desire applied research[4] for industrial innovations, and universities basic research for academic publications (Lee, 2000; Nelson, 2004; D'Este and Perkmann, 2011). Third, communications are found to be critical for UIC development and success (Plewa et al., 2013a), but differences in terminology and communication styles are likely to inhibit collaboration (Barnes et al., 2002; Galán-Muros and Plewa, 2016). It is often claimed that academic researchers lack training in communicating their findings outside of academia (Mittion et al., 2007; Galán-Muros and Plewa, 2016).

7.4 IMPLICATIONS FOR PRACTICE

This chapter has presented a comprehensive overview of UIC, exploring different types of UIC, and four drivers and barriers of UIC. In conclusion, three areas seem particularly important.

First, the success factors in the UIC literature are quite general (Thune, 2011; Plewa et al., 2013a), which relates to the general gap in the UIC literature, namely, that the collaborative dynamics and processes that reveal how successful outcomes are developed are less explored (Perkmann and Walsh, 2007; Thune and Gulbrandsen, 2014; Boardman and Bozeman, 2015). Hence, it is challenging for firms, universities, scholars and policymakers in UIC to assess and evaluate the success and impact of UICs (Thune, 2011, Perkmann et al., 2011). Especially due to the uncertain nature of basic research, which makes it difficult to set clear objectives and evaluate outcomes that often materialize many years after the UIC has ended (Perkmann et al., 2011). However, because limited insight has been shed on firms' innovation portfolio (Nagji and Tuff, 2012), an important realization is that long-term research partnerships (e.g., research centers) may lead to short-term research services, where research partnerships, in, for example, research centers, often lead to research services that are conducted "outside" of the research center (Lauvås, 2017; Iglebæk et al., 2018), illustrating the additionality effects of UICs.

Second, while most research on UIC has been conducted in science-based industries such as pharmaceuticals and biotechnology (Lundvall, 2007; Broström, 2012), this chapter has pointed toward some nuances regarding the firm's size, characteristics and industrial affiliation, and the importance of different dimensions of proximity. Geographical proximity toward universities is found to be particularly important for the establishment of UICs for both small and engineering-based-firms (Slavtchev, 2013; Dornbusch and Neuhäusler, 2015; Steinmo and Rasmussen, 2016). Hence, efforts to increase UIC can rely on collaborations with local universities as an important starting point for these types of firms (Steinmo and Rasmussen, 2016; Fitjar and Gjelsvik, 2018). This could be a starting point for increasing the levels of UICs, because former collaboration is one of the key factors for explaining successful UICs (Bruneel et al., 2010; Bishop et al., 2011; Núñez-Sánchez et al., 2012).

Third, firms' involvement and commitment in UICs are found to be important in building firms' absorptive capacity and to generate research results that are relevant for both university and firm partners (Santoro and Chakrabarti, 2002; Knudsen et al., 2017; Lauvås and Steinmo, 2019). This is also indicated in the recent report by the OECD (2019), which recommend focusing on knowledge co-creating between universities and industry, rather than a knowledge-transfer approach. For further research, it would thus be important to study such collaboration processes in-depth, to illustrate how different types of firms can both initiate and participate in UICs, and where the learning could be progressed into new UICs.

NOTES

1. UIC is often used as an umbrella term including all types of "public research organizations," which are predominantly government-funded, such as universities, research institutes and research centers (e.g., Perkmann and Walsh, 2007; Estrada et al., 2016).
2. This development has spurred additional concepts such as the triple helix (Etzkowitz and Leydesdorff, 1997) and Mode 2 (Gibbons et al., 1994), which is more concerned with greater demands for the practical relevance of research and science.
3. By dynamics and processes in UIC, this chapter refers to the activities, events, practices and interactions enacted by university and industry partners and how they affect the development of UICs.
4. Although some science-intensive sectors may prefer basic research, such as pharmaceuticals, this general statement, and the rest of the chapter, corresponds to the R&D preferences of most industrial sectors (Meyer-Krahmer and Schmoch, 1998; Cohen et al., 2002; Perkmann et al., 2011).

REFERENCES

Adler, N., Elmquist, M. and Norrgren, F. 2009. The challenge of managing boundary-spanning research activities: Experiences from the Swedish context. *Research Policy,* 38, 1136–49.

Alexander, A., Martin, D.P., Manolchev, C. and Miller, K. 2018. University–industry collaboration: Using meta-rules to overcome barriers to knowledge transfer. *Journal of Technology Transfer,* 45(2), 371–92.

Ambos, T.C., Mäkelä, K., Birkinshaw, J. and D'Este, P. 2008. When does university research get commercialized? Creating ambidexterity in research institutions. *Journal of Management Studies,* 45, 1424–47.

Ankrah, S. and Al-Tabbaa, O. 2015. Universities–industry collaboration: A systematic review. *Scandinavian Journal of Management,* 31, 387–408.

Balland, P.-A. 2011. Proximity and the evolution of collaboration networks: Evidence from research and development projects within the Global Navigation Satellite System (GNSS) industry. *Regional Studies,* 46, 741–56.

Banal-Estañol, A., Jofre-Bonet, M. and Lawson, C. 2015. The double-edged sword of industry collaboration: Evidence from engineering academics in the UK. *Research Policy,* 44, 1160–75.

Barnes, T., Pashby, I. and Gibbons, A. 2002. Effective university–industry interaction: A multi-case evaluation of collaborative R&D projects. *European Management Journal,* 20, 272–85.

Belderbos, R., Carree, M., Lokshin, B. and Sastre, J.F. 2015. Inter-temporal patterns of R&D collaboration and innovative performance. *The Journal of Technology Transfer,* 40, 123–37.

Bishop, K., D'Este, P. and Neely, A. 2011. Gaining from interactions with universities: Multiple methods for nurturing absorptive capacity. *Research Policy,* 40, 30–40.

Bjerregaard, T. 2010. Industry and academia in convergence: Micro-institutional dimensions of R&D collaboration. *Technovation,* 30, 100–8.

Boardman, P.C. 2012. Organizational capital in boundary-spanning collaborations: Internal and external approaches to organizational structure and personnel authority. *Journal of Public Administration Research and Theory,* 22, 497–526.

Boardman, P.C. and Bozeman, B. 2015. Academic faculty as intellectual property in university-industry research alliances. *Economics of Innovation and New Technology,* 24, 403–20.

Boschma, R. 2005. Proximity and innovation: A critical assessment. *Regional Studies,* 39, 61–74.

Bozeman, B., Fay, D. and Slade, C. 2013. Research collaboration in universities and academic entrepreneurship: The-state-of-the-art. *The Journal of Technology Transfer,* 38, 1–67.

Broström, A. 2012. Firms' rationales for interaction with research universities and the principles for public co-funding. *The Journal of Technology Transfer,* 37, 313–29.

Bruneel, J., D'Este, P. and Salter, A. 2010. Investigating the factors that diminish the barriers to university–industry collaboration. *Research Policy,* 39, 858–68.

Calderini, M., Franzoni, C. and Vezzulli, A. 2007. If star scientists do not patent: The effect of productivity, basicness and impact on the decision to patent in the academic world. *Research Policy,* 36, 303–19.

Caloghirou, Y., Tsakanikas, A. and Vonortas, N.S. 2001. University-Industry cooperation in the context of the European Framework Programmes. *The Journal of Technology Transfer,* 26, 153–61.

Carayol, N. 2003. Objectives, agreements and matching in science–industry collaborations: Reassembling the pieces of the puzzle. *Research Policy,* 32, 887–908.

Chesbrough, H.W. 2003. *Open Innovation: The New Imperative for Creating and Profiting from Technology.* Cambridge, MA: Harvard Business School Press.

Cohen, W., Nelson, R.R. and Walsh, J.P. 2002. Links and impacts: The influence of public research on industrial R&D. *Management Science,* 48, 1–23.
Cyert, R.M. and Goodman, P.S. 1997. Creating effective university-industry alliances: An organizational learning perspective. *Organizational Dynamics,* 25, 45–57.
D'Este, P. and Patel, P. 2007. University–industry linkages in the UK: What are the factors underlying the variety of interactions with industry? *Research Policy,* 36, 1295–313.
D'Este, P. and Perkmann, M. 2011. Why do academics engage with industry? The entrepreneurial university and individual motivations. *The Journal of Technology Transfer,* 36, 316–39.
Dahlander, L. and Gann, D.M. 2010. How open is innovation? *Research Policy,* 39, 699–709.
Davids, M. and Frenken, K. 2017. Proximity, knowledge base and the innovation process: Towards an integrated framework. *Regional Studies,* 52(1), 23–34.
Dornbusch, F. and Neuhausler, P. 2015. Composition of inventor teams and technological progress – the role of collaboration between academia and industry. *Research Policy,* 44, 1360–75.
Estrada, I., Faems, D., Martin Cruz, N. and Perez Santa, P. 2016. The role of interpartner dissimilarities in industry-university alliances: Insights from a comparative case study. *Research Policy,* 45, 2008–22.
Etzkowitz, H., and Leydesdorff, L. (1997). *Universities and the Global Knowledge Economy: A Triple Helix of University-Industry-Government Relations*. London: Pinter.
Fitjar, R.D. and Gjelsvik, M. 2018. Why do firms collaborate with local universities? *Regional Studies,* 52, 1525–36.
Fitjar, R.D. and Rodríguez-Pose, A. 2013. Firm collaboration and modes of innovation in Norway. *Research Policy,* 42, 128–38.
Fontana, R., Geuna, A. and Matt, M. 2006. Factors affecting university-industry R&D projects: The importance of searching, screening and signalling. *Research Policy,* 35, 309–23.
Galán-Muros, V. and Plewa, C. 2016. What drives and inhibits university-business cooperation in Europe? A comprehensive assessement. *R&D Management,* 46, 369–82.
Gibbons, M., Limoges, C., Nowotny, H., Schwartzman, S., Scott, P., and Trow, M. (1994). *The New Production of Knowledge: The Dynamics of Science and Research in Contemporary Societies.* London: Sage.
Giuliani, E. and Arza, V. 2009. What drives the formation of "valuable" university–industry linkages?: Insights from the wine industry. *Research Policy,* 38, 906–21.
Gulbrandsen, M. and Smeby, J.-C. 2005. Industry funding and university professors' research performance. *Research Policy,* 34, 932–50.
Gulbrandsen, M., Mowery, D. and Feldman, M. 2011. Introduction to the special section: Heterogeneity and university–industry relations. *Research Policy,* 40, 1–5.
Gulbrandsen, M., Thune, T., Borlaug, S.B. and Hanson, J. 2015. Emerging hybrid practices in public–private research centres. *Public Administration,* 93, 363–79.
Hagedoorn, J. 2002. Inter-firm R&D partnerships: An overview of major trends and patterns since 1960. *Research Policy,* 31, 477–92.
Hall, B.H., Link, A.N. and Scott, J.T. 2001. Barriers inhibiting industry from partnering with universities: Evidence from the advanced technology program. *The Journal of Technology Transfer,* 26, 87–98.
Hayter, C.S., Rasmussen, E. and Rooksby, J.H. 2020. Beyond formal university technology transfer: Innovative pathways for knowledge exchange. *The Journal of Technology Transfer,* 45, 1–8.
Hessels, L.K., Wardenaar, T., Boon, W.P.C. and Ploeg, M. 2014. The role of knowledge users in public–private research programs: An evaluation challenge. *Research Evaluation,* 23, 103–16.

Huber, F. 2012. On the role and interrelationship of spatial, social and cognitive proximity: Personal knowledge relationships of R&D workers in the cambridge information technology cluster. *Regional Studies,* 46, 1169–82.
Iglebæk, F., Nygård, G., Bruvoll, A. and Christian, G. 2018. *Effekter av energiforskningen (Effects of the energy research)* (ISBN:978-82-691556-1-7). Retrieved from https://www.regjeringen.no/contentassets/816c63dcb0ea49768ec03cd64828af5a/effekter_av_energiforskningen.pdf.
Jakobsen, S. and Steinmo, M. 2016. The role of proximity dimensions in the development of innovations in competition: A longitudinal case study. *International Journal of Technology Management,* 71, 100–22.
Jakobsen, S., Lauvås, T.A. and Steinmo, M. 2019. Collaborative dynamics in environmental R&D alliances. *Journal of Cleaner Production,* 212, 950–9.
Jarvenpaa, S.L. and Valikangas, L. 2016. From governance void to interactive governing behaviors in new research networks. *Academy of Management Discoveries,* 2, 226–46.
Johansson, M., Axelson, M., Enberg, C. and Tell, F. 2011. Knowledge integration in inter-firm R&D collaboration: How do firms manage problems of coordination and cooperation. In: Berggren, C., Bergek, A., Bengtsson, L., Hobday, M. and Söderlund, J. (eds), *Knowledge Integration & Innovation: Critical Challenges Facing International Technology-based Firms.* Oxford: Oxford University Press, pp. 148–69.
Johnston, A. and Huggins, R. 2018. Partner selection and university-industry linkages: Assessing small firms' initial perceptions of the credibility of their partners. *Technovation,* 78, 15–26.
Kaiser, U. and Kuhn, J.M. 2012. Long-run effects of public–private research joint ventures: The case of the Danish Innovation Consortia support scheme. *Research Policy,* 41, 913–27.
Knockaert, M., Spithoven, A. and Clarysse, B. 2014. The impact of technology intermediaries on firm cognitive capacity additionality. *Technological Forecasting and Social Change,* 81, 376–87.
Knudsen, M.P., Tranekjer, T.L. and Bulathsinhala, N. 2017. Advancing large-scale R&D projects towards grand challenges through involvement of organizational knowledge integrators. *Industry and Innovation,* 26(1), 1–30.
Lauvås, T.A. 2017. The dynamics of university-industry collaboration: A longitudinal case study of research centers, no. 63–2017, Nord University Business School.
Lauvås, T. and Steinmo, M. 2019. The role of proximity dimensions and mutual commitment in shaping the performance of university-industry research centres. *Innovation,* 23(2), 182–208.
Lee, Y.S. 2000. The sustainability of university-industry research collaboration: An empirical assessment. *The Journal of Technology Transfer,* 25, 111–33.
Lhuillery, S. and Pfister, E. 2009. R&D cooperation and failures in innovation projects: Empirical evidence from French CIS data. *Research Policy,* 38, 45–57.
Lind, F., Styhre, A. and Aaboen, L. 2013. Exploring university-industry collaboration in research centres. *European Journal of Innovation Management,* 16, 70–91.
Lundberg, H. and Andresen, E. 2012. Cooperation among companies, universities and local government in a Swedish context. *Industrial Marketing Management,* 41, 429–37.
Lundvall, B.Å. 2007. National innovation systems – analytical concept and development tool. *Industry and Innovation,* 14, 95–119.
Mansfield, E. 1991. Academic research and industrial innovation. *Research Policy,* 20, 1–12.
McKelvey, M., Zaring, O. and Ljungberg, D. 2015. Creating innovative opportunities through research collaboration: An evolutionary framework and empirical illustration in engineering. *Technovation,* 39–40, 26–36.
Meyer-Krahmer, F. and Schmoch, U. 1998. Science-based technologies: University–industry interactions in four fields. *Research Policy,* 27, 835–51.

Mittion, C., Adair, C.E., McKenzie, E., Patten, S.B. and Perry, B.W. 2007. Knowledge transfer and exchange: Review and synthesis of the literature. *The Milbank Quarterly,* 85, 729–68.
Monteiro, F. and Birkinshaw, J. 2017. The external knowledge sourcing process in multinational corporations. *Strategic Management Journal,* 38, 342–62.
Mora-Valentin, E.M., Montoro-Sanchez, A. and Guerras-Martin, L. A. 2004. Determining factors in the success of R&D cooperative agreements between firms and research organizations. *Research Policy,* 33, 17–40.
Mowery, D.C. 1983. The relationship between intrafirm and contractual forms of industrial research in American manufacturing, 1900–1940. *Explorations in Economic History,* 20, 351–74.
Muscio, A. and Pozzali, A. 2013. The effects of cognitive distance in university-industry collaborations: Some evidence from Italian universities. *Journal of Technology Transfer,* 38, 486–508.
Nagji, B. and Tuff, G. 2012. Managing your innovation portfolio. *Harvard Business Review,* 66–73.
Nelson, R.R. 2004. The market economy, and the scientific commons. *Research Policy,* 33, 455–71.
Nilsen, T. and Lauvås, T.A. 2018. The role of proximity dimensions in facilitating university-industry collaboration in peripheral regions: Insights from a comparative case study in Northern Norway. *Arctic Review,* 9, 312–31.
Núñez-Sánchez, R., Barge-Gil, A. and Modrego-Rico, A. 2012. Performance of knowledge interactions between public research centres and industrial firms in Spain: A project-level analysis. *The Journal of Technology Transfer,* 37, 330–54.
OECD 2019. *University-Industry Collaboration: New Evidence and Policy Options.* Paris, OECD Publishing.
Okamuro, H. and Nishimura, J. 2018. Whose business is your project? A comparative study of different subsidy policy schemes for collaborative R&D. *Technological Forecasting and Social Change*, 127, 85–96.
Perkmann, M. and Walsh, K. 2007. University–industry relationships and open innovation: Towards a research agenda. *International Journal of Management Reviews,* 9, 259–80.
Perkmann, M. and Walsh, K. 2008. Engaging the scholar: Three types of academic consulting and their impact on universities and industry. *Research Policy,* 37, 1884–91.
Perkmann, M. and Walsh, K. 2009. The two faces of collaboration: Impacts of university-industry relations on public research. *Industrial and Corporate Change,* 18, 1033–65.
Perkmann, M., Neely, A. and Walsh, K. 2011. How should firms evaluate success in university–industry alliances? A performance measurement system. *R&D Management,* 41, 202–16.
Perkmann, M., Tartari, V., McKelvey, M. et al. 2013. Academic engagement and commercialisation: A review of the literature on university–industry relations. *Research Policy,* 42, 423–42.
Petruzzelli, A.M. 2011. The impact of technological relatedness, prior ties, and geographical distance on university–industry collaborations: A joint-patent analysis. *Technovation,* 31, 309–19.
Pfeffer, J. and Salancik, G.R. 2003. *The External Control of Organizations: A Resource Dependence Perspective.* Stanford, CA: Stanford University Press.
Plewa, C., Korff, N., Baaken, T. and Macpherson, G. 2013a. University-industry linkage evolution: An empirical investigation of relational success factors. *R&D Management,* 43, 365–80.
Plewa, C., Korff, N., Johnson, C., Macpherson, G., Baaken, T. and Rampersad, G.C. 2013b. The evolution of university–industry linkages – a framework. *Journal of Engineering and Technology Management,* 30, 21–44.

Polt, W., Gassler, H., Schibany, A., Rammer, C. and Schartinger, D. 2001. Benchmarking industry – science relations: The role of framework conditions. *Science and Public Policy,* 28, 247–58.

Robin, S. and Schubert, T. 2013. Cooperation with public research institutions and success in innovation: Evidence from France and Germany. *Research Policy,* 42, 149–66.

Roper, S. and Hewitt-Dundas, N. 2013. Catalysing open innovation through publicly-funded R&D: A comparison of university and company-based research centres. *International Small Business Journal,* 31, 275–95.

Sampson, R.C. 2007. R&D alliances and firm performance: The impact of technological diversity and alliance organization on innovation. *Academy of Management Journal,* 50, 364–86.

Santoro, M.D. 2000. Success breeds success: The linkage between relationship intensity and tangible outcomes in industry–university collaborative ventures. *The Journal of High Technology Management Research,* 11, 255–73.

Santoro, M.D. and Chakrabarti, A.K. 1999. Building industry–university research centers: Some strategic considerations. *International Journal of Management Reviews,* 1, 225–44.

Santoro, M.D. and Chakrabarti, A.K. 2002. Firm size and technology centrality in industry–university interactions. *Research Policy,* 31, 1163–80.

Santoro, M.D. and Gopalakrishnan, S. 2001. Relationship dynamics between university research centers and industrial firms: Their impact on technology transfer activities. *The Journal of Technology Transfer,* 26, 163–71.

Slavtchev, V. 2013. Proximity and the transfer of academic knowledge: Evidence from the spatial pattern of industry collaborations of East German professors. *Regional Studies,* 47, 686–702.

Spithoven, A., Clarysse, B. and Knockaert, M. 2011. Building absorptive capacity to organise inbound open innovation in traditional industries. *Technovation,* 31, 10–21.

Steinmo, M. 2015. Collaboration for innovation: A case study on how social capital mitigates collaborative challenges in university–industry research alliances. *Industry and Innovation,* 22(7), 597–624.

Steinmo, M. and Rasmussen, E. 2016. How firms collaborate with public research organizations: The evolution of proximity dimensions in successful innovation projects. *Journal of Business Research,* 69, 1250–9.

Steinmo, M. and Rasmussen, E. 2018. The interplay of cognitive and relational social capital dimensions in university-industry collaboration: Overcoming the experience barrier. *Research Policy,* 47, 1964–74.

Takanashi, C. and Lee, K.-J. 2018. Boundary spanning leadership, resource mobilisation, and performance of university-industry R&D projects: A study in a Japanese university. *Technology Analysis & Strategic Management,* 31(2), 140–54.

Thune, T. 2011. Success factors in higher education-industry collaboration: A case study of collaboration in the engineering field. *Tertiary Education and Management,* 17, 31–50.

Thune, T. and Gulbrandsen, M. 2014. Dynamics of collaboration in university–industry partnerships: Do initial conditions explain development patterns? *The Journal of Technology Transfer,* 39, 977–93.

Villani, E., Rasmussen, E. and Grimaldi, R. 2017. How intermediary organizations facilitate university–industry technology transfer: A proximity approach. *Technological Forecasting and Social Change,* 114, 86–102.

Wirsich, A., Kock, A., Strumann, C. and Schultz, C. 2016. Effects of university–industry collaboration on technological newness of firms. *Journal of Product Innovation Management,* 33, 708–25.

Wry, T., Cobb, J.A. and Aldrich, H.E. 2013. More than a metaphor: Assessing the historical legacy of resource dependence and its contemporary promise as a theory of environmental complexity. *The Academy of Management Annals,* 7, 441–88.

8. Contextualizing technology transfer: a review of university-industry transfer in the construction industry

Laís Bandeira Barros, Mirjam Knockaert and Laura Lecluyse

8.1 INTRODUCTION

The transfer of knowledge and technology from university to industry is a process fraught with difficulty (Shane, 2004). Over the past decades, the entrepreneurship, innovation, strategy, management and technology transfer (TT) literatures have extensively discussed barriers and facilitators to TT and have provided insights into the different channels through which TT takes place (such as spin-offs, patenting and licensing, etc.), the intermediaries that support TT (such as science parks, TT offices, incubators, etc.) and TT outcomes. The strong interest in TT has resulted in an abundance of studies on these topics, with a number of bibliometric analyses and review articles recently presenting an outline of the work published. Such recent reviews deal with university-industry collaborations (Skute et al., 2019), TT in general (Bengoa et al., 2021), academic entrepreneurship ecosystems (Hayter et al., 2018), TT intermediaries (such as science parks) (Lecluyse et al., 2019) and incubators and accelerators (Hausberg and Korreck, 2021), amongst others. While such reviews have provided interesting overviews of the state of the literature, they have often considered TT as a homogeneous phenomenon, hereby typically considering studies from any region, context or technology. This is somewhat in contrast to calls in the entrepreneurship and management literatures to take the context in which phenomena are studied into consideration. In management research, context (such as circumstances, conditions, situations or environments) is external to the phenomenon and may enable or constrain it. Contexts can be business-related, social, institutional or spatial, with the industry or market representing an important element of a business-related context (Welter, 2011). Important elements of industries and markets, for instance, include their industry life stage and the size and nature of the competition (Welter, 2011). In this chapter, we take a business-related contextual perspective into the TT literature and focus specifically on understanding what we know and what is still to be explored in terms of TT in the construction industry.

The construction industry represents an interesting setting for a review with a contextual dimension. In Europe, the construction industry has often been considered a mature, traditional industry (Pries and Janszen, 1995) that is conservative and attached to familiar technologies (Beerepoot and Beerepoot, 2007). At the same time, the construction industry is challenged by increased global competition, urging firms in this industry to engage in innovation (Ahmad et al., 2005). Furthermore, research has identified inefficiencies in the construction process (Holley and Farrow, 2013), which has spurred the demand for innovation in the sector. Also, while the construction industry has traditionally, to a larger extent, engaged in incremental innovations and less in radical innovations, the emerging importance of environmental issues and informatics has affected firm R&D policies (Pries and Janszen, 1995; Uusitalo and Lavikka, 2021). Specifically, while firms in this industry have predominantly focused on cost leadership, they have to a larger extent, engaged in a differentiation strategy (competition on quality instead of price), in which they more closely connect their offer to market demand (Pries and Janszen, 1995). As such, new construction technologies need to demonstrate not only cost reduction potential but also a potential increase in productivity and optimal resource usage (Ganesan and Kelsey, 2006). However, the extent to which firms are able to embrace and adopt new technologies is largely contingent on their dynamic capabilities, enabling them to adapt, integrate and reconfigure their organizational competences and resources in order to reach a fit with the changing business environment (Pinske and Dommisse, 2009; Teece et al., 1997). Furthermore, construction firms are challenged by the integration of new technologies into their project-based structure, which is typical for the construction industry (Pinske and Dommisse, 2009). Following these challenges, university-industry collaboration and TT have become an essential part of innovation and growth in the construction industry, resulting in new product, service and process inventions, new patents and licenses (Al-Gasim et al., 2021). By consequence, in this chapter, we provide an overview of what we know on TT in the construction industry based upon the literature, taking into consideration the specific nature of this industry. Furthermore, alongside synthesizing current knowledge and contrasting it to the TT literature in general, we also present a research agenda for future research into this intriguing field. Specifically, in what follows, we first present our research method, followed by the results and our recommendations for future research.

8.2 METHOD

Our study aims at providing a review of the current literature on university-industry TT in the construction sector. In order to reach our research aim, we conducted a systematic literature review following the recommendations of White and Schmidt (2005). In what follows, we present the different steps recommended by this review method, followed by an overview of the selected papers.

8.2.1 Article Search and Classification

We performed a six-stage search procedure including the following steps, which are also illustrated in Figure 8.1: (1) basic literature search, (2) title, keyword and abstract screening, (3) screening of the full-text publications, (4) prior literature search: reviewing of the titles in the reference lists of the selected articles, (5) screening of the full texts of articles in the reference lists, and (6) analysis of the qualified articles for the review.

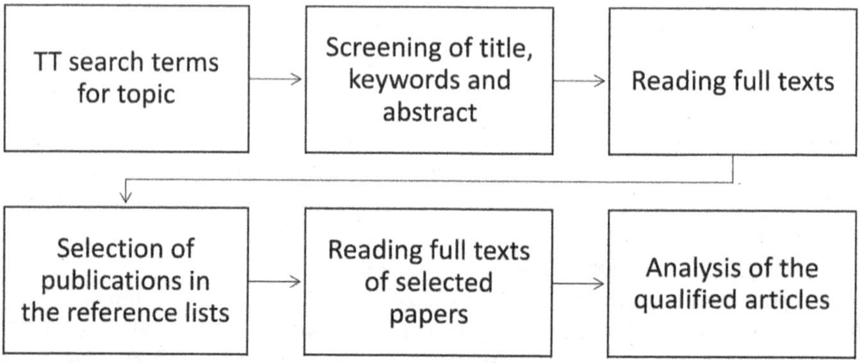

Figure 8.1 *Literature review steps*

In the first step, we used the Web of Science database to search for articles and reviews. We included papers published until October 2020. Conference papers, books and book chapters were excluded from the analysis. Search terms include terms for technology transfer and channels, as shown in Table 8.1. The search was done by topic comprising title, abstract and author keywords. We conducted our search in the following categories of the Web of Science: management, business, economics, engineering, materials and environmental sciences, and construction building technology. This first step resulted in 1,649 articles.

The second step of this process was the screening of articles by analyzing the title, keywords and abstract. We subsequently considered the scope of the title, keywords and abstract and only selected papers that studied the transfer of technology from university to industry. We excluded papers on entrepreneurship education, the impact of cultural differences in technology transfer, gender bias and policy proposals for innovation because they are out of scope. Duplicate articles were also removed. This step yielded 73 relevant publications.

Table 8.1 *Web of Science search terms and their respective number of articles after steps 1 and 2*

Web of Science search terms	Number of articles – step 1	Selected articles – step 2
("technology transfer" OR "TT" OR commerciali?ation) AND (universit* OR academi*) AND (construction OR building OR engineering)	369	21
("spin-off" OR "spin-out" OR spinoff OR spinout) AND (universit* OR academi*) AND (construction OR building OR engineering)	75	3
("start-up*" OR startup*) AND (universit* OR academi*) AND (construction OR building OR engineering)	197	2
patent* AND (universit* OR academi*) AND (construction OR building OR engineering)	218	7
licens* AND (universit* OR academi*) AND (construction OR building OR engineering)	117	0
("contract research" OR "research and development project" OR "R&D project" OR "R&D" or "research and development") AND (universit* OR academi*) AND (construction OR building OR engineering)	490	24
(consultan* OR "consultan* project") AND (universit* OR academi*) AND (construction OR building OR engineering)	183	16

Note: ? is to allow for –s and –z in the middle of a word; * is used to allow for multiple extra letters (e.g., universit* returns university, universities).

The third step was the screening of full texts using the screening strings "technology transfer," "commercialization," "university" and "construction". Papers only focusing on the development of technologies, without any aspect of transferring these to industry or bringing them onto the marketplace were considered out of scope. In addition, studies investigating international TT or simply not mentioning the involvement of universities were also excluded. This resulted in 28 studies that were of relevance to our research aim.

The fourth step was the selection of relevant titles in the reference lists of the 28 articles, which yielded 18 articles, which were then screened in a similar way as explained above, resulting in six additional papers fitting the scope of our review. Consolidating the two lists led to a final list of 34 papers which are further analyzed.

8.2.2 Coding Frame and Characteristics

In this section, we present the information transferred from the papers to the coding frame used in our analyses and provide deeper insights into the sample characteristics. The 34 papers were coded along the following dimensions: author, year, journal, title, author keywords, research questions/objectives, definitions, construction sector link, university-industry link, type of technology, methodology, units observed, country of study, key findings and future research.

Table 8.2 Number of papers per journal

Journal	Number of articles
3D Printing and Additive Manufacturing	1
Advanced Engineering Materials	1
Architectural Engineering and Design Management	1
Climatic Change	1
Composites Engineering	1
Construction Economics and Building	2
Construction Innovation	1
Construction Management and Economics	2
Environmental Science and Pollution Research	1
Informes de la construcción	1
International Journal of Construction Management	1
Journal of Civil Engineering and Management	1
Journal of Cleaner Production	2
Journal of Construction Engineering and Management	2
Journal of Engineering and Technology Management	1
Journal of Hazardous Materials	1
Journal of Professional Issues in Engineering Education and Practice	2
Journal of Technology Transfer	1
Proceedings of the Institution of Civil Engineers: Civil Engineering	1
Processes	1
Renewable and Sustainable Energy Reviews	2
Smart and Sustainable Built Environment	1
Sustainability	2
Sustainable Cities and Society	1
Technovation	1
Transportation Management and Public Policy 2003: Planning and Administration	1
World Journal of Microbiology and Biotechnology	1

The total of 34 papers is published in 27 different journals. Table 8.2 gives an overview of the distribution of the final list of papers across the different journals. This overview shows that the construction industry has mainly been addressed in construction and engineering journals, with some interest in technology management and innovation journals.

Figure 8.2 provides an overview of the years in which these papers were published, showing an increase in interest over the past decade.

We observed that many papers address new technologies in the energy-efficiency field to reduce carbon emissions in construction. This is aligned with the concern about global warming and circular economy goals towards sustainability, which seems to spur innovation in the construction industry.

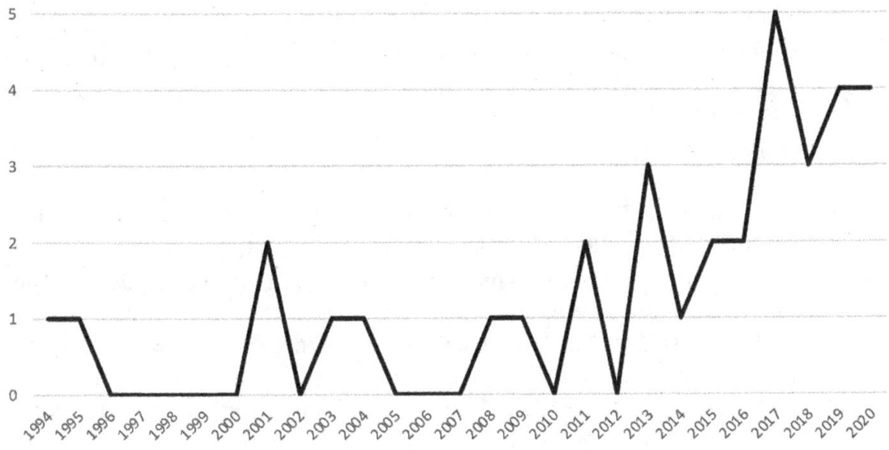

Figure 8.2 Number of articles per year of publication

8.3 RESULTS OF THE LITERATURE REVIEW

Our review points to three overarching themes which have been discussed in the literature. First, research has studied the TT process. Second, it has looked into factors influencing this process or its implications, and finally, it has studied the outcomes of the TT process. We discuss the main take-aways from prior studies in what follows. We also summarize the current state of the literature in Figure 8.3.

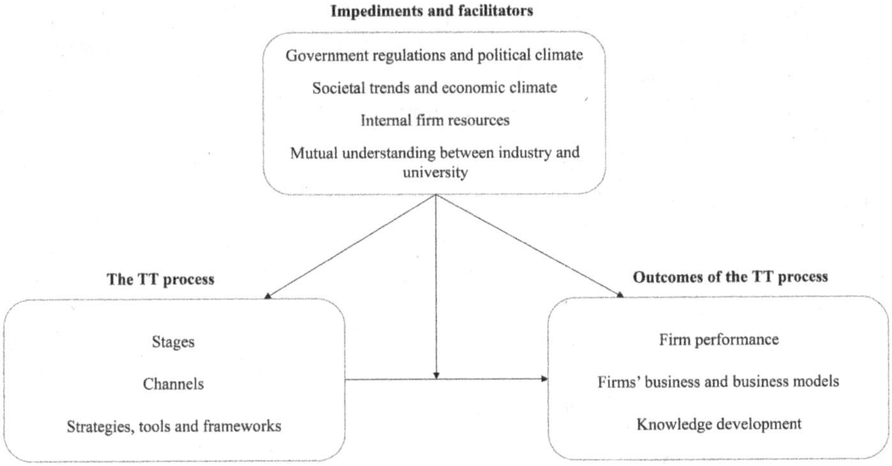

Figure 8.3 Overview of the literature on TT in the construction industry

8.3.1 The TT Process

As to what the TT process is concerned, the literature has so far studied three main topics, namely, stages, channels and strategies, tools and frameworks.

8.3.2 Stages

A traditional university TT process consists of several stages. It ranges from technology discovery by the university scientist, through the evaluation of intermediaries such as TT offices, to patent and technology licensing (Bradley et al., 2013). These stages are not specific to any industry and not linear in practice. In what follows, we discuss the TT stages that have been documented in the construction sector. While studies that looked into the stages (Bazan, 2019; Bossink, 2020; Laborde and Sanvido, 1994; Shapira and Rosenfeld, 2011) show some overlap, they also present their own view on the TT process.

Bazan (2019) considers basic research, applied research, new product development, and commercialization as the four consecutive stages of the TT process for engineering research. At the same time, the author also identifies three "efforts" that need to be made simultaneously during the four stages, through a constant feedback mechanism, in order to successfully commercialize university research. These three efforts are related to the market (i.e., assuring and validating that the technology is aligned with the market needs), the product (i.e., designing the product's attributes), and the science (i.e., technology development). Bossink (2020) indicates that successful commercialization of new sustainable energy technologies takes place through a series of consecutive demonstration projects. These projects typically start with the development and improvement of prototypes in labs and are followed by the development of operational processes on production sites at which these prototypes are applied. The projects typically end with gaining knowledge regarding market demand. The case study of Shapira and Rosenfeld (2011) discusses the promotion and implementation of a construction technology that is developed through joint research by academia and industry. In doing so, they identify seven stages in the TT process: problem identification, formulation of a solution, prototype development, system test, initial implementation, commercialization, and capitalization on success. For each stage, the authors describe how the university and the firm cooperate. This model of continuous cooperation at each stage of the process implies that, in order to be effective, a continuous combination of the transfer of tacit knowledge and explicit knowledge of the technology itself is needed. The work by Laborde and Sanvido (1994) then focuses on a particular type of TT, namely, one in which construction companies involve university students through a program to foster construction innovation. The TT process they identify consists of technology identification and motivation for implementation, cost analysis and the development of a business plan as part of the technology evaluation, followed by the technology implementation and feedback.

We can conclude that there are different descriptions of the stages of the TT process in the construction sector. However, while the building blocks of the TT processes described here are distinct, there are some communalities. First and foremost, the studies describe the stages of the TT process in construction, from research to commercialization, which does not differ from the traditional TT process described by Bradley et al. (2013). In contrast, there are quite some differences between the studies, and each study has a slightly different emphasis. While Bazan (2019) proposes a framework with feedback loops, the other studies (Bossink, 2020; Laborde and Sanvido, 1994; Shapira and Rosenfeld, 2011) mainly consider the TT process to be linear and consecutive, with technology implementation to test and eventually prove its efficiency to reach the market. Also, in comparison with the TT literature in general, the TT process in the construction sector seems to rely on a much larger involvement in demonstration projects and prototypes. Also, TT in this sector seems to rely on a tight and continuous collaboration between industry and science, which already starts at the early stages of development, namely, at the stage of technology discovery. This seems to be somewhat different from the general TT literature, in which industry is often only considered later in the TT process (e.g., when the technology is already patent-protected).

8.3.3 Channels

There are different channels through which technology from university can be transferred to industry. Wright et al.'s (2008) study in a general, non-construction-specific context identifies the following channels: spin-off creation, patenting and licensing, contract research, consulting, and graduate and researcher mobility. Our review on channels of TT in the construction sector points to a more limited number of channels used in this industry, namely, university spin-off creation, joint R&D, and patenting and licensing.

First, *university spin-off creation* is one of the commercialization routes through which researchers bring their ideas to market. Several studies have highlighted the suitability of establishing university spin-offs to successfully commercialize university-developed construction technologies, such as geosynthetic technology (Jones et al., 2017), nanotechnology (Ozin et al., 2013) and technology for water quality monitoring (Bazan, 2019). Moreover, spin-off creation seems to be particularly relevant for the commercialization of sustainable solutions because this approach allows one to extensively conduct experiments, test the value proposition and the market potential in the university context before making a significant investment (Bazan, 2019). Indeed, the work of Nejabat and Geenhuizen (2019) analyzes 107 spin-offs in Northwest Europe, including Denmark, Finland, Norway, Sweden and the Netherlands, and indicates that 60 percent of sustainable energy spin-offs are able to reach the market and most of them do so within the first five years of their existence. On the downside, however, they do find that the entrepreneurs in these spin-offs face difficulties in scaling their ventures due to insufficient capital and constraints from regulations, and are sometimes forced by market forces to select strategies that are cost-focused instead of sustainability-focused. In contrast, some authors are less convinced that establishing university spin-offs is the best commercialization route for construction innovation. For instance, Bröchner and Lagerqvist (2016) investigate the most important university-industry types of interaction through a questionnaire with industry and university respondents. They find that academic spin-offs are not considered the mainstream type of interaction. Instead, they point to the importance of informal contacts, joint research projects and staff mobility.

Second, the literature on TT in the construction sector has discussed *joint R&D* as a commercialization channel. Watson et al. (2015), for instance, document that in the sector of low-carbon technologies such as energy-efficient cement production, Chinese firms largely benefit from engaging in joint R&D projects with local universities. This is because the jointly developed technologies are cheaper than foreign-made equipment, the after-sales service is more convenient, further improvements are easier to implement, and there is a desire to support the local industry. Along the same lines, Lee and Win (2004) consider joint research to be efficient as it ensures high commitment of the industry to increase the transferability and, as such, enhances the facilities and expertise of both university and industry. The work by Hall (2003) and Shapira and Rosenfeld (2011) also studies joint R&D projects and points to the importance of organizing professional conferences and seminars for both academics and practitioners to boost the transfer of technology developed in joint R&D projects. Alwan et al. (2017) consider the involvement of academia important as it provides research and design expertise, as well as technical innovation and skills training. In their case study, joint research is adopted to develop a model that can facilitate strategies to reduce environmental impacts in construction. The authors state that the interaction between academia and industry already exists in the construction sector, but that it is often informal and uncoordinated. With the aim of fostering industry innovation, Laborde and Sanvido (1994) propose a program to establish partnerships between university and industry to achieve excellence in construction. Apart from educating students as future leaders and improving their skills, the program has the objective to study and develop new construction methods and processes, as well as to provide a forum to share the knowledge among industry, students and faculty. In addition, Bröchner and Lagerqvist (2016) suggest that industry and university consider joint research as the preferred TT route while pointing to the importance of combining transfer of tacit knowledge with acquiring intellectual property rights.

Third, *patenting and licensing* is another way in which technology can be transferred from university to industry. The study by Zhong et al. (2009) analyzes almost 40,000 Chinese patents in the construction field from 2000 to 2015 and indicates that the majority of patents deal with the preparation, handling or processing of building materials or building components on the construction site. They find that universities have increasingly engaged in patent applications since 2007. The important role of universities in patenting activity is also highlighted in the study by Dapurkar and Telang (2017) in which they investigate 228 patents on the application of microorganisms used in the construction industry as cement and concrete additives. The authors find that universities or research institutes are the second-largest major player in the field (holding 28 percent of all patents), while patents stemming from university-industry collaborations only account for 4 percent of the patents. Interestingly, there seem to be important regional differences in terms of patenting. For instance, Bröchner (2013) indicates that in the Nordic region (Denmark, Finland, Norway and Sweden), industry-science collaborations seldomly result in patents, while in the US, many patents stem from university-industry collaborations. The case study conducted by Holley and Farrow (2013) presents a collaborative interdisciplinary product development program at a university in the US to improve productivity and safety in the construction sector. As a result of the program and the partnership with a manufacturer, the developed products were submitted for patenting and licensing. However, there is little unanimity in the literature on the use of patents in construction as a measure to indicate the success of university-industry collaboration (Bröchner, 2013).

In summary, the literature has discussed three TT channels, namely, spin-off creation, joint R&D, and patenting and licensing. Our review indicates that joint research is the most studied TT channel in the construction industry, which is in line with our observation that industry-science collaboration is rather common in all stages of the TT process in the construction industry. The advantages of joint research are addressed in several studies (Laborde and Sanvido, 1994; Lee and Win, 2004; Shapira and Rosenfeld, 2011; Watson et al., 2015). Joint research seems to facilitate the introduction of innovation in the market. Despite the contributions that the discussed studies have brought to the TT field, the current state of the literature is largely descriptive and provides few insights into the contingencies under which specific channels are more or less effective. Neither do these present the decision factors that are taken into account when choosing the channel nor do they present the advantages and disadvantages of the selected channel. In addition, the transfer channels in the selected articles are restricted to spin-offs, joint R&D, and patenting and licensing, with few insights presented into the (potential) role of other TT channels, such as consulting or contract research.

8.3.4 TT Strategies, Tools and Frameworks

Effective strategies are required to overcome challenges in the TT process and ensure its success. For instance, according to Owusu-Manu et al. (2015), there is a need to synthesize sustainable methodologies to improve organizational efficiency of construction consulting firms and innovations into a business framework that enables monitoring TT developments (Gann, 2001). In this regard, the literature has explored strategies, business management tools and frameworks that have the purpose of achieving a sustainable planning process and reaching the market. Specifically, studies have been undertaken to evaluate and identify optimal management practices in TT in the construction industry (Alwan et al., 2017; Muslemani et al., 2020; Shakeel et al., 2017; Suprun et al., 2022; Yin and Li, 2018a, 2018b).

For instance, to improve the energy efficiency for cleaner production in the construction industry, Yin and Li (2018a) propose and implement a decision-making model to match supply and demand of Green Building Technologies (GBT) to promote TT from academic research institutes to construction enterprises. Along the same lines, Yin and Li (2018b) study the TT process based on a dynamic framework through a stochastic differential game approach to investigate strategies that work better in the transfer of GBT. These strategies are related to incentive mechanisms and revenue distribution of GBT. The findings show that the effort level in terms of time of academic research institutes and construction enterprises is proportional to the government subsidy of GBT innovation, pointing to government subsidies as potential long-term incentives to promote TT for these technologies. Alwan et al. (2017) apply a framework for strategic sustainable development in the UK construction industry to assist with the transfer and implementation of new technologies that aim to reduce negative environmental impacts of construction. The framework is tested by use of a case study to evaluate its potential, in which the academic institution provided research and design expertise, as well as technical innovation and skills training. Through this case study, ineffective strategies, policies and leadership were highlighted, which have prevented the full exploitation towards sustainable production. Next, by exploring the theoretical underpinnings of low-carbon technology business models and through semi-structured interviews with industry experts and academics, Muslemani et al. (2020) point to revenue as the central element in creating value for these business models. However, despite the promising potential of low-carbon technologies for energy efficiency in the construction sector, the lack of a viable business model has deterred the commercialization of these technologies developed within universities (Muslemani et al., 2020). Shakeel et al. (2017) present a comprehensive framework for effective commercialization of renewable energy technologies in Finland, highlighting the importance of university-industry collaboration. The framework comprises barriers related to commercialization and measures to overcome these barriers, such as encouraging university-industry collaboration to reduce costs. Uusitalo and Lavikka (2021) develop a platform strategy as a technology in which they aim to standardize construction materials and processes used in the TT process. The platform strategy entails identifying predictable supply chains to provide technological solutions, and it helps companies to overcome uncertainties associated with the TT process.

In sum, the studies presented strategies and frameworks for effective technology commercialization of specific technologies in the construction industry. These studies present a rather fragmented view as they are often focused on a single technology. Also, they have typically focused on a particular strategy or methodology, as such providing few insights into the extent to which these strategies or methodologies could be applicable to other technologies or markets. While the current state of the literature provides some insights into strategies and methodologies currently used in specific cases, it does not present specific solutions regarding the most suitable strategy for a certain purpose. Furthermore, neither contingencies at technology or contextual level affecting the effectiveness of strategies and frameworks have been studied so far.

8.3.5 Impediments and Facilitators to TT

Several studies have elucidated and examined factors that influence the TT process in the construction sector. These factors can either spur or constrain the TT process, its efficiency and outcomes. In what follows, we identify these factors and discuss how they can affect the TT process and its outcomes.

As the TT process is embedded in the regional context, many studies discuss *government regulations and the political situation* that positively or negatively affect TT and its outcomes. Indeed, it is often argued that the government can impose restrictive regulations and barriers, for instance, related to safety, which may impede TT and innovation in the construction sector (Laborde and Sanvido, 1994; Suprun et al., 2022). Nejabat and Geenhuizen (2019) even found that many university spin-offs had to close down partly as a result of the severe policy restrictions they were confronted with. However, the government can also adapt policy and provide incentives in support of TT and innovation in the construction sector (Jones et al., 2017; Suprun et al., 2022). These government incentives can, for instance, include providing loans, fiscal concessions, subsidies or funding. By doing so, the government can promote collaboration between academia and the construction industry (Bazan, 2019; Suprun et al., 2022), which has been shown to enhance innovative activity in the construction sector (Suprun et al., 2022). Relatedly, the government can provide incentives to increase the demand for products and services that may be developed and commercialized through TT (Suprun et al., 2022; Tassicker et al., 2016). Finally, the political situation may affect TT. Suprun et al. (2022) indicate that the political situation in some countries, such as Russia, may limit international TT, requiring increased national R&D efforts with efficient collaboration between university and industry (Suprun et al., 2022).

Some studies have also considered the impact of *larger societal trends and the economic climate* when studying TT in the construction sector. The larger trends in society, such as the transition to sustainability and concerns regarding climate change, represent pressing issues in society and can affect the TT process and its outcomes. Specifically, the construction sector is required to reduce its environmental footprint (Richner et al., 2017), which represents opportunities to develop new energy-efficient materials, use renewable energy, and so on. Yet, these trends require much additional research, and environmental considerations will likely impact the TT and innovation process (Jones et al., 2017). Furthermore, the broader (economic) climate and situation may also influence the TT process and its outcomes (Suprun et al., 2022). For instance, Shapira and Rosenfeld (2011) refer to the impact of the Israeli recession on a successful university-industry collaboration in the construction sector, and argue that, in general, an industry is driven to innovation in the face of a crisis.

Another critical element that affects TT in the construction sector and its success is the *internal resources* of the parties involved in the process. Specifically, engaging the right people, having sufficient internal R&D capabilities and money to invest appear to be crucial for firms and universities to successfully go through the TT process. First, firms aiming to adopt and commercialize university research require internal R&D capabilities, suitable infrastructure and skilled staff members to be able to absorb and utilize these new ideas (Gann, 2001). However, calculations by Gann (2001) show that only 1 percent of the construction firms in the UK have sufficient internal R&D capabilities to successfully absorb and act upon academic research results. Moreover, the author argues that, even when construction firms have the technical competencies to absorb academic research, they often do not have the internal structure nor the culture to internalize these results, learn from them in the long run, and capitalize on them. Along the same lines, for university spin-offs to be successful, it is recommended that experienced business managers join the founding team (Nejabat and Geenhuizen, 2019). Second, having access to investment capital (with acceptable refunding periods) is also essential for successful TT (Nejabat and Geenhuizen, 2019). This is especially important for firms in hardware, machinery and new materials, which face large investment needs, quick depreciation, and require testing at large scale. Particularly, innovations in the construction sector often require a large upfront investment, with effects taking a longer time span to mature. As such, the development cost is often considered the main barrier for the implementation of new technologies (Addy et al., 2020; Dapurkar and Telang, 2017; Peña et al., 2014; Sorrell and Hoffman, 1994). However, the costs can be compensated by savings on maintenance and the subsequent favorable life cycle (Tassicker et al., 2016). The literature indicates that it is often difficult to get innovations accepted into civil engineering practice (Jones et al., 2017), and a variety of restrictive codes and regulations contribute to this (Laborde and Sanvido, 1994). Furthermore, prior research in the construction sector suggests that university spin-offs are often not sufficiently attractive for investors (Nejabat and Geenhuizen, 2019).

Another factor that undeniably affects the TT and its outcomes, especially since the prior section shows that firms and universities often do not have sufficient internal resources to bring TT to a successful conclusion, is the *mutual understanding and collaboration between the university and the industry*. Good collaboration can facilitate the TT process and lead to positive outcomes. Shapira and Rosenfeld (2011) studied a case of a highly successful academic-practitioner cooperation that led to a successful innovation, and summarize several important lessons learned. The lessons include that construction academics should visit construction sites and meet regularly with practitioners and laborers to fully understand the issues and needs of the industry. Also, in the testing and implementation phases, close cooperation between practitioners and academics with daily feedback loops appears to be decisive. The lack of a mutual understanding and collaboration often originates from the differences between university and industry. A first issue relates to the differences in the ways in which they generate and manage knowledge (Gann, 2001). For instance, while university knowledge production encompasses a wide range of disciplines and typically takes a long time, industrial knowledge production generally entails combining limited specialized and generalist knowledge and is developed more rapidly. Relatedly, academic researchers and practitioners may not speak the same language because of different training and experience (Bazan, 2019). These distinctions, along with differences in research objectives, may not only impede collaborations between the two parties, but may already discourage them from considering to collaborate in the first place. Additionally, universities typically value openness and produce knowledge as a public good. Yet, while some construction engineering and design firms also value that their employees publish in prestigious journals, the industry typically aims at protecting its knowledge and intellectual property as much as possible to secure private profits (Gann, 2001; Hairstans and Smith, 2018). These tensions between the university and industry are again difficult to reconcile and may hinder TT. However, it is important to note that many universities have also started to realize the benefits related to commercializing university research. Therefore, many have introduced general TT policies encouraging academics to engage in TT and research commercialization (Bazan, 2019). Such policies can, for instance, state that academics will be rewarded for the contributions they make to R&D projects with commercial purposes. Also, many universities have set up Technology Transfer Offices (TTOs) that provide a range of integrated services, from managing intellectual property to helping with funding applications, all to support TT.

In summary, the literature has pointed to a number of factors that can impede or stimulate TT in the construction industry. Specifically, it has pointed to the role of government regulations, societal trends and the economic climate, internal resources at the firm level and a mutual understanding and collaboration between industry and university. Overall, these factors seem to coincide to a large extent with the factors that have been considered relevant for TT in general (see, e.g., Bruneel et al., 2010).

8.3.6 Outcomes of the TT Process

A number of studies have considered the outcomes of TT. In what follows, we provide an overview of studies that have discussed the impact of university-industry TT for firm performance, the firm's business and business models and knowledge development. First, a new technology may affect *performance and efficiency* (Jones et al., 2001; Jones et al., 2017). In addition to the optimization of construction processes, the interaction between university and industry can reduce costs, provide support for companies to globalize their operations and offer infrastructural support (Shakeel et al., 2017). In general, increasing the efficiency of the process implies cost savings, but the outcome of TT includes furthering efforts for technology's consistency and standardization (Peña et al., 2014). Construction companies expect to increase profitability and productivity with the TT process (Suprun et al., 2022). At the same time, the involvement of researchers provides better solutions to industry (Gambatese and Hallowell, 2011; Jones et al., 2017; Ryńska et al., 2019), either through new technologies or decision support models. The current studies have mainly made such claims, often supported by theory. However, we did not find any studies that empirically studied the relationship between TT and firm performance or other firm outcomes in the construction industry.

Second, construction innovation has an impact for a firm's *business and business models*. By transferring new technology to the market and promoting it, construction enterprises can improve their competitiveness (Yin and Li, 2018b), whereby technological innovation acts as an important marketing element for the company (Shapira and Rosenfeld, 2011). Innovation is likely to attract new customers. This is particularly the case when more sustainable technologies are being developed, not only because of demands following climate change but also due to the growing awareness of people pushing policies towards sustainability. Consequently, these technologies can shape the industry as it has been a pressing issue to reduce carbon dioxide emissions. New construction materials, processes and designs have been developed to improve energy efficiency and meet the building sector requirements to reduce its environmental footprint (Richner et al., 2017). TT is an essential part of business models, as well as a way to gain competitive advantage through market disruption (Uusitalo and Lavikka, 2021). Specifically, it is argued that TT can be an effective way for construction companies to move away from the current economic model towards a more resource-efficient one, in line with increased demand for sustainability. Next to these efficiency gains, CEOs of construction companies are also interested in TT as it could benefit their employees and the community (Uusitalo and Lavikka, 2021). Bridging the gap between academic research and university through TT helps firms in bringing new construction products and services to the market, driving the development of the sector (Richner et al., 2017). The industry then benefits from the opportunities that a new technology brings by providing value to clients (Owusu-Manu et al., 2015).

Third, the literature has discussed benefits of TT in terms of *knowledge development*. Specifically, by engaging in academia-industry cooperation, the construction industry benefits from the synergic outcome (Shapira et al., 2008), improving its absorptive capacity by developing better feedback and lifelong learning mechanisms, education and training (Gann, 2001). Ryńska et al. (2019) consider the fact that TT provides better knowledge to industry as one of the main implications. The interaction between educational institutions and industries provides vital resources to the companies in the form of human capital and knowledge that can be useful in improving the technical aspects of technology (Shakeel et al., 2017). At the same time, TT brings research opportunities for the university, allows for expanding relationships with industry partners and provides research validation (Holley and Farrow, 2013).

In summary, studies into the outcomes of the TT process in the construction industry have focused on firm performance, the firm's business and business models and knowledge development. Despite the insights that these studies have generated, a number of important gaps remain in the literature. First, the focus of studies into the implications of TT has mainly been case-based. While cases provide important insights, they often do not allow for generalization. Second, the literature has remained vague on how TT effectiveness can be measured and assessed. Third, the predominant focus on the outcomes for firms and the industry has led to scant insights into the outcomes and implications of TT for the involved universities and research institutions.

8.4 A FUTURE RESEARCH AGENDA FOR TT STUDIES IN THE CONSTRUCTION INDUSTRY

In what follows, we propose a future research agenda, guided by our analysis on the current state of the literature on TT in the construction industry. Specifically, we highlight opportunities for qualitative and quantitative research to further contribute to the advancement of knowledge regarding the TT process, its influencing factors and outcomes in the construction industry.

8.4.1 TT Process

First, our study indicated that *prototyping and demonstration projects* are rather common in TT in the construction industry. Interestingly, the general TT literature has not considered these as typical elements of the TT process or as part of TT channels (Wright et al., 2008). Therefore, we are convinced that both the general and the construction-specific TT literature could benefit from more in-depth studies into prototyping and demonstration projects. This is because prototypes are not only common in construction but may also be particularly relevant in other industries, such as software development and car manufacturing (Elverum et al., 2016). Prototypes and demonstration projects are relevant to verify and validate assumptions and calculations and are particularly relevant in contexts in which large upfront investments are required. They can also help in gaining institutional support for new technologies and convincing stakeholders such as investors and customers (Elverum et al., 2016). Furthermore, the case studies of prototypes and demonstration projects addressed in this review are mainly focused on incremental innovations. Thus, further studies could discuss radically new technologies and shed light on the role of prototypes and demonstration projects in reaching the market.

Second, our review also indicates that some TT channels, such as spin-offs, joint R&D, patenting and licensing have been frequently studied. Future research could purposefully *focus on less frequently studied channels*, such as consulting, contract research and graduate mobility, and shed light on whether or not they are used in the construction sector. Furthermore, future studies could investigate the advantages and disadvantages of the chosen channel for this specific industry. More insights into these channels could also provide important indications on which is the most suitable channel and TT strategy for a certain purpose, and it could help in understanding the factors that affect the effectiveness of specific channels. As such, future research could develop a framework that could guide the choice for specific channels. Finally, we also urge future research to investigate the later stages of the TT process, in which new technologies are brought to market, in contrast to many studies which have considered the earlier stages of the process, namely, that of technology development.

Finally, although our review indicates that the literature has discussed the stages, channels and strategies, frameworks and tools of the TT process in the construction industry, we still have scant insights into how TT in the construction has emerged and what the main *best practices* during the evolution were. While the general TT literature refers to the Bayh-Dole Act as a major milestone for TT in general (Shane, 2004), the literature on TT in the construction industry provides few insights into when and how TT in the construction industry emerged and what the major achievements of TT in this industry are. Future research could provide deeper insights into the *history of TT in the construction industry* and would allow us to learn from best practices in this field.

8.4.2 Impediments and Facilitators

First, studies into TT in the construction industry have either used a case-based approach or have considered the construction industry as a whole (e.g., studies into patents). However, the construction industry likely consists of subindustries with *diverse market dynamics*. For instance, while the construction industry is typically considered conservative (Beerepoot and Beerepoot, 2007), some subindustries such as environmental engineering may be less conservative and more open to radical innovations and TT collaborations. Future research could assess the extent to which the subindustry a firm belongs to affects its engagement in TT. In order to do so, a research design that unites firms or TT collaborations in different subindustries should be developed. Taking such a perspective would allow for a stronger contextualization of TT within and beyond the construction industry.

Second, our review indicates that there are regional differences in the use of specific TT channels. However, the literature provides few insights into the *factors that affect the choice of specific TT channels*. Future research could assess these factors, which could be situated at the regional or national (e.g., regulations, market dynamics, political and economic situation) or firm level (management experience, available absorptive capacity, etc.).

8.4.3 Outcomes

First, following the largely case-based nature of studies into the TT process in the construction industry, future research could use *quantitative, large-scale studies* to provide insights that are more generalizable. In order to do so, the identification or development of performance indicators that are of relevance to this industry or subindustry would help to provide such generalizable and quantifiable insights. As the construction industry is a conservative one and is resistant to the entry of new technologies and innovation (Shapira et al., 2011), such generalizability and quantification may have important practical implications, as it may encourage the transfer of technology.

Second, future research could also *identify business models of TT* that work and conditions under which they do. For instance, does the seminal model on commervcialization strategies for new technologies developed by Gans and Stern (2003), based upon Teece's work (1986), also hold for the construction industry? Does it allow for differentiating between construction technologies that should be brought to the market for ideas or the market for products? Are there factors that the model does not incorporate but that are of particular relevance to this industry?

Finally, as much of the research into outcomes of the TT process has taken the perspective of the firm or industry, future studies can *evaluate how and when TT in the construction sector* may or may not be *beneficial to universities and research institutions*, and may identify barriers and stimuli to such transfer from the academic perspective. Here again, the identification or development of performance indicators would be an important step forward.

8.5 CONCLUSION

This study analyzed 34 papers on TT in the construction industry, which is considered a rather conservative industry, experiencing a number of pressures towards the implementation of new technologies and innovation. We found that the literature has mainly studied the TT process, impediments and facilitators and outcomes. We summarized these insights from the literature and provided a research agenda for future research.

ACKNOWLEDGEMENT

This project has received funding from the European Union's Horizon 2020 research and innovation programme under the Marie Skłodowska-Curie grant agreement No 860006.

REFERENCES

Addy, M., Adinyira, E., Danku, J.C., and Dadzoe, F. (2020). Impediments to the development of the green building market in sub-Saharan Africa: The case of Ghana. *Smart and Sustainable Built Environment.*

Ahmad, I.U., Russell, J.S., and Abou-Zeid, A. (1995). Information technology (IT) and integration in the construction industry. *Construction Management and Economics*, *13*(2), 163–71.

Al-Gasim, Z., Senin, A.A., and bin Yusoff, M.E. (2021). A review and comprehensive analysis of the performance of university–construction industry collaboration. *Civil Engineering Journal*, *7*(4), 763–74.

Alwan, Z., Jones, P., and Holgate, P. (2017). Strategic sustainable development in the UK construction industry, through the framework for strategic sustainable development, using Building Information Modelling. *Journal of Cleaner Production, 140*, 349–58.

Bazan, C. (2019). "From lab bench to store shelves:" A translational research & development framework for linking university science and engineering research to commercial outcomes. *Journal of Engineering and Technology Management, 53*, 1–18.

Beerepoot, M., and Beerepoot, N. (2007). Government regulation as an impetus for innovation: Evidence from energy performance regulation in the Dutch residential building sector. *Energy Policy, 35*(10), 4812–25.

Bengoa, A., Maseda, A., Iturralde, T., and Aparicio, G. (2021). A bibliometric review of the technology transfer literature. *The Journal of Technology Transfer, 46*, 1514–50.

Bossink, B. (2020). Learning strategies in sustainable energy demonstration projects: What organizations learn from sustainable energy demonstrations. *Renewable and Sustainable Energy Reviews, 131*, 110025.

Bradley, S., Hayter, C.S., and Link, A. (2013). Models and methods of university technology transfer. *Foundations and Trends in Entrepreneurship, 9*(6).

Bröchner, J. (2013). Construction patents and university-industry research interaction. *Construction Innovation, 13*(4), 410–23.

Bröchner, J., and Lagerqvist, O. (2016). From ideas to construction innovations: Firms and universities collaborating. *Construction Economics and Building*, 16(1), 76–89.

Bruneel, J., d'Este, P., and Salter, A. (2010). Investigating the factors that diminish the barriers to university–industry collaboration. *Research Policy, 39*(7), 858–868.

Dapurkar, D., Telang, M. (2017). A patent landscape on application of microorganisms in construction industry. *World Journal of Microbiology and Biotechnology, 33*(7), 1–12.

Elverum, C. W., Welo, T., and Tronvoll, S. (2016). Prototyping in new product development: Strategy considerations. *Procedia CIRP, 50*, 117–122.

Gambatese, J. A., Hallowell, M. (2011). Factors that influence the development and diffusion of technical innovations in the construction industry. *Construction Management and Economics, 29*(5), 507–517.

Ganesan, S., Kelsey, J. (2006). Technology transfer: international collaboration in Sri Lanka. *Construction Management and Economics, 24*(7), 743–753.

Gann, D. (2001). Putting academic ideas into practice: technological progress and the absorptive capacity of construction organizations. *Construction Management & Economics, 19*(3), 321–330.

Gans, J., Stern, S. (2003). The product market and the market for "ideas": commercialization strategies for technology entrepreneurs. *Research Policy, 32*(2), 333–50.

Hairstans, R., and Smith, R.E. (2018). Offsite HUB (Scotland): establishing a collaborative regional framework for knowledge exchange in the UK. *Architectural Engineering and Design Management, 14*(1–2), 60–77.

Hall, K.D. (2003). Arkansas Superpave® Seminars: model technology transfer partnership. *Transportation Research Record, 1848*(1), 101–5.

Hausberg, J.P., and Korreck, S. (2021). Business incubators and accelerators: A co-citation analysis-based, systematic literature review. In Sarfraz A. Mian, Magnus Klofsten, and Wadid Lamine (eds), *Handbook of Research on Business and Technology Incubation and Acceleration* (pp. 39–63). Cheltenham, UK and Northampton, MA, USA: Edward Elgar Publishing.

Hayter, C.S., Nelson, A.J., Zayed, S., and O'Connor, A.C. (2018). Conceptualizing academic entrepreneurship ecosystems: A review, analysis and extension of the literature. *The Journal of Technology Transfer, 43*(4), 1039–82.

Holley, P.W., and Ben Farrow, C. (2013). Expanding collaboration in academia: Case study of the development of construction products. *Journal of Professional Issues in Engineering Education and Practice, 139*(2), 139–47.

Jones, C., Lamont-Black, J., Huntley, D., Alder, D., and Glendinning, S. (2017, April). Electrokinetic geosynthetics: From research to hype to practice. In *Proceedings of the Institution of Civil Engineers-Civil Engineering* (Vol. 170, No. 3, pp. 127–34). Thomas Telford.

Jones, K.W., Feng, H., Stern, E.A., Lodge, J., and Clesceri, N.L. (2001). Dredged material decontamination demonstration for the port of New York/New Jersey. *Journal of Hazardous Materials, 85*(1–2), 127–43.

Laborde, M., and Sanvido, V. (1994). Introducing new process technologies into construction companies. *Journal of Construction Engineering and Management, 120*(3), 488–508.

Lecluyse, L., Knockaert, M., and Spithoven, A. (2019). The contribution of science parks: A literature review and future research agenda. *The Journal of Technology Transfer, 44*(2), 559–95.

Lee, J., and Win, H.N. (2004). Technology transfer between university research centers and industry in Singapore. *Technovation, 24*(5), 433–42.

Muslemani, H., Liang, X., Kaesehage, K., and Wilson, J. (2020). Business models for carbon capture, utilization and storage technologies in the steel sector: A qualitative multi-method study. *Processes, 8*(5), 576.

Nejabat, R., and Geenhuizen, M.V. (2019). Entrepreneurial risk-taking in sustainable energy: University spin-off firms and market introduction in northwest Europe. *Sustainability, 11*(24), 6952.

Owusu-Manu, D.G., Badu, E., Holt, G.D., and Edwards, D. (2015). Determinants of management innovation in the Ghanaian construction consulting sector. *Construction Economics and Building, 15*(1), 75–88.

Ozin, G.A. (2013). Spin of a nanotech spin-off. *Advanced Engineering Materials, 15*(1–2), 8–18.

Peña, V., Lal, B., and Micali, M. (2014). US federal investment in the origin and evolution of additive manufacturing: A case study of the National Science Foundation. *3D Printing and Additive Manufacturing, 1*(4), 185–93.

Pinkse, J., and Dommisse, M. (2009). Overcoming barriers to sustainability: An explanation of residential builders' reluctance to adopt clean technologies. *Business Strategy and the Environment, 18*(8), 515–27.

Pries, F., and Janszen, F. (1995). Innovation in the construction industry: The dominant role of the environment. *Construction Management and Economics, 13*(1), 43–51.

Richner, P., Heer, P., Largo, R., Marchesi, E., and Zimmermann, M. (2017). NEST – a platform for the acceleration of innovation in buildings. *Informes de la Construcción, 69*(548), e222.

Ryńska, E., Koźmińska, U., and Rucińska, J. (2019). Effectivity–ecosphere–economics in nZEB retrofit procedures. *Environmental Science and Pollution Research, 26*(29), 29544–59.

Shakeel, S.R., Takala, J., and Zhu, L.D. (2017). Commercialization of renewable energy technologies: A ladder building approach. *Renewable and Sustainable Energy Reviews, 78*, 855–67.

Shane, S. (2004). Encouraging university entrepreneurship? The effect of the Bayh-Dole Act on university patenting in the United States. *Journal of Business Venturing, 19*(1), 127–51.

Shapira, A., and Rosenfeld, Y. (2011). Achieving construction innovation through academia-industry cooperation – keys to success. *Journal of Professional Issues in Engineering Education & Practice, 137*(4), 223–31.

Shapira, A., Rosenfeld, Y., and Mizrahi, I. (2008). Vision system for tower cranes. *Journal of Construction Engineering and Management, 134*(5), 320–32.

Skute, I., Zalewska-Kurek, K., Hatak, I., and de Weerd-Nederhof, P. (2019). Mapping the field: A bibliometric analysis of the literature on university–industry collaborations. *The Journal of Technology Transfer, 44*(3), 916–47.

Sorrell, C.A., and Hoffman, P.A. (1994). Innovative composite materials research at the US Department of Energy, Office of Energy Efficiency and Renewable Energy. *Composites Engineering*, *4*(8), 857–82.

Suprun, E., Sahin, O., Anthony Stewart, R., and Panuwatwanich, K. (2022). Examining transition pathways to construction innovation in Russia: A system dynamics approach. *International Journal of Construction Management*, *22*(4), 556–78.

Tassicker, N., Rahnamayiezekavat, P., and Sutrisna, M. (2016). An insight into the commercial viability of green roofs in Australia. *Sustainability*, *8*(7), 603.

Teece, D.J. (1986). Profiting from technological innovation: Implications for integration, collaboration, licensing, and public policy. *Research Policy*, 15, 285–305.

Teece, D.J., Pisano, G., and Shuen, A. (1997). Dynamic capabilities and strategic management. *Strategic Management Journal*, *18*(7), 509–33.

Uusitalo, P., and Lavikka, R. (2021). Technology transfer in the construction industry. *The Journal of Technology Transfer*, *46*, 1291–320.

Watson, J., Byrne, R., Ockwell, D., and Stua, M. (2015). Lessons from China: Building technological capabilities for low carbon technology transfer and development. *Climatic Change*, *131*(3), 387–99.

Welter, F. (2011). Contextualizing entrepreneurship – conceptual challenges and ways forward. *Entrepreneurship Theory and Practice*, *35*(1), 165–84.

White, A., and Schmidt, K. (2005). Systematic literature reviews. *Complementary Therapies in Medicine*, *13*(1), 54–60.

Wright, M., Clarysse, B., Lockett, A., and Knockaert, M. (2008). Mid-range universities' linkages with industry: Knowledge types and the role of intermediaries. *Research Policy*, *37*(8), 1205–23.

Yin, S., and Li, B. (2018a). Matching management of supply and demand of green building technologies based on a novel matching method with intuitionistic fuzzy sets. *Journal of Cleaner Production*, *201*, 748–63.

Yin, S., and Li, B. (2018b). Transferring green building technologies from academic research institutes to building enterprises in the development of urban green building: A stochastic differential game approach. *Sustainable Cities and Society*, *39*, 631–8.

Zhong, B., Hei, Y., Li, H., Rose, T., and Luo, H. (2019). Patent cooperative patterns and development trends of Chinese construction enterprises: A network analysis. *Journal of Civil Engineering and Management*, *25*(3), 228–40.

9. The role of Universities of Applied Sciences in technology transfer: the case of Germany[1]

Alexander Starnecker and Katharine Wirsching

9.1 INTRODUCTION

In the mid-1960s, German public universities were no longer able to supply the human capital needed by industry. Higher educational qualifications for non-scientific jobs were needed and, moreover, met the interests of the young generation. Thus, the German government developed a new institution for the higher education sector, the University of Applied Sciences *(Fachhochschule)*. By applying a well-known concept used in industry for a long time, the division of labor, the new concept can be interpreted as a direct answer to the demand of industry (German Science Council, 2002; Krause, 2005; Kulicke and Stahlecker, 2004), providing basic researchers graduating from public universities on the one hand and graduates specialized in applied sciences from Universities of Applied Sciences (UAS) on the other hand (Lehmann and Starnecker, 2012; Pasternack, 2009). The goal was to educate experts who were able to "change from a white collar to a blue collar," as required (Spiegel, 1969, p. 97). Looking at the regional distribution of public universities, it is noticeable that they are mainly located in major German cities. Thus, they are primarily orientated towards cooperation with large companies, whereas UAS, specialized in applied research, tend to be located in rural areas and are more attractive to small and medium-sized enterprises (SMEs). The contribution of UAS to the economy, particularly as a reliable partner to SMEs, is still generally underestimated. Although UAS have improved in research, traditionally the preserve of public universities, they are still not regarded as equal partners in the higher education sector (Osel, 2012). The persistent impression that education and research at UAS are less valuable than that carried out at public universities is deeply rooted in the history of UAS. One minor example may serve to illustrate this: in the beginning, engineers graduating from public universities received a company-paid cup of coffee with their lunch, while engineers from UAS did not (Spiegel, 1969).

In the last decade, this division of labor between UAS and public universities has faced an additional objective – the technology transfer process. While the role of public universities and their role within the technology transfer processes is intensively studied (Acs et al., 2016; Hülsbeck et al., 2013), the impact of UAS remains little investigated. Although they are quite successful in their cooperation with the industry and are nevertheless another backbone in the university-industry relationship, only anecdotal evidence on this type of university exists (Lehmann and Starnecker, 2012). However, UAS could be a role model, in particular for countries and regions where SMEs dominate the industrial landscape. Besides Germany, this is also shown by the success of comparable UAS in the Netherlands, Finland and Switzerland (Lepori and Kyvik, 2010). This chapter tries to shed some light on this particular type of university that is one of the major success factors of German economy (Audretsch and Lehmann, 2016). The following section provides an overview of the German higher education landscape, introduces the concept of the UAS and highlights their importance for the higher education sector as well as the economy. Section 9.3 focuses on technology transfer at UAS before the Section 9.4 points out parallels to similar concepts in other European countries. The chapter concludes in Section 9.5 with a research outlook.

9.2 THE UNIVERSITIES OF APPLIED SCIENCES IN GERMANY

The German Federal Statistical Office counted 422 higher education institutions in 2020/21 (Table 9.1), whereof about half are UAS, educating one-third of all German students. Around 2.8 million students were enrolled in German higher education institutions, of which 0.75 million (27 percent) visited state-owned UAS and 1.7 million (61 percent) attended public universities.

Table 9.1 *Distribution of higher education institutions in 2020/21*

Universities	108
Universities of Applied Science	210
Colleges of Art & Music	52
Colleges of Public Administration	30
Colleges of Education	6
Colleges of Theology	16
Total Number of Higher Education Institutions	422

Source: German Federal Statistical Office (2021a).

There are two major reasons why public universities outweigh state-owned UAS in the number of students, although there is an overall higher number of UAS institutions. First, public universities tend to be located in major cities and thus can potentially attract more students. Second, public universities have a higher capacity due to major structural differences (Fritsch et al., 2007). The average public university has almost 22,000 students and 260 professors, whereas the average UAS consists of 7,000 students and 150 professors (German Statistical Office, 2021a; Figure 9.1). Obviously, German public universities are larger than UAS. However, the intensity of student care at UAS is almost twice as high (2.3 per 100 students) compared to public universities (1.3). Taking research fellows into account, the intensity of care shifts in favor of public universities. The average UAS employs only 270 research fellows compared to 1,500 in public universities. Historically, only public universities have kept the right to award doctoral degrees, which was one major motivation for research fellows to stay at the university after a Master's degree. In the last decade, Germany underwent a change of policy, based on a judgment of the German Federal Constitutional Court in 2010 granting equal rights to UAS professors. In 2016, Hessen was the first German state to allow UAS the right to award doctorate degrees, followed by Saxony-Anhalt in 2020. Some states, for example, North Rhine-Westphalia, decided to allow only doctoral study courses in cooperation with a public university. The equalization of UAS and public universities, although legally possible, is elusive, explained by three major structural differences between these two types of institutions: (1) teaching hours of UAS professors are twice as long, leaving less time for research; (2) the objective of UAS is not to educate research fellows and young researchers, but is a central mission of public universities; (3) small UAS sometimes only have one professor per field, restricting scientific discourse within the institutions and making it seem impossible to create a qualified doctoral committee (Pautsch, 2020).

Public universities are usually focused on basic research in different scientific areas, whereas the specialization of UAS is clearly in applied fields and disciplines, such as all of the engineering sciences. Two-thirds of all German engineers graduate from UAS (German Federal Statistical Office, 2021a). The more applied approach in the educational system also allows a shorter education time, which reduces the time to graduation and provides skilled workers at a younger age (Krause, 2005, p. 53). In the initial founding phase of UAS, the German Minister of Research, Gerhard Stoltenberg, between 1965 and 1969, gave the target that a student at a UAS should be educated at 25–30 percent of the cost required for a student at a public university, due to considerably lower investment in research equipment and a shorter period of study (Spiegel, 1969). In 2019, 8,140 euros per year was spent on each student at a public university, while only an average of 4,330 euros per year was spent at UAS. The expense per professor is nearly three times higher at a public university (665,670 euros per year) compared to UAS (238,560 euros per year) (German Federal Statistical Office, 2021b). Regards the time to graduation, students at UAS need ten semesters on average to complete their degrees (Bachelor and Master), whereas students at public universities finish after 12.8 semesters (German Federal Statistical Office, 2021b). Since education is considered a public good, UAS are intended to make higher education more affordable. The importance of UAS for the German higher education system is continuously rising. In the last 15 years, growth rates of UAS were almost three times that of public universities (Osel, 2012). In 2021, almost 45 percent of all first-semester students enrolled in UAS (German Federal Statistical Office, 2021b).

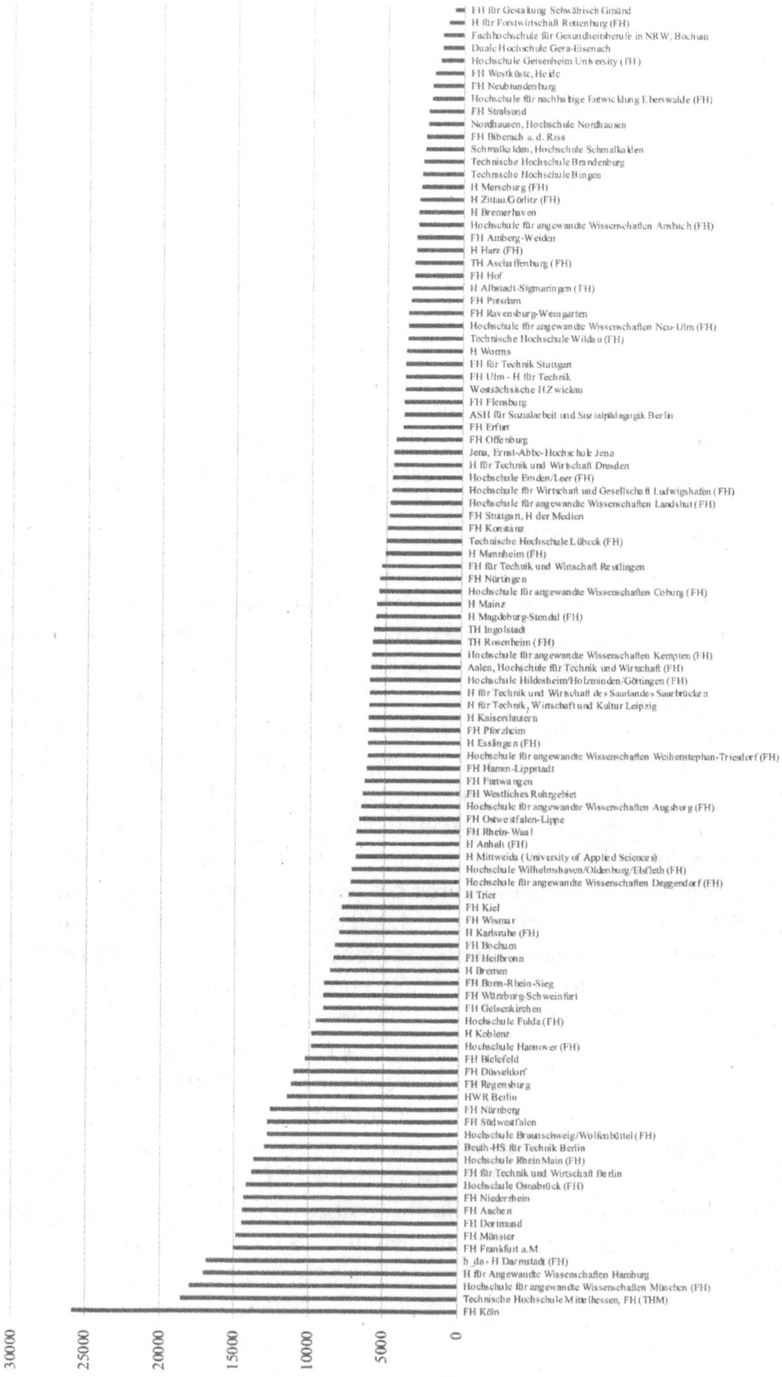

Source: German Federal Statistical Office (2021a).

Figure 9.1 *Number of students per UAS in 2018*

Additionally, the contribution of UAS to applied research is also growing (Figure 9.2). Consequently, research is developing to be the second pillar of UAS' profile. The core competence of UAS is short-term and implementation-oriented research, while research at public universities is more basic and long-term oriented (Schmoch et al., 2000). In conclusion, the most important aim of public policy (when founding UAS) was to create higher education institutions specialized in applied research and education as well as to facilitate knowledge spillovers to industry predominantly located in the more rural areas of Germany.

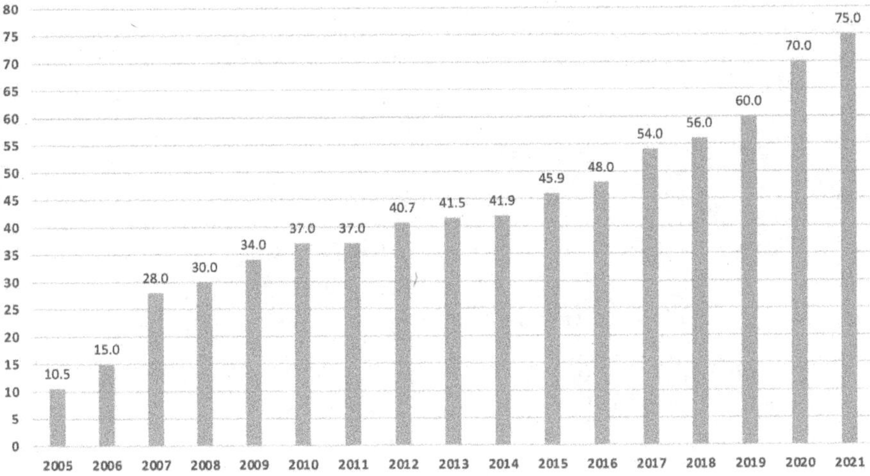

Source: BMBF (2021).

Figure 9.2 *Funding for research projects of UAS by the German Ministry of Education & Science (in million €)*

Approximately one-third of the UAS existed before 1967. They had their origin in Technical Schools, Academies for Engineers, and other similar institutions (Kulicke and Stahlecker, 2004). The first period yielding a considerable number of UAS was in 1971 when more than a fifth of the UAS were founded. Then again between 1991 and 1996 – after the Reunion of Germany – another 33 were set up, particularly in Eastern Germany, to foster a rapid improvement of the educational level in the former communist country (Kulicke and Stahlecker, 2004). Only very few UAS have been established outside these two key periods (Figure 9.3).

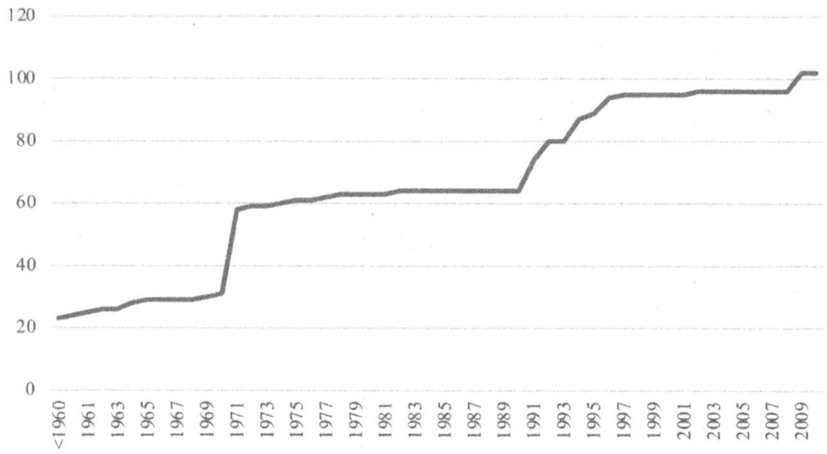

Source: Hochschulkompass (2021).

Figure 9.3 Years in which UAS were founded

The organization and structure of UAS encourage applied research and education. The right to grant a doctorate degree is almost exclusively the preserve of public universities, making it very difficult for the UAS to attract research fellows (Lehmann and Starnecker, 2012). Thus, UAS are mainly made up of professors, administrators and students. Not surprisingly, examination of performance indicators such as third party funding, publications and patents show dramatically lower success rates for UAS (Einhaeupl, 2005). Compared to public universities, the time available for research is significantly lower at UAS. UAS professors are expected to teach about 18 hours a week (in contrast to professors at public universities who only teach nine hours). Consequently, 55 percent of the available time of professors at public universities is scheduled for research compared to just 5–10 percent at UAS (Pasternack, 2009). Czarnitzki et al. (2000) carried out a comparison of major German research institutions. They found that about 90 percent of the research activities of professors at UAS could be considered as applied research and development, compared to only 40 percent at public universities.

Table 9.2 Comparison of the division of research activities of professors at public universities and AUS

	Basic Research (%)	Applied Research (%)	Development (%)	No. R&D (%)
Public Universities	57	33	10	0
Universities of Applied Science	6	50	39	5

Source: Czarnitzki et al. (2000).

By contrast, basic research activities at public universities account for almost 60 percent as opposed to just 6 percent at UAS (Czarnitzki et al., 2000). The approximate values for public universities and UAS are shown in Table 9.2.

This circumstance is not surprising since, with the foundation of UAS, policy did not pursue the goal of improving access to human capital and applied knowledge for industry in general, but rather to specifically promote medium-sized companies in rural areas and thus stimulate economic performance in these regions. While public universities were predominantly concentrated in major cities such as Berlin, Munich or other areas of high population density like the Ruhr valley, UAS tend to be located in less populated areas (Figure 9.4). This means that relatively small cities have UAS, giving them closer access to higher education opportunities for cooperation. Most of the studies analyzing the role of universities in the regional innovation system come to the conclusion that proximity to a higher education institution has a positive influence on different facets of a region as well as on the performance of its companies (e.g., Anselin et al., 1997; Audretsch and Lehmann, 2005; Audretsch and Stephan, 1996; Audretsch et al., 2013; Beise and Spielkamp, 1996; Glaeser et al., 1992).

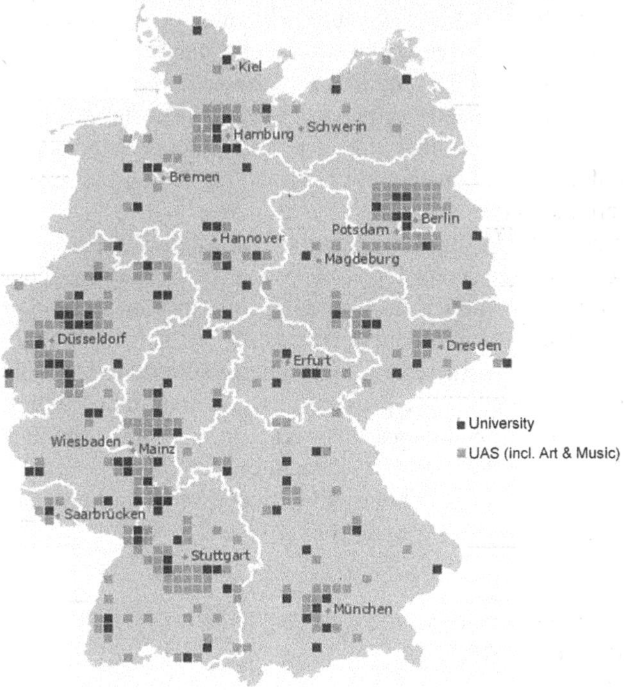

Source: Kulicke and Stahlecker (2004).

Figure 9.4 *Distribution of public universities and UAS in Germany*

Geographical proximity always leads to positive effects when scientists and people working in industry make personal contacts, leading to an informal technology transfer (Audretsch et al., 2012; Audretsch et al., 2014; Fritsch and Slavtchev, 2007; Grimpe and Fier, 2010; Link et al., 2007). The local distribution of UAS leads to close geographical proximity, particularly to smaller companies, which is generally supposed to enhance technology transfer, especially to SMEs (Audretsch and Lehmann, 2015; Boettcher, 2004; Link et al., 2007).

Proximity is one of the important advantages of UAS to enhance regional knowledge transfer (Beise and Stahl, 1999; Einhaeupl, 2005; Schmoch et al., 2000). Moreover, UAS respond to the needs of industry. Czarnitzki et al. (2000) found that the main drivers behind research and development (R&D) activities at UAS were SMEs. In addition to regional proximity, UAS also offer cognitive proximity. Overcoming cultural barriers and thus "boundary spanning" between scientists as suppliers and firms as customers of research outcomes is one of the key factors for successful technology transfer (Siegel et al., 2003). This cooperation between SMEs and UAS is very successful in Germany and reflected in the high number of so-called Hidden Champions. Hidden Champions are market leaders in their continent or number one, two or three worldwide in their market niche and generate revenues up to 4 billion euros (Audretsch and Lehmann, 2016, pp. 30–5). Additionally, hidden champions are largely unknown by the public because they often operate in business-to-business (B2B) markets and are settled in the regions (mostly rural) where they were established (Lehmann et al., 2019). These companies are highly innovative, re-engineering their products and processes is a continuous process. Therefore, hidden champions benefit from the regional proximity to UAS and their applied research activities.

Another reason for the applied approach in research and education at UAS can be found in a special requirement for professors at UAS: they must obtain at least five years of practical job experience outside higher education institutions. These practical experiences then lead to the fact that existing contacts to companies in the region are further exploited and, therefore, SMEs located closed to UAS in particular profit from the regional and cognitive proximity (Beise and Stahl, 1999; German Science Council, 2000, pp. 23–4). Against this backdrop, it is not surprising that UAS have a significant impact on the economy since they were introduced to enhance technology transfer to SMEs, the backbone of German economy (Audretsch and Lehmann, 2016).

9.3 TECHNOLOGY TRANSFER AT UNIVERSITIES OF APPLIED SCIENCE

Important knowledge for the German industry is not only generated at public universities (Audretsch et al., 2012; Audretsch et al., 2014) but also at UAS, focusing on implementation-oriented R&D as well as education. Krause (2005) highlights the importance of UAS, particularly to SMEs. The geographical proximity, as well as

Table 9.3 Third party funding of UAS in 2018

	Third Party Funding (in 1,000 €)	**Percentage (%)**
Engineering	311 948	52
Economics	78 529	13
Computer Science	61 834	10
Agricultural Science	34 435	6
Natural Science	33 810	6
Social Services	32 722	5
Art	10 978	2
Others	37 064	6
Total	601 320	100

the focus on applied science, makes it easier for smaller firms to cooperate with UAS than with public universities. Applied research has become essential for industry, especially for SMEs (German Science Council, 2000, pp. 22–3). In particular, those SMEs located nearby UAS profit from the regional and cognitive proximity (Beise and Stahl, 1999; German Science Council, 2000, pp. 23–4). Until now, the contribution of UAS to the technology transfer process has barely been looked at. The high spin-off activity of UAS (Krause, 2005) is an additional hint at the technology transfer potential of UAS.

Beise and Stahl (1999) were among the first to examine the impact of UAS and highlight the significance of the proximity between UAS and companies. In their study, representatives of companies were asked to name useful knowledge resources, for example, public universities, UAS and several other research institutes. Around 80 percent of the enterprises that considered UAS as a useful source of knowledge were located within 75 kilometers of a UAS (Beise and Stahl, 1999). By comparison, only around 40 percent of company representatives naming public universities as a useful source of knowledge were within the same catchment area (Beise and Stahl, 1999). Einhaeupl (2005, p. 36) and Schmoch et al. (2000) confirm the importance of UAS for regional knowledge transfer. In addition, UAS are responsive to the needs of industry. Czarnitzki et al. (2000) found that the principal drivers behind R&D activities at UAS are SMEs. Dziatzko et al. (2011) explain the need for an innovation manager in SMEs, with the responsibility to put the innovation process into practice. Firms that cannot afford to employ an innovation manager could find a reliable partner in the UAS (Lehmann and Starnecker, 2011). Overall, three major sources of knowledge and technology transfer exist (Czarnitzki et al., 2000). Most important are final-year theses, which address current problems existing in companies. The second major source derives from previous associations with industry. The third very important type of technology transfer from UAS is research cooperation with firms. All three sources are fundamentally due to their close relations with industry and their main focus on applied research. In conclusion, there can be no doubt about the importance of the close connection between UAS and industry. Companies, especially SMEs, are highly dependent on the knowledge and human capital generated at UAS for their success.

The focus of technology transfer in UAS is, beyond doubt, the field of engineering. Besides the large number of students graduating in the field of engineering, the third party funding, a proxy for technology transfer, is also significantly high. Compared to other scientific fields, more than half of all third party funding received by UAS is associated with engineering (Table 9.3). Within the field of engineering, the most successful segment in technology transfer is mechanical engineering.

As already mentioned above, besides regional proximity, UAS also show closer cognitive proximity to companies, especially SMEs. Most SMEs are not willing to hire their own researchers and scientists due to their small financial capabilities or to the fact that there is no need for basic research. This fact may increase the complications in cooperation between public universities and SMEs because of a missing common understanding. At this point, UAS have the advantage that their professors mandatorily have to gain at least five years of work experience outside institutions of higher education and thus have lower cognitive barriers. At public universities, professors pursue a traditional academic career, often without any practical experience outside the higher education sector. The work experience of professors at UAS equips them with a better understanding of how companies work, thus lowering the cultural barriers between higher education and industry. Furthermore, Beise and Stahl (1999) state that UAS tend to specialize in line with the needs of the local economy, making it easier for industry to access knowledge not only due to regional but also cognitive proximity.

Due to the fact that Germany is well known for its skills in mechanical engineering throughout the world, the education of engineers is crucial to the continuing success of the German economy. UAS exhibit the highest level of spin-off activities of any research institutes, even if UAS are smaller than public universities (Krause, 2005; Heumann and Schmude, 2008, pp. 52–5). UAS tend to be successful in supporting start-ups based on their knowledge and expertise, despite structural disadvantages like fewer research fellows and a lower budget (Krause, 2005). To compensate for the lack of non-tenured support staff and their lesser ability to attract and retain suitably qualified employees, UAS began setting up so-called "An-Institutes" (Kulicke and Stahlecker, 2004, p. 39). An "An-Institute" or "Adjunct Institute" is legally separated from the UAS, essentially by creating a private limited company (GmbH) or registered association (Kulicke and Stahlecker, 2004). Besides improving the quality of teaching and the diversification of research, the possibility of facilitating project cooperation with industry – due to reduced administrative overheads – is one of the most important reasons for channeling funding via an Adjunct Institute (Kulicke and Stahlecker, 2004). The probability that an Adjunct Institute is managed by professors from the UAS is more than 90 percent, a fact that highlights the close connection between UAS and Adjunct Institutes (Kulicke and Stahlecker, 2004).

Technology transfer projects between UAS and industry are very diverse. They range from solving new challenges in mechanical engineering to applied material research and from cooperation with small and local companies to global players. When it comes to applied knowledge, companies favor the cooperation with UAS because of their fast development cycles and orientation towards feasible results. Three examples explain the value of such cooperation. First, the LipoPep project, a cooperation between two of the ten most successful UAS (Cologne and Aix-la-Chapelle) and two German global market leaders (BASF and Henkel). The goal of the applied natural science departments of the UAS and their industry partners was to develop Tensid made from renewable raw materials (TH Cologne, 2019). Tensid is an important ingredient of washing lotion and powder, usually made in mineral oil. The results of this project led to organic-based products that contribute not only to more sustainability but also improve skin compatibility. The second example is the cooperation between UAS Darmstadt and Werkzeugbau Weidemann GmbH and Co. KG. In a very ambitious project, their goal is to build a fourth dimension in 3D printing of thermoplastics (Hochschule Darmstadt, 2020). With approximately 15 employees, Werkzeugbau Weidemann can be considered a small but highly innovative company. The fourth manufacturing axle will increase the coating adhesion and consequently improve mechanical properties, which guarantees Werkzeugbau Weidemann GmbH and Co. KG a huge competitive advantage. Third, the cooperation between UAS Berlin and the Berlin-based start-up MOWEA to develop wind farms, which are also effective in cities, is evolutionary since, in Berlin, rooftops are easier to find than mountains (HTW Berlin, 2021). The challenge is to find a balance between technical efficiency, cost-effectiveness and discreetness by reducing the vibrations of the rotor blades. Therefore, this applied research cooperation will contribute highly to the energy revolution in Germany.

In conclusion, UAS are an important source of knowledge needed by industry. Since UAS are not intended to do basic research, which is carried out at public universities and private research institutes such as the Fraunhofer Institute, UAS have to be successful in more practical education. Due to the division of labor in the higher education sector, UAS can specialize in applied research and education, thus providing the more professional training needed by employees in industry. UAS provide the largest number of engineers (Krause, 2005), who can be considered the backbone of the German industry. The regional and cognitive proximity lowers barriers to knowledge transfer compared to public universities. One core competence of UAS is short-term and implementation-oriented research (Schmoch et al., 2000). Since research at public universities is more long-term oriented (Schmoch et al., 2000), this emphasizes the division of labor in the German higher education sector. In the 50 years since the founding of UAS, applied research has come to play an essential role in the economy, especially for SMEs in general (German Science Council, 2000) and particularly for those located close to UAS (Beise and Stahl, 1999; Einhaeupl, 2005; German Science Council, 2000). Summing up, the UAS are fulfilling their task not only by offering a practical education but also by facilitating technology transfer.

9.4 THE CONCEPT OF UNIVERSITIES OF APPLIED SCIENCES AND ITS ROLE IN HIGHER EDUCATION AROUND EUROPE

The division of higher education in different organizational institutions is quite common around the world. Especially in Europe, colleges and universities were established to provide highly skilled academics focused on applied knowledge within a short education period (Lepori and Kyvik, 2010). In three European countries, the Netherlands, Finland and Switzerland, UAS that are comparable to German UAS have been established. The Netherlands founded the "Vereniging Hogescholen" in the mid-1980s, almost 20 years after Germany. The main principle of the establishment of these institutions was also to install a binary education system. In contrast to the universities, the Hogescholen are less research- but more profession-oriented (Lepori and Kyvik, 2010). The focus on higher professional education is quite successful since 95 percent of students find employment within one year after graduating. Because of the applied knowledge students gain from problem-based projects, identifying and solving real global challenges in a "learning by doing" concept (Vereniging Hogescholen, 2021). Like UAS in Germany, the professional orientation of the Hogescholen has a specific role in the innovation process by intensifying collaboration with industry, in particular with SMEs (de Weert and Leijnse, 2010). This approach also enables new ways of knowledge transfer due to the specific research (de Weert and Leijnse, 2010).

In the 1990s, Finland established the "Ammattikorkeakoulu," a comparable higher education institution to the Fachhochschule and the Hogescholen. The focus of the UAS in Finland is to enhance regional development with multidisciplinary and applied higher education. Additional key elements are applied research and close relationships with regional SMEs to facilitate technology transfer. In contrast to the institutions in Germany and the Netherlands, the UAS in Finland are independent legal entities (Ministry of Education and Culture, 2021). Nevertheless, Finnish UAS are not completely independent from politics. An operating license granted by the government is necessary to establish a UAS. Comparable to Germany, engineering is a highly important field of studies and research, with roughly one-fifth of overall students (Statistics Finland, 2017). UAS are also found to be highly important for the support of regional development via technology transfer, especially to Finnish SMEs (Lepori and Kyvik, 2010).

Compared to the other countries, the binary system in Switzerland is even further divided, having two clearly separated sectors (Lepori and Kyvik, 2010). UAS are depending on a different law, have their own system of funding and differing career paths (Lepori, 2008). Similar to the UAS in Finland and, to some extent, the UAS in Germany, the main goal of UAS in Switzerland is regional development. For a long time, the seven UAS in Switzerland have been almost exclusively teaching-oriented (Arvanitis et al., 2008). However, since the mid-1990s, applied research activities have been growing, setting their focus on engineering and management. In the meantime, compared to federal institutions and cantonal universities, the UAS in

Table 9.4 *UAS in Europe: an international comparison*

Country	Name	No.	Type	Students	Degrees	Primary Goal	USP
The Netherlands	Vereniging Hogescholen	36	state-owned	500 000	Bachelor (98%) Master (2%)	profession-oriented study programs	problem-based projects to identify and solve real global challenges ("learning by doing")
Finland	Ammattik-orkeakoulu	22	public limited companies in the Ministry of Education	150 000	Bachelor (83%) Master (17%)	regional development	independent legal entities
Switzerland	Fachhochschule	7	state-owned	80 000	Bachelor (81%) Master (19%)	technology transfer & regional development	most successful higher education institution in technology transfer
Germany	Fachhochschule/ Hochschule	105	state-owned	600 000	Bachelor (82%) Master (18%)	applied knowledge for SME and regional development	efficient partner for applied solutions

Source: Own summary.

Switzerland are significantly more successful in knowledge and technology transfer (Arvanitis et al., 2008). Especially, Swiss SMEs benefit from the applied research and education at UAS (Lepori and Kyvik, 2010).

UAS in the above-mentioned European countries seem to have a comparable perspective to their German counterparts. The main focus is to educate a profession-oriented workforce and use technology transfer and applied research to enhance regional development (Table 9.4). In conclusion, the German UAS are not alone in their success. UAS in the Netherlands, Finland and Switzerland also provide an important contribution to the application of applied knowledge in their countries. Despite this success, other countries in Europe have chosen different approaches. In the United Kingdom, for instance, the UAS ("polytechnics") were assimilated to the universities in 1992. Other European countries have established a binary higher education system, but with colleges in addition to universities (Lepori and Kyvik, 2010). These colleges focus on education but neglect the impact of technology transfer. Therefore, future research might explain the impact of differences in the higher education systems on economic performance of European countries.

9.5 CONCLUSION

Due to their focus on applied research, UAS are a perfect partner for industry, particularly for SMEs (Audretsch and Lehmann, 2016; Einhaeupl, 2005; German Science Council, 2000; Lehmann and Starnecker, 2011, 2012; Schmoch et al., 2000). The differences between UAS and public universities (Lehmann and Starnecker, 2012; Pautsch, 2020) explain the challenges and the opportunities of UAS. First, research and teaching at public universities follow a theoretical approach, while UAS use a more practical concept. In other words, the objective of UAS is not to educate research personnel but the workforce and to facilitate technology transfer. This is supported by the fact that two-thirds of all German engineers graduate from UAS (German Federal Statistical Office, 2021a) and that the more applied approach used in the educational system allows a shorter education time (Lehmann and Starnecker, 2012). Second, in contrast to public universities, UAS are not exclusively located in major cities, which decreases the distance to companies in more rural parts of Germany, notably SMEs. Third, due to a lower number of research assistants compared to public universities, research staff is limited at UAS; thus applied research is mainly conducted by the UAS professors. Among other structural differences, this is caused by the absence of the right to grant doctorate degrees. If one considers in addition that teaching hours of UAS professors are twice as long compared to those of professors at public universities, the great number and quality of applied research at UAS is impressive.

Despite their importance and success, the potential of UAS is still largely underestimated (Hamm and Wenke, 2002; Lehmann and Starnecker, 2012). This is especially surprising since there is no doubt about the significant impact UAS have on regional development (Audretsch and Lehmann, 2016; Audretsch et al., 2012; Lehmann and Starnecker, 2012). The research cooperation with small as well as big companies on applied knowledge has already led to important technologies that impact the energy revolution and a more sustainable world. Not surprisingly, politics are focused on enhancing UAS, which is reflected in the increasing financial support for applied research at UAS. Other European countries, the Netherlands, Finland and Switzerland, have developed a binary system of higher education similar to Germany. In line with research on German UAS (Beise and Stahl, 1999; Einhaeupl, 2005; Lehmann and Starnecker, 2012; Schmoch et al., 2000), these countries also demonstrate the significant importance of UAS for technology transfer and regional development (Lepori and Kyvik, 2010). There is no doubt about the positive impact of UAS on the economy, particularly as a reliable partner to SMEs. Future research should therefore continue to focus on elucidating UAS' success factors for regional economic performance to find additional potential to increase the success of universities of applied sciences, not only for the higher education sector but also for regions and national economies.

NOTE

1. This chapter is partly based on the joint work of Erik E. Lehmann and Alexander Starnecker, published in the edited volume *Technology Transfer in a Global Economy* (Lehmann and Starnecker, 2012).

REFERENCES

Acs, Z.J., Audretsch, D.B., Lehmann, E.E., and Licht, G. (2016). National systems of entrepreneurship. *Journal of Small Business Economics, 46*, 527–35.

Anselin, L., Varga, A., and Acs, Z. (1997). Local geographic spillovers between university research and high technology innovations. *Journal of Urban Economics, 42*(3), 422–48.

Arvanitis, S., Kubli, U., and Woerter, M. (2008). University-industry knowledge and technology transfer in Switzerland: What university scientists think about co-operation with private enterprises. *Research Policy, 37*(10), 1865–83.

Audretsch, D.B., and Lehmann, E.E. (2005). Does the knowledge spillover theory of entrepreneurship hold for regions? *Research Policy, 34*(8), 1191–202.

Audretsch, D.B., and Lehmann, E.E. (2015). The emergence of the Mittelstand Company: A German perspective. Working Paper, University of Augsburg.

Audretsch, D.B., and Lehmann, E.E. (2016). T*he Seven Secrets of Germany: Economic Resilience in an Era of Global Turbulence*. New York: Oxford University Press.

Audretsch, D.B., and Stephan, P.E. (1996). Company-scientist locational links: The case of biotechnology. *The American Economic Review, 86*(3), 641–52.

Audretsch, D.B., Lehmann, E.E., Link, A.N., and Starnecker, A. (2012). *Technology Transfer in the Global Economy*. International Handbook Series on Entrepreneurship. New York: Springer.

Audretsch, D.B., Hülsbeck, M., and Lehmann, E.E. (2013). Families as active monitors of firm performance. *Journal of Family Business Strategy, 4*(2), 118–30.

Audretsch, D.B., Lehmann, E.E., and Wright, M. (2014). Technology transfer in a global economy. *Journal of Technology Transfer, 39*, 301–12.

Beise, M., and Spielkamp, A. (1996). Technologietransfer von Hochschulen: Ein Insider-Outsider Effekt. ZEW Discussion Paper No. 96–10, Mannheim.

Beise, M., and Stahl, H. (1999). Public research and industrial innovations in Germany. *Research Policy, 28*(4), 397–422.

BMBF. (2021). Forschung an Fachhochschulen. Accessed May 2021, https://www.bmbf.de/bmbf/de/forschung/das-wissenschaftssystem/forschung-an-fachhochschulen/forschung-an-fachhochschulen.html.

Boettcher, M. (2004). Der regionale Wissens- und Technologietransfer einer neuen Fachhochschule. Das Beispiel des RheinAhrCampus Remagen Bonn: Rheinische Friedrich-Wilhelms-Universität.

Czarnitzki, D., Rammer, C., and Spielkamp, A. (2000). Interaktion zwischen Wissenschaft und Wirtschaft in Deutschland. Ergebnisse einer Umfrage bei Hochschulen und öffentlichen Forschungseinrichtungen. ZEW Documentation No. 00–14.

De Weert E., and Leijnse F. (2010). Practice-Oriented Research: The Extended Function of Dutch Universities of Applied Sciences. In Kyvik S. And Lepori B. (eds), *The Research Mission of Higher Education Institutions outside the University Sector*. Higher Education Dynamics, Vol. 31, 199–217. Dordrecht: Springer.

Dziatzko, N., Kielkopf, M., Schittenhelm, F., and Streinwandt, A. (2011). Die Bedeutung des Innovationsmanagements in mittelständischen Unternemen – eine empirische

Untersuchung. In Haubrock, A., Rieg, R., and Stiefl, J. (eds), *Zweite Aalener KMU-Konferenz - Beiträge zum Stand der KMU Forschung*, 35–51. Aachen: Shaker Verlag.

Einhaeupl, K. (2005). Forschung als differentia specifica von Fachhochschulen? Gegenwart und Perspektiven, Beiträge zur Hochschulpolitik, 5/2005, 29–46. Dokumentation der 34. Jahrestagung des Bad Wieseer Kreises.

Fritsch, M., and Slavtchev, V. (2007). Universities and innovation in space. *Industry and Innovation, 14*(2), 201–18.

Fritsch, M., Henning, T., Slavtchev, V., and Steigenberger, N. (2007). Hochschulen, Innovation, Region: Wissenstransfer im räumlichen Kontext, 82, 264. Edition Sigma.

German Federal Statistical Office. (2021a). Data from Genesis-Database. Accessed October 2021, https://www-genesis.destatis.de/genesis/online?language=de&sequenz=tabellen&selectionname=213*#abreadcrumb.

German Federal Statistical Office. (2021b). Destatis Statistisches Bundesamt. Bildung und Kultur. Monetäre hochschulstatistische Kennzahlen 2019. Accessed December 2021, https://www.destatis.de/DE/Themen/Gesellschaft-Umwelt/Bildung-Forschung-Kultur/Hochschulen/_inhalt.html#sprg233706.

German Science Council. (2000). Thesen zur künftigen Entwicklung des Wissenschaftssystems in Deutschland. Accessed August 2021, https://www.wissenschaftsrat.de/download/archiv/4594–00.pdf?__blob=publicationFile&v=1.

German Science Council. (2002). Empfehlungen zur Entwicklung der Fachhochschulen Accessed August 2021, http://www.wissenschaftsrat.de/download/archiv/5102–02.pdf.

Glaeser, E., Kallal, H., Scheinkman, J., and Schleifer, A. (1992). Growth in cities. *Journal of Political Economy, 100*(6), 1126–52.

Grimpe, C., and Fier, H. (2010). Informal university technology transfer: A comparison between the United States and Germany. *Journal of Technology Transfer, 35*(6), 637–50.

Hamm, R., and Wenke, M. (2002). Die Bedeutung von Fachhochschulen für die regionale Wirtschaftsentwicklung. *Raumforschung und Raumordnung, 60*(1), 28–36.

Heumann, S., and Schmude, J. (2008). Rahmenbedingungen und Ausprägungsformen der akademischen Gründungsförderung an 100 deutschen Fachhochschulen. BMWi Forschungsbericht 576. Accessed October 2021, http://www.exist.de/imperia/md/content/studien/studie_fhs.pdf.

Hochschule Darmstadt. (2020). Forschung und Entwicklung – Abgeschlossene Projekte. Accessed June 2021, https://fbmk.h-da.de/forschung/institut-fuer-kunststofftechnik/aktuelle-projekte-am-ikd/abgeschlossene-projekte.

Hochschulkompass (2021). Übersicht mit allen Hochschulen. Accessed October 2021, https://www.hochschulkompass.de/hochschulen/downloads.html.

HTW Berlin. (2021). Campus Stories – Windräder in die Stadt und auf hohe Berge. Accessed June 2021, https://campus-stories.htw-berlin.de/jahr/2021/windraeder-in-die-stadt-und-auf-hohe-berge/.

Hülsbeck, M., Lehmann, E.E., and Starnecker, A. (2013). Performance of technology transfer offices in Germany. *Journal of Technology Transfer, 38*(3), 199–215.

Krause, P. (2005). Fachhochschulen und KMU – Innovationspartner für die Region. Beiträge zur Hochschulpolitik 5/2005, 47–68. Dokumentation der 34. Jahrestagung des Bad Wieseer Kreises.

Kulicke, M., and Stahlecker, T. (2004). Forschungslandkarte Fachhochschulen – Potenzialstudie (189). Berlin, Bonn: Bundesministerium fuer Bildung und Forschung, Fraunhofer Institut für Systemtechnik und Innovationsforschung. Accessed October 2021, https://forschungslandkarte.isi-projekt.de/pdf/endbericht.pdf.

Lehmann, E.E., and Starnecker, A. (2011). Wissenstransfer aus Hochschulen und regionale Entwicklung. In Haubrock, A., Rieg, R., and Stiefl, J. (eds), Zweite Aalener KMU-Konferenz – Beiträge zum Stand der KMU Forschung, 115–30. Aachen: Shaker Verlag.

Lehmann, E.E., and Starnecker, A. (2012). Introducing the University of Applied Science in the technology transfer process. In Audretsch, D.B., Lehmann, E.E., Link, A.N., and Starnecker, A. (eds), *Technology Transfer in the Global Economy*, 99–115. New York: Springer.

Lehmann, E.E., Schenkenhofer, J., and Wirsching, K. (2019). Hidden champions and unicorns: A question of the context of human capital investment. *Small Business Economics, 52*, 359–74.

Lepori, B. (2008). Research in non-university higher education institutions. The case of the Swiss universities of applied sciences. *Higher Education, 56*(1), 45–58.

Lepori, B., and Kyvik, S. (2010). The research mission of universities of applied sciences and the future configuration of higher education systems in Europe. *Higher Education Policy, 23*(3), 295–316.

Link, A.N., Siegel, D.S., and Bozeman, B. (2007). An empirical analysis of the propensity of academics to engage in informal university technology transfer. *Industrial and Corporate Change, 16*(4), 641–55.

Ministry of Education and Culture. (2021). Higher education institutions and science agencies. Accessed May 2021, https://minedu.fi/en/heis-and-science-agencies.

Osel, J. (2012). Fachhochschulen begehren gegen Unis auf. Accessed September 2021, http://www.sueddeutsche.de/bildung/streit-um-promotionsrecht-fachhochschulen-begehren-gegen-unis-auf-1.1453804.

Pasternack, P. (2009). Differenzierung des Hochschulsystems – Aktueller Stand und künftige Entwicklungen Beiträge zur Hochschulpolitik 3/2009, 47–65. Dokumentation der 38. Jahrestagung des Bad Wieseer Kreises.

Pautsch, A. (2020). Warum das FH-Promotionsrecht ein Irrweg ist. Forschung & Lehre. Accessed May 2021 https://www.forschung-und-lehre.de/politik/warum-das-fh-promotionsrecht-ein-irrweg-ist-3027/.

Schmoch, U., Licht, G., and Reinhard, M. (2000). Wissens- und Technologietransfer in Deutschland – Executive Summary.

Siegel, D.S., Waldman, D., and Link, A.N. (2003). Assessing the impact of organizational practices on the relative productivity of university technology transfer offices: An exploratory study. *Research Policy, 32*(1), 27–48.

Spiegel, D. (1969). Mit dem Latein am Ende. Spiegel Serie über die Krise und Zukunft der deutschen Hochschulen, 41/1969, 92–105.

Statistics Finland. (2017). Universities of Applied Sciences Education 2016. Accessed October 2021, https://www.stat.fi/til/akop/2016/akop_2016_2017–04–19_en.pdf.

TH Cologne. (2019). Lipopeptid-Tenside aus Lupinen, Sonnenblumen und Raps. Accessed October 2021, https://www.th-koeln.de/hochschule/lipopeptid-tenside-aus-lupinen-sonnenblumen-und-raps_68180.php.

Vereniging Hogescholen. (2021). Website. Accessed May 2021, https://www.verenigingh ogescholen.nl/english.

10. The role of university in a time of crisis: learn from the past to shape the future

Henry Etzkowitz, Chunyan Zhou and Rosa Caiazza

10.1 INTRODUCTION

In a "normal" era, under-utilizing latent innovation potential may be tolerable, even if it is wasteful. Within these conditions, such shortfall only comes to the forefront of attention when a dissatisfied constituency agitated for change makes its displeasure known and provides an action plan and way forward. For example, the Stanford student government organized the StartX Accelerator in 2010, activating the latent capacity of alumni networks by creating an entrepreneurial support structure and raising the university's startup rate from seven to thirty firms per annum, proving the existence of a gap by filling it (Etzkowitz, 2013).

Facing economic and social crises with the loss of its traditional industries in the early 20th century, New England mobilized its unique concentration of academic resources, using a not inconsiderable regional and federal government political resources and remaining industry resources, creating the groundwork for a knowledge-based economy (Etzkowitz, 1993). At the advent of the Second World War, this novel regional innovation model was scaled up to the national level, producing a national innovation system that materially contributed to Allied victory during the Second World War (Bush, 1945). In the post-war period the US triple helix devolved into a "hidden innovation practice," submerging the role of government due to ideological strictures (Etzkowitz and Gulbrandsen, 1999).

Nevertheless, the US maintained international leadership in innovation and entrepreneurship, despite and perhaps due to the 1957 Soviet Sputnik challenge inspiring the formation of the defense advanced research project, recuperating in part the integrated wartime triple helix. In recent decades the DARPA model has been extended to additional topics, including energy and health, and research programs have been extended into innovation at the margins, creating public venture capital at the federal and state levels without daring to use the label (Etzkowitz, Gulbrandsen and Levitt, 2000).

Facing multiple and interrelated climate, energy, health and innovation crises, the US National Science Foundation is mounting a macroscopic program to raise the level of under-resourced regions by linking them to highly successful regions like Silicon Valley. In effect, the US is revisiting the original intent of the Bush Endless Frontier proposal to address the entire scope of military and civilian innovation at the scale necessary to transform the whole society, not merely favored pockets of high-level research universities. The lateral Triple Helix U-I-G model, exemplified by the rise of Silicon Valley (Etzkowitz, 2022), despite periodic recession and crises, has only become more salient, both to greenfield and declining regions (Etzkowitz and Beddows, 2020).

10.2 THE EVOLUTION OF UNIVERSITY TECHNOLOGY TRANSFER

The evolution of the university technology transfer office (TTO) has gone through several stages from a legal orientation to protect intellectual property created within the university, to a marketing arm whose purpose is to identify potential users or licensees for the university's intellectual property, and finally, a business development arm, to help create start-up firms that can develop the university's inventions, especially in cases where existing firms are not interested in licensing them. These developments all extend the reach of the transfer office outside of the university, changing it from a relatively passive to an increasingly active actor in moving technology out of the university.

A second set of developments has extended the reach of the TTO within the university. The traditional initial step of an office's interaction with inventors begins when they contact the office, at their initiative, to file a disclosure statement, an official form outlining an invention that they have made. Increasingly, offices are making efforts to identify potential inventions and even the potential for invention at earlier stages. Some offices keep in touch with their counterparts at the university's Office of Sponsored Research in order to identify proposals that may have technological implications. TTOs may then contact the faculty member and discuss the proposal with them in order to increase their sensitivity to the technology as well as the science that may emanate from the research process. Other TTOs simply try to increase faculty and student awareness of the technological potential of their research by making presentations at department meetings, through discussions with new and prospective faculty and by sponsoring events and contests.

A third set of developments focused on research goes in both directions, externally and internally. The TTO becomes involved in projects to gain research funds. This activity takes place from two standpoints: to increase the research base of the university and thus the disclosure rate of inventions; and to provide funds to further develop the potential of inventions, to reduce them to practice and create a prototype, thereby making them more marketable. Sometimes, the university uses its own funds for this purpose, especially from earnings of the TTO.

More commonly, the university seeks funds from external sources for this purpose. It may attempt to interest regional governments seeking to promote economic development as well as investors interested in technology opportunities. Sometimes, the effort to translate the invention into practice may take the route of creating a start-up firm as a vehicle both for accessing government funds and garnering investor support. While most start-ups are intended to create fully functioning businesses, the start-up firm is also utilized as a vehicle to move a technology a stage or two through the development process to the point at which a larger firm might be willing to take it on. Even though a technology is more expensive at a later stage, a working prototype, and better yet, customers and sales, provide greater assurance to potential licensees or purchasers. Thus, the start-up firm is increasingly the method of choice for university technology transfer.

Sometimes, the only asset a start-up has to support its ability to raise capital is its protected intellectual property. Absent an exclusive license, the start-up process as an engine of innovation and economic growth will be inhibited. Since a patent is the prerequisite force behind the exclusive license as an incentive to economic growth, we come full circle back to the issue of private versus public knowledge, hopefully exposed above as an inherently false dichotomy. With academic patenting, however, public and private interests are even more intensely intertwined, and, given the academic-industry conflicts over exclusive licenses, one might even say inverted, since the private interest here protecting its intellectual property is the university, pre-eminent among institutions serving the public interest, whether land grant or Ivy League, and the entities clamoring for freer access are corporations, the very paradigm of private interest.

10.3 THE BAYH-DOLE ACT

An amendment to US patent law, the Bayh-Dole Act was passed in 1980, turning over the rights to intellectual property generated from federally funded research at universities and gave title to the universities with the provision that the universities had to make an effort to put it to use.

The Bayh-Dole Act altered the regulatory infrastructure that determined how the intellectual property rights arising from research funded by the federal government at universities should be handled. By changing the rules of the game and placing responsibility for technology transfer in the university where it was being conducted rather than with government (the funder) potentially useful research was moved closer to users. A business model that emerged in the early 20th century at a relatively few universities was generalized throughout academia from 1980.

In the context of federal research budgets that were not growing at a fast enough rate to meet academic researchers' needs, universities were given the opportunity to earn monies from the royalties and equity that they could generate from federally sponsored research on their campuses. Universities that were not previously involved in technology transfer typically hired a scientist from industry to organize a TTO. TTOs grew over time, sometimes into full-scale intellectual property and business development support structures. The overall trend, such as at Columbia University and the University of Virginia, is toward assisting firm formation as well as sale of intellectual property rights (Acs et al., 2009). At other schools, TTOs remained pro-forma units, taking out patents but doing little to valorize them.

A significant number, especially at less research-intensive universities, view their effort as a service to their university and their region, rather than as a moneymaker. Unfortunately, most technology transfer metrics focus on licensing income and start-ups formed rather than faculty assisted to collaborate with local firms or pursue research funding, so the broader contribution of university technology transfer is often neglected (Etzkowitz and Göktepe-Hultén, 2010). The potentially contentious issue of division of proceeds appears to have been settled within the context of the Bayh-Dole Act. The Act guarantees the inventor a share through a more or less equal three-way division of proceeds among the university administration on behalf of the institution as a whole, the faculty members' department and the individuals themselves.

The Act provided a stable ground for university technology transfer, although no funds were appropriated for its implementation or administration. Nevertheless, universities were able to charge for some of the costs of their patent licensing offices to their administrative overhead rates for federal grants, thus receiving an indirect subsidy. More important, initially, was the need to meet a requirement to develop the capacity to assist the commercialization of research to remain eligible for federal funding, which rapidly expanded the universe of academic institutions involved with technology transfer.

Nevertheless, it was very controversial for many academics, industrialists and politicians. Should universities take this action or should government make the intellectual property that it had paid for with billions of dollars of federally funded research in the post-war era freely available to all interested parties? It was argued that if university originated technologies were made freely available there was a free-rider problem: Firm A, believing that if they took title to the intellectual property and had great success based on it, firm B would come along and demand access to it on the grounds that it had been publicly funded. Although the Act provided a framework for resolving this blockage to academic involvement in technology transfer, other issues remained.

Student-generated rights were traditionally left entirely with the individual, perhaps on the grounds that they paid tuition fees and did not have employee status. However, since most graduate students in the sciences are on fellowships, often from a government or university source, graduate students have increasingly been defined as officers of the university subject to the same intellectual property rules as faculty. Whatever the administrative definition, student status is quite different in reality. Student work is typically conducted under the control of faculty members, and it can become a matter of faculty discretion on including students in disclosure statements unless a university has clear rules and procedures. Failure to include students as inventors has led to later legal challenges when a patent was realized to have considerable value.

Access of academic researchers to patented discoveries has also become an issue. In a previous era, when there was a clear distinction between basic and applied research, this was not a great problem since patenting largely took place away from the research frontier. However, at present, when basic discoveries in molecular biology produce practical results simultaneously with advances in research, the issue of access comes to the fore. The Bayh-Dole Act places control over this problem in the hands of the universities. Universities can protect access of academic researchers by including provisions in licensing agreements authorizing access.

Only a few research universities, like the University of Wisconsin, had developed highly professional technology transfer practices since the early 20th century. Most entered the field after 1980 when patent law made receipt of federal research funds, which had become the mainstay of academic research support during the post-war era, conditional on a commercialization effort. The reorientation of US universities toward a commercial role was not intervention in the sense of specific government measures requiring targeting of particular areas of R&D for support, as in Japan, or requiring enterprises and research institutes to make research contracts with each other, as in the Eastern European socialist model. Instead, incentives were built into the research funding system to move the universities closer to industry, in their motivation and structure. Government went through the universities to reach industry, resulting in an "indirect industrial policy" that was more fruitful than many more direct schemes.

10.4 THE NATURE OF KNOWLEDGE

Knowledge can be a public and private good at one and the same time, with degrees of publicity and privacy. Secrecy has been identified as part of the discovery strategy of investigators, who may not wish to notify competitors of their preliminary findings even from research that they eventually intend to disseminate solely in classic academic fashion through journal publication. Unless one is a member of an organization that pays for a license allowing access to its members, individuals have to rent or buy articles. Articles may be published in an open access journal, but then authors are typically expected to pay for the privilege. Making knowledge freely available still requires public or philanthropic sponsorship, in one form or another.

The patent system is an exemplar of organizing knowledge as public and private at one and the same time, with a temporary monopoly available to cover costs. An alternative to secrecy, a patent is a grant of a temporary monopoly in exchange for disclosing the workings of an invention. Concepts and technologies are made accessible to others in the patent literature available at public depositories and on the Internet even as rents are accrued. Operating on a different axis than the academic journal peer review system based on honor and prestige, the patent system is, nevertheless, in both principle and practice, congruent with the academic value placed on disseminating knowledge.

In addition to providing a temporary monopoly, patents are an information system and disclosure process as the invention's workings and at least basic knowledge of its means of achievement become publicly available upon issuance. What better definition is there of the "universal" attribute of the university than the ability to create a knowledge community and collective knowledge; may not the same be said of the patent system? Even if a university wishes to maintain a hands-off position, it may be forced to become involved in issues of patent management, both to protect its reputation and to fulfill expectations raised by academia's own justification for government research funding: that it will contribute to economic and social advance.

10.5 PUBLIC VERSUS PRIVATE KNOWLEDGE

The Bayh Dole regime, with its provisions for transparency and sharing between the academic institution and the actual inventors (faculty, students and staff) provides a creative framework for invention. Although ownership is legally and formally with the US university, the individual academic inventor, especially if a professor, retains significant influence over the disposition of their invention and can, almost as a default informal understanding, expect to receive a license if they wish to form a firm. Thus, the formal US system of university ownership is halfway between the traditional European model of individual academic ownership or "Professor's Privilege" and firm control through employment contract.

Indeed, in practice, the actually existing US model is a "partial professor's privilege" in contrast to corporate practice whereby firms from the late 19th century assumed total ownership and control of their employee's invention made on and even off the premises. In an earlier era, railroad employees, for example, owned the inventions they made during their work hours and their employers had to purchase them. Through the legal stratagem of employment contracts, this relationship has been reversed. Thus, firms have largely superseded the constitutional intent of patents: to encourage individual valorization of inventorship. In contrast to the academic inventor, the firm inventor is entirely at the mercy of their employer to receive more than token remuneration.

Much of the controversy over academic patenting and technology transfer arises from differing conceptions of the role of the university in society. If the university is seen as having an explicit economic and social role, the transfer of technology and utility of patenting becomes an issue of means, how best to organize this activity in relation to the other missions of the university. If the role of the university is believed to be limited to the production and dissemination of knowledge, then the issue of patenting research becomes a matter of principle. It is fought on the grounds of whether the university should play an explicit economic and social role in society and whether such a mission adds to or detracts from its other missions.

Nevertheless, irrespective of ideological positions, the research mission of the university inevitably brings with it the potential for technology development, with economic and social implications that must be taken into account. Even if a university wishes to maintain a hands-off position, it may be forced to become involved in issues of patent management, both to protect its reputation and to fulfill expectations raised by academia's own justification for government and philanthropic research funding: that it will contribute to economic and social advance.

In the early 1980s, most US research universities merely hoped to earn enough to pay for running a TTO so that they could satisfy the new government requirements of playing a role in supporting industry in exchange for receiving government research funds. That level of income was achieved relatively quickly at many universities, almost to their own surprise. US universities earned a total of one billion dollars, mostly in royalty income, in 2000. However, a few large returns came from infringement proceedings in which a company was found to be utilizing patented research without authorization and from sale of equity in a firm founded with university intellectual property.

Nevertheless, as larger numbers of universities have established offices to arrange patent licensing and technology transfer, critics have questioned both the propriety and efficacy of the trend (Mowery and Sampat, 2004). Mowery and Sampat (2004) argue that the Bayh-Dole Act did not cause the increase in patenting by universities, which was already on the rise during the 1970s. From this perspective, the law is seen as symptomatic of changes that were already underway as much as causative. Equally significant was a body of changes to patent law extending the scope of what could be patented. The authors also argue that structural aspects of the US higher education system – including a long history of regional engagement through the land grant system – are key to the exchange of knowledge between academia and industry. Mowery and Sampat (2004) thus warn other countries against adopting Bayh-Dole type legislation in the belief that this will single-handedly spur the kind of innovation systems seen in the US.

In a comprehensive study, Owen-Smith (2005) investigates trends at 87 research universities in the US over the period 1981–95, specifically focusing on the relationship between patenting and publishing. He found a major transition in the institutional environment in the early 1990s, such that high-volume patenting begins to be associated with a decrease in publications. The exception was those institutions that were able to secure high-impact patents and saw gains in both overall publications and high-impact publications. In conclusion, universities engaged in high-volume but indiscriminate patenting may see declines in publications, but those with the experience and competencies to pursue high-impact patents may be able to parlay their success across academic and commercial efforts.

He concludes that the current environment strongly favors first-movers, and universities that have entered later into commercialization may be at a permanent disadvantage. Skepticism comes from an assessment of the efficacy of technology transfer as highly path-dependent so that the success of certain institutions may not be generalizable to the rest (Krücken, 2003; O'Shea et al., 2005; Owen-Smith, 2005). On the other hand, highly technology transfer adverse universities, like Johns Hopkins and Yale, have successfully executed a complete reversal of culture and previous policy and become leaders in the field. Common elements in these cases are strong leadership from above in response to the exigencies of the surrounding economically depressed area (Breznitz, 2014).

In a review of the literature on technology transfer, Caiazza et al. (2015) argue that organizational, individual and institutional context of technology transfer must be considered together as they are interrelated.

10.6 TECHNOLOGY TRANSFER

There are three essential elements of academic technology transfer: heightened awareness among faculty, staff and students of the potential of knowledge to have useful properties that make it amenable to translation into intellectual property and/or a firm as well as publications, lectures and degrees; inventor's interest in further developing the technology and/or collaborating with others to do so as academic inventions are often insufficiently developed to be transferred even if a patent can be obtained; and identification of potential users of protected intellectual property outside the university in the form of a license to practice a patent, either among existing firms or a start-up formed by the inventor, on their own or in collaboration with others.

The first step is dependent upon the culture of the university, awareness of faculty and students, and university policy. A TTO can influence these preconditions through an active outreach and education program. While the third of these steps in fact constitutes the actual transfer of the technology because of the very early stage of academic technology, it is not usually worthwhile making the investment unless cooperation of the inventor is also achieved, also suggesting the importance of the underlying premise stated in step one.

10.7 DIFFERENTIAL CAPACITIES

The stage of development of the university in which a TTO is situated affects its organizational capacity and strategy. The literature classifies TTOs based on their organizational structure (for example, whether the organization is centralized or has multiple decentralized units), finding that different forms correlate with strengths in different areas. The literature also classifies TTOs as traditional university structure, research foundation, and for-profit private venture extraction. They find that for-profit strategies are best at new firm creation and that most TTOs overemphasize royalty income and underemphasize entrepreneurship.

The following cross section of university and region types may be identified:

- high level of research at the university and high level of business development services in the region (high/high)
- high level of research at the university and low level of business development services in the region (high/low)
- low level of research at the university and low level of high-tech and business development services in the region (low/low)

- low level of research at the university and high level of business development services in the region (low/high).

The above variables relate to organizational capacity. In case 1, a TTO can be highly successful by providing the interface between research outcomes and business development services, including potential licenses and venture capitalists. In cases 1 and 2, the TTO may play a role in stimulating the development of business services in the region. This is especially likely in case 2; in case 1, the TTO will likely be occupied with licensing technology. Case 3 requires an entrepreneurial TTO director both to be involved in efforts to raise the level of research at the university and services outside. When the university and the region are more highly developed, the TTO director can operate in an administrative, support and networking capacity.

Table 10.1 *TTO organizational capacity*

University research		
High	1	2
Low	4	3
Business development services in region	High	Low

Cultural orientation is an additional factor that is both affected by organizational capacity and helps to create it. Some authors have found that cultural and informational differences between universities and firms impeded technology transfer. They also found that faculty reward systems were one of the most significant factors relating to technology transfer (a third factor is structure and staffing practices – that is, mix of science, law and business backgrounds – of the TTO). Incentive structures can be changed to help induce and/or reward entrepreneurial behavior. For example, Stanford University, whose presidency is currently occupied by a computer scientist who participated in firm formation from his research earlier in his academic career, recently took the step of crediting patents as well as publications toward tenure and promotion, bringing the university's reward structure into alignment with its culture.

On the other hand, Johns Hopkins University, known for its aversion to an economic orientation until quite recently, is taking steps to alter its culture toward an active interest in technology transfer. As one university official put it, changing a university's culture takes time; like turning a tanker takes time, there's a lot of inertia to overcome. In contrast to many other universities that give preference to local licenses and start-ups, Stanford's Office of Technology Licensing (OTL) performs a stringent due diligence to see that an invention is licensed to the firm best suited to develop it. A campus inventor, who wants to go the firm formation route, has to make the case for a license. Nevertheless, the weight of presumption is in favor of the faculty member, as discussed above.

10.8 TTOS AT LESS RESEARCH-INTENSIVE UNIVERSITIES

Whereas the lack of business development services leads the TTO to focus externally, a low level of research funding encourages the TTO to expand its focus to build up research support. As a director in this situation said, "An important objective of the office is to build up research at the university. This may take the form of assisting in the organization of research centers to capture larger funded research projects." The relative lack of research also leads the TTO to focus more closely on the potential of the research that is underway. Instead of waiting for disclosures to come in and seeing that event as the start of the technology transfer process, a TTO at a less research-intensive university may focus on the potential for invention in projects just getting underway.

Under these conditions, the TTO may develop a close relationship to the Sponsored Research Office, the administrative arm of the university charged with facilitating and monitoring funded research projects in order to identify projects with technology transfer potential at the earliest stage. As one such director put it, "We work very closely with sponsored research, lot of places don't do, and they really should." In this mode, the research proposals are the starting point rather than invention disclosures. The TTO also takes a proactive stance. Instead of waiting to be contacted by the researcher with an invention; the TTO director (and it is typically the director since the office will likely be a one- or two-person operation at a less research-intensive university) contacts the faculty member with a research project in order to discuss the potential for invention. As a director said, "when they get grants, as soon as we are aware of funding, we let them know we are available." The objective is to sensitize the researcher to issues of intellectual property and invention, so that they will be aware of what they are doing and contact the TTO if they happen to invent something.

Even more research-intensive universities find it advantageous to maintain close contact with potential faculty inventors. As a director said, "We try to make sure faculty are aware and thinking about invention opportunities." The director regularly makes presentations at department meetings and maintains "one-on-one" contact on a regular basis with especially inventive faculty members. Another director described how "We maintain close contact with faculty we think are close to invention. We are constantly working with them, discussing their work. The objective is to go with the flow of the 'meandering stream' of basic research, not divert its flow."

10.9 OUTSOURCING TECHNOLOGY TRANSFER

An academic scientist who confronted these issues in the early 21st century at the University of California, Berkeley came up with a novel solution: an independent non-profit-corporation that would aggregate academic patents nationwide, license them to industry and distribute research grants to faculty who applied, not necessarily those who generated the funds. Professor Frederick Cottrell, the physical chemist who invented the electrostatic precipitator that could control industrial pollution, turned over his patent rights that were earning $1,000 per day to the Research Corporation (RC), and the University of California encouraged other academic inventors to do likewise. Most patent-active universities during the interwar era developed similar relationships to RC and it became a significant source of support to novel projects like the Cyclotron.

A flaw in the model was its distance from researchers with potential inventions. Headquartered in New York, RC sent visitors to campuses to seek out and assess inventions for patentability. A virtue in the model was that it had the potential to amalgamate intellectual property from various campuses into larger packages more suitable for licensing. Campus-based TTOs have addressed the first issue but opened up the second. RC gradually declined as more universities undertook to manage their own patents. MIT's withdrawal was an especially sharp blow. Some suggest that the model could usefully be renewed to perform the aggregation function. A variety of private firms, some with deep pockets, have emerged for this purpose.

The creation of a research institute with a permanent staff, capable of working on the early stages of development of several technologies at the same time, at a distance from academia, is an alternative to the start-up firm as a means of moving very early-stage technology forward. This is the premise of the German Fraunhofer Institute system, which helps nascent university instituted technologies take further steps forward to development, on the one hand, and pursues contracts with firms to produce technologies to order, on the other. While the institute model has the attractive feature of stability, continuity of staff and does not require going through the firm formation process each time a new technology is identified, it is also perceived as lacking the momentum and impetus that the start-up process, with an entrepreneur, can provide for a new invention. Moreover, the system works best for incremental improvements in existing technologies rather than the creation of new platforms.

10.10 INDUSTRY'S VIEW OF UNIVERSITY TECHNOLOGY TRANSFER

Academics and corporate employees have a different legal framework with respect to intellectual property.

This imbalance has changed somewhat in recent years as spin-off firms abscond with their employers' intellectual property and, like the founders of Compaq Computer in departing from Texas Instruments, make it the basis of their new firm, with payback settled years later in the courts. The increasing ease of starting-up and a loose employment law tradition in California has redressed some of the balance in favor of the individual inventor. Nevertheless, it was not without reason that large corporations opposed the passage of the Bayh-Dole Act, fearing that its favorable provisions for academic inventors might someday be extended to their employees.

Academics and corporate employees have a different legal framework with respect to intellectual property!

During the 1980s industry's view of academia began to shift, initially among pharmaceutical companies and then across a broader range of industries interested in discontinuous innovation, which brought with it an increased willingness to source technology outside of the firm. Increased international competition also led the US to decrease the barriers between firms as well as encourage university-industry interaction. Nevertheless, some large corporations view the rise of academic patenting and licensing to start-ups rather than to major companies as a threat and press for universities to offer free access to their technologies as a condition of close working relationships and establishment of joint projects. This stance is especially attractive to engineering schools, with little expectation of producing patentable research but rather smaller increments of useful knowledge that are best exchanged informally.

Likewise, some large firms question the university granting exclusive licenses, joining with the ivory tower critics of university patenting. Unlike those critics, however, these companies' real concern is that if a university grants an exclusive license to a start-up, then they might have to pay more for innovative university-originated technology. On the other hand, such firms increasingly realize that they are not well suited to develop early-stage technology internally and prefer to monitor and perhaps invest in start-ups as an academic interface strategy. Given that relatively few firms are willing to license early-stage technology, the inventor's projected firm is often the only real choice.

10.11 THE MARKETING MODEL

Stanford's OTL was founded in 1969 as the result of a proposal from an employee of the university's contracts and grants office who believed that an organized effort could produce better results than the paltry amount that the university had earned happenstance from a few patents. OTL's hallmark has been a marketing model of technology transfer. The literature evidences that the founding director of OTL held that TTOs were too focused on patenting rather than getting inventions into use. He instituted procedures to identify a wide range of firms that might be interested in an invention, followed up by contacts, traditionally by telephone and more recently via email.

Although marketing typically starts from contacts generated by the inventor, licensing associates, with responsibility for an invention, typically brainstorm additional uses that are beyond those foreseen by the inventor and tap their own contacts as part of the marketing process. The marketing model was transferred to MIT by Reimers, whom MIT hired as a consultant after that school experienced difficulties with a venture capital model in which the university was deciding among various candidates to invest its funds, angering faculty who did not receive an investment. It has since become the prevalent US model.

The existence of a parallel universe of TTOs in industry and government is assumed as part of an institutional culture of porous boundaries, encouraging interface within and across institutional spheres. For example, the TTO as an internal search mechanism within the university and as an external search mechanism outside to identify customers is paralleled by the establishment of similar intelligence and transfer capacities within the firm. Such units identify technology developed within the firm that may be utilized in other parts of the company, or be marketed outside, as well as technology that it needs to import.

Many universities patent their potential intellectual property only when they can be reasonably assured of finding a market for it. Niels Reimers, founding director of Technology Licensing at Stanford, originated the shift from a legal to a marketing approach. Every technology transfer officer is on the lookout for a "blockbuster patent" and is keenly aware of the transience and inherently obsolescent nature of invention. Indeed, the impetus to establishing an intellectual property regime has often been the cautionary story of a significant invention not patented, either because the inventor was unaware of its potential or because the university had not yet established a TTO.

Technology transfer has been professionalized, with specialists devoted to its practice, relocated from independent intermediaries between university and industry, internalized within universities and firms and decentralized from central units on multi-campus universities and occasionally to the school level within individual campuses. It evolved from an occasional initiative of individual entrepreneurially oriented faculty members to a professional intermediary organization (Research Corporation) serving the entire academic community selectively, to a more intensive format as an arm of individual universities, whether as an administrative entity or a nominally independent foundation. More recently, independent consulting firms have been formed, partially retaking the intermediary role between university and industry, hired by either side as they see fit. Professional associations, primarily representing practitioners on both the university and industry sides of transfer, provide training and opportunities for interchange. Professional credentialing procedures have been made available to complete the contemporary technology transfer universe. Perhaps the only missing element is a degree offering bespoke professional training programs at the master's level, as in law and business, to complete the framework of an independent profession.

10.12 REGIONAL FACTORS

Licensing to the highest bidder tends to move technology out of the university's region unless it is a highly innovative region like Silicon Valley, while start-ups tend to be based close to where an invention is made. The most common mechanism of technology transfer, licensing-for-cash, is negatively correlated with new venture formation, suggesting that universities may be sacrificing long-term growth for short-term gains. On the other hand, Apple's owing the Wisconsin Alumni Research

Foundation more than $600 million, assuming the courts uphold its claim, would appear to put the bloom back on the licensing model. In fact, Wisconsin supports 2.5 percent of its budget with earnings from technology licensing, a relatively small but not negligible proportion.

An entrepreneurial orientation at the university can substitute for business development services in the region, especially by creating them internally within the university as a substitute. Located in a greenfield site beyond the outflow of business activity from New York City, Stonybrook University took this approach in creating a biotechnology industry on Long Island during the 1990s. On the other hand, a university with a high level of research and business development services in the region provides the ideal substrate for entrepreneurial activities from the university to take root and grow. The response of the TTO to these different situations will, to a great degree, influence its success.

In a region without much previous high-tech development, the TTO director may be the first person with official responsibility for this topic. Even though his or her remit is focused on the university, an entrepreneurial director will soon expand it to include helping create the conditions for high-tech development in the area. Once local economic development considerations are taken into account, the issue broadens from the difficult enough one of finding a licensee to one of identifying a local source to develop the technology.

In addition to retaining the technology, the human capital goal is to encourage inventors and entrepreneurs who graduate to remain in the area, or if they have left, to find ways to attract them back. As a director explained, we "didn't want students to take technology and leave." As another director put it, "We would like to see spin-offs to retain students in the area, rather than seeing the best move to Boston." Skilled people can also be attracted back to the region as entrepreneurs to run start-ups. Another director noted that "There are a group of expatriates who formerly lived in this area; they have an affinity for this area. There are events where they come home at holidays. The director takes the opportunity to set up events where we describe what we are doing, including the opportunity to take a position as head of a firm founded to commercialize a technology developed at the university."

Regional economic development is an increasingly important objective for university funders that is sometimes at odds with university administrators who wish to earn from patent licensing. As another director of a TTO at a leading university in a large city said, "80% of startups situate around the discovery source, not by mandate, by result." Indeed, an entrepreneurial university surrounded by clusters of high-tech firms is the emerging ideal for knowledge-based economic development in every part of the world. There is also an advantage to the university of having a group of high-tech firms nearby, offering collaboration possibilities to its researchers. As a director noted, "they are better than us at some things." Thus, having companies surrounding the university with complementary capabilities becomes an academic as well as an economic advantage.

10.13 INTERNATIONALIZATION OF UNIVERSITY TECHNOLOGY TRANSFER

Although this has changed in recent years, in many European countries and Japan, ownership of intellectual property arising from academic research traditionally resides with the professor. Nevertheless, actually taking out a patent, especially across national boundaries, can be an expensive proposition. As efforts to commercialize academic research get underway in Europe and Japan, offers to pay for the taking out of a patent, in exchange for part of the rights, resulted in a division between the university and the individual professor. Under these conditions, individual negotiations are soon superseded by precedent as the basis for a division of proceeds between the individual, the research unit or department and the university.

As precedent takes hold, a third rule is elaborated internationally, whether by law as in Denmark or through purchase of a share of the rights as in Sweden, where the individual academic retains ownership. The relatively autonomous position of the professor in the university, in comparison to the scientist as an employee of a corporation, is the fundamental basis for this shared intellectual property regime. Whether speeded by law or developed more slowly through precedent, the outcome is basically the same; so are the issues that must be resolved. The production of new knowledge with technological and economic implications inevitably drew the university toward utilization of the patent system as a support structure for firm formation and regional development as well as a means to enhance the university's income.

10.14 THE CULT OF NUMEROLOGY IN TECHNOLOGY TRANSFER

Traditional measures do not capture the full extent of these developments. The analysis of university patenting and technology transfer has suffered from a misplaced focus on numbers without context, a phenomenon that Pitrim Sorokin (1956), the founder of Harvard's Sociology Department, identified as Quantophrenia in his classic work *Fads and Foibles in Modern Sociology*. Statistics themselves, not merely unreliable data, contribute to creating misplaced objectives. This is the principle that *every measure which becomes a target becomes a bad measure* by inducing a fixation on the metric rather than on the creativity and initiative that any practice requires. Quantophrenia is the competitive tendency inherent in statistics to become a basis of rating comparisons that may counterproductively skew activity.

Some innovation researchers take patents produced by academics as a proxy for university contribution to economic development since they are subject to quantification and the data is often provided by technology transfer organizations. In 2001, for example, 626 licenses from US university TTOs formed the basis of 494 start-up firms, increasing to 550 in 2005, a low-end indicator of academic firm formation involving intellectual property, flowing through official channels. Patents, licenses and even number of start-up firms created, by themselves, are inadequate metrics of contribution to economic development. For example, it has been noted that while Japan has, in recent years, witnessed a significant increase in start-up activity, many of these firms, lacking either entrepreneurial intent, support structure or ecosystem, have not yet had significant economic effects (Tiaki and Ito, 2014).

Co-authorship between academic and industrial scientists provides a seemingly "hard measure" of university-industry linkages. New metrics, focusing on academic entrepreneurship needed to supplement existing university-industry cooperation metrics such as co-authorship and co-inventorship of patents as well as metrics of university invention licensing and licensing income: for example, a regional involvement index, a strategic direction capability, availability of university venture capital, leave policies for firm formation, contribution to economic and social development rewarded in promotion, entrepreneurship training at undergraduate, postgraduate and faculty levels.

Moreover, not all outputs are included in the official statistics. For example, AUTM spin-off data includes what is reported by university TTOs but misses those start-ups that have gone out from under the radar of the transfer office or were begun by undergraduates, or others, who are not employees required to disclose. Some universities keep close watch, but others do not consider themselves to be intellectual property police. The problem is not only with quality of data: "the garbage in – garbage out" phenomenon and with the overemphasis on quantification, but also with the shortening of time frames that may cut off significant potential results that have a long gestation period.

However, even conventional measures show that university technology transfer has generated an increase in revenues from $7.3 million in 1981 to $3.4 billion in 2008, and from 390 patents in 1980 to 3,088 by 2009 (Loise and Stevens, 2010). Nevertheless, technology transfer modalities that can be turned into numerical indicators, like licensing income, disclosures or patents attained, tend to overwhelm valuable activities that are not easily subject to quantification. Statistical analysis of TTOs emphasize income generation at the expense of mentoring services to faculty members in developing the commercial implications of their research. Judged according to quantitative criteria, TTOs at emerging research universities may appear as failures due to relative lack of income even as they perform an important academic and business development function that will only pay off in the longer term). Their work may become visible if they are allowed sufficient gestation period rather than being prematurely aborted as failures. Qualitative criteria are required to distinguish these proactive mentoring offices, with their outreach activities, from the failed TTO, disconnected from its clientele, waiting patiently in its offices for lightning to strike.

Some observers project an impending shake-out of university patent offices, unable to meet their costs (Mowery et al., 2001). Even at a less intensive level, academic technology transfer efforts have been found to be successful, if the research generation and regional outreach efforts of emerging research universities are taken into account (Etzkowitz and Göktepe-Hultén, 2010). Many observers were quite skeptical a decade ago, and some still are, that universities could even make enough money from their technology transfer efforts to cover costs. The rule of thumb is that it takes approximately seven years to reach break-even.

The university's economic base begins to be transformed as it generates some of its income from the sale of inventions. Although the funds are rarely large, they are monies that do not come with prescribed uses, whether from government grants or private donors and thus afford a flexibility that is an attractive attribute even if the scale is not yet great. Although most universities do not yet earn from patent licensing and technology transfer, a few have gained significantly. In 1992 Columbia University earned $24 million from its intellectual property rights, the equivalent of the income on almost half a billion dollars of endowment. This income rose to $100 million by the end of the decade, most of it derived from a single patent on the transgenic mouse that would soon expire. In recognition of this eventuality, the university is expanding its efforts into additional areas such as software. The Axel patents were so financially rewarding to Columbia University that it made an attempt, which ultimately failed, to change the law and extend their life. Nevertheless, the Cohen-Boyer patent around the same time produced more than $200 million in royalties for the University of California and Stanford, the home institutions of the researchers.

The techniques were broadly introduced to a wide variety of firms, with the academic institutions requesting relatively modest payments from each firm (Feldman et al., 2007). A decade or so later, Rockefeller University announced the receipt of $20 million in payment for a patent license for the obesity gene from Amgen, a biotechnology company, with an agreement to pay many times that amount if the protein proves useful in treating fat people. In this case, the result did not prove useful in the end, and no further funds were transferred.

10.15 CONCLUSION

The current crisis has evidenced some problems in knowledge creation and diffusion all around the world. Firms are focused on their economic problems and institutions are focused on health and social problems. Thus, the huge technology transfer gap that has emerged with such crisis has to be filled by universities through several kinds and levels of support. The main approaches are intensified search to enhance the disclosure rate and providing entrepreneurial assistance to the inventor to improve innovation chances. Several cases of good practices can be used as a benchmark to improve other universities' activities. A half-time position in technology transfer and an academic department has been instituted in the Columbia University Medical School as a unique arrangement for an individual faculty member. This dual-life scheme formalized the scouting function of ARD, the original venture capital firm that served as an informal TTO and incubator for MIT in the early post-war era. A serial entrepreneur, working at OTL as a part-time licensing officer in between start-ups, frustrated with the paucity of licensing opportunities for a technology that he strongly believed in, formed a firm with special permission. This one-off instance of a de facto entrepreneur in residence program may regularly be found in Swedish university incubators. The two modes may also be combined. Thus, the Chalmers Innovation System includes a master's program in innovation and entrepreneurship to which student teams apply with commercialization ideas that they often source in academic research groups. A Swedish hospital encourages nurses to be aware of the commercialization potential of devices they have invented and pairs them with "idea pilots" and advisors to speed the innovation process. These experiments may be synthesized into an intensive transfer regime to encourage other universities, including those not traditionally thought of as innovators, to raise the innovation rate.

REFERENCES

Audretsch, D.B. (2009). The entrepreneurial society. *The Journal of Technology Transfer*, 34(3), 245–54.

Breznitz, S.M. (2014). *The Fountain of Knowledge: The Role of Universities in Economic Development*. Stanford, CA: Stanford University Press.

Caiazza R., Richardson A., and Audretsch D.B., 2015, Knowledge effects on competitiveness: From firms to regional advantage. *The Journal of Technology Transfer*, 40(6), 899–909.

Etzkowitz, H. (1993) Enterprises from Science: The Origins of Science-based Regional Economic Development. *Minerva*, 31(3).

Etzkowitz, H. (2013). StartX and the "Paradox of Success": Filling the gap in Stanford's entrepreneurial culture. *Social Science Information*, 52(4).

Etzkowitz, H. (2022). Entrepreneurial university icon: Stanford and Silicon Valley as innovation and natural ecosystem. *Industry and Higher Education*. August

Etzkowitz, H., and Gulbrandsen, M. (1999). Public entrepreneur: the trajectory of United States science, technology and industrial policy. *Science and Public Policy*, 26(1), 53–62.

Etzkowitz, H., and Beddows, J. (2020). From Smart Specialization to Insightful Diversification. *Helice*, webinar series. *www.triplehelixassociation.org*

Etzkowitz, H., and Göktepe-Hultén, D. (2010). Maybe they can? University technology transfer offices as regional growth engines. *International Journal of Technology Transfer and Commercialisation*, 9(1–2), 166–81.

Feldman, M.P., Colaianni, A., and Liu, K. (2007). Lessons from the commercialization of the Cohen-Boyer patents: The Stanford University licensing program. In A. Krattiger, R.T. Mahoney, L. Nelsen, J.A. Thomson, A.B. Bennett, K. Satyanarayana, G.D. Graff, C. Fernandez, and S.P. Kowalski (eds), *Intellectual Property Management in Health and Agricultural Innovation: A Handbook of Best Practices*, Vols 1 and 2, 1797–1807. Oxford: MIHR.

Krücken, G. (2003). Mission impossible? Institutional barriers to the diffusion of the "third academic mission" at German universities. *International Journal of Technology Management*, 25(1–2), 18–33.

Loise, V., and Stevens, A.J. (2010). The Bayh-Dole Act turns 30. *Science Translational Medicine*, 2(52), 185–94.

Mazzucato, M. (2013). *The Entrepreneurial State: Debunking the Public vs. Private Myth in Risk and Innovation*. London, UK: Anthem

Mowery, D.C., and Sampat, B.N. (2004). The Bayh-Dole Act of 1980 and university–industry technology transfer: A model for other OECD governments?. *The Journal of Technology Transfer*, 30(1), 115–27.

Mowery, D.C., Nelson, R.R., Sampat, B.N., and Ziedonis, A.A. (2001). The growth of patenting and licensing by US universities: An assessment of the effects of the Bayh–Dole Act of 1980. *Research Policy*, 30(1), 99–119.

O'Shea, R.P., Allen, T.J., Chevalier, A., and Roche, F. (2005). Entrepreneurial orientation, technology transfer and spinoff performance of US universities. *Research Policy*, 34(7), 994–1009.
Owen-Smith, J. (2005). Trends and transitions in the institutional environment for public and private science. *Higher Education*, 49(1), 91–117.
Sorokin, P. (1956). *Fads and Foibles in Modern Sociology and Related Sciences* (357). Chicago, IL: H. Regency.
Tiaki, S., and Ito, C. (2014). University start-up or technology transfer; Japan–US comparison on commercialization of new technology at the dawn of the computer age. http://www.lib.kobe-u.ac.jp/handle_kernel/9002474 (accessed in 2014).

PART IV

COUNTRIES

11. Academic entrepreneurship in Italy

Alice Civera, Michele Meoli and Silvio Vismara

11.1 INTRODUCTION

Science-based entrepreneurial firms play a key role in the modern knowledge economy since they contribute to the creation and enhancement of regional high-tech clusters and the transformation of investments in basic science into economic growth, employment, and competitive advantage (Murray, 2004). Among science-based entrepreneurial firms, academic spinoffs have been considered especially important for national economies. Academic spinoffs are companies created by academic personnel (Colombo et al., 2010) to exploit technological knowledge that originated within universities (Fini et al., 2011). They have received great attention from scientists and policymakers (Perkmann et al., 2013). Indeed, although they are unlikely to generate major short-term shifts in macroeconomic performance on their own, their indirect effects on technology-using sectors are significant (Ljungberg and McKelvey, 2012). Therefore, these firms are expected to act as engines of development, resulting in economic and social spillovers above the private returns (Griliches, 1992).

National governments have shown interest in academic spinoffs, implementing laws to foster their establishment (Clarysse et al., 2007). The European Commission recommends research institutions set incentives for researchers who take part in such knowledge transfer activities. The growing interest of policymakers is paralleled in the high number of research publications seeking to better understand and address the forces that shape this activity. The literature is divided into three main streams: the characteristics of individual academics who become entrepreneurs (Landry et al., 2006), the factors in the university environment that facilitate the creation of business activities (O'Shea et al., 2005), and the institutional conditions under which spinoffs are incubated (Lockett et al., 2005).

In this chapter, we first review the literature and discuss the regulatory aspects. The review of the literature presents the determinants of the establishment of academic spinoffs at three levels, namely, the individual, the institutional, and the contextual level. Then, we focus on the Italian context, which has been extensively studied in the academic entrepreneurship literature. The academic spinoff phenomenon has indeed increased significantly starting in Italy from the 1990s (Fini et al., 2009), following the introduction of a new dedicated regulatory framework focused on supporting scientific and technological research, knowledge transfer, and researchers' mobility

(Law 297/1999 and Ministerial Decree 593/2000). Since then, public researchers can be involved in technology transfer projects while keeping their university positions and wages. We discuss the main characteristics of the Italian academic spinoffs by shedding light on the rationales behind their diffusion and their configuration. In particular, we include information on the sector, the technological level, as well as the survival rate of these companies.

11.2 LITERATURE REVIEW

11.2.1 Individual-level Determinants of the Establishment of Academic Spinoffs

Several studies emphasize the impact of individual attributes and dispositions on the creation of academic spinoffs. Landry et al. (2006) by considering a sample of 1,554 Canadian researchers in natural sciences and engineering, finding that 16.8 percent of the sample exhibits entrepreneurial behavior through attempts to create spinoffs, whereas 32 percent use diverse types of mechanisms to protect their intellectual property. Estimates show that a complementary set of resources needs to be mobilized by researchers to launch university spinoffs.

The multifaceted nature of the resources and assets researchers need to become entrepreneurs is confirmed by D'Este and Perkmann (2011). They examine what motivates academics to engage with industry using both informal collaboration and formal models of interaction by surveying a sample of 4,337 research grant holders from the UK's Engineering and Physical Sciences Research Council (EPSRC) covering 1999 to 2003. Factor analysis identifies four factors as motivations: commercial exploitation of technology or knowledge; learning intended as expected benefits from gaining new insights, receiving feedback on research, and accessing new knowledge through engagement with industry; access to private funding complementing public research monies; and access to in-kind resources such as industry-provided equipment, materials, and data for research. Only one of these factors (commercialization) is related to the intention to be entrepreneurial. Link et al. (2007) similarly present empirical evidence on the determinants of three types of informal technology transfer by faculty members: knowledge transfer, joint publications with industry scientists, and consulting. They find that male and tenured faculty members are more likely to engage in all three forms of informal technology transfer. They also find that academics who allocate a relatively higher percentage of their time to grants-related research are more likely to engage in informal commercial knowledge transfer. In fact, when the firm involves a university in its funded project, more scientific papers result (Audretsch, Link et al., 2019). Scientific papers submitted for publication from US Small Business Innovation Research (SBIR)-funded research projects have a higher level of innovativeness if they involve an individual from the university who originally developed the technology being pursued by the firm in its SBIR project. In other words, the project-specific technical human capital knowledge from the university that spills over to the firm's project begets additional knowledge in the form of scientific papers submitted for publication (Audretsch, Link et al., 2019).

Scientists may thus start a business for different reasons (Hayter, 2011; Lam, 2011). Within this set of factors, the enhancement of academic status is perceived to be the most important incentive for fostering the creation of a new company. By establishing a spinoff, researchers expect to achieve desired outcomes such as the generation of further stimuli for research activities, gaining prestige and reputation as leading academics, the creation of funding opportunities (grants) for students or research assistants, and the possibility of obtaining new infrastructure and facilities for research activities.

The existence of non-financial motivations for the creation of firms is highlighted by the entrepreneurship literature's classification of "push" and "pull" factors. In the pull approach, or "opportunity" entrepreneurship, incentives for the creation of a firm are driven directly or indirectly by the desire to increase the founder's wealth (Baumol, 1996); in the push approach, also known as "necessity" entrepreneurship, individuals are moved to create a firm to escape an unsatisfactory situation. Among push factors, unemployment is typically the most important (Ritsilä and Tervo, 2002; Storey, 1991). Different motivations correspond to a different performance, as necessity-oriented academic spinoffs are associated with higher survival profiles, while opportunity-oriented spinoffs exhibit higher post-entry growth rates (Civera, Meoli et al., 2020).

The founding team of the academic spinoffs needs to be developed according to the motivations and the desired performance. If the ultimate goal is creating a financially successful company, a team composed of scientists only may have downsides, as the academic founders of a company may not have enough business and commercial experience to properly exploit their innovations. Hmieleski and Ensley (2005) underline this deficiency by examining differences in the composition of the top management team (TMT) and performance between matched samples of 256 university-based and 500 independent high-technology startups. They demonstrate that university support tends to focus on technology and marketing instead of spending adequate time developing the TMTs of the new ventures. Zahra et al. (2007) provide further evidence of a lack of commercial experience within universities through a comparative study between 91 corporate spinoffs and 78 academic spinoffs. The results indicate that corporate spinoffs benefit from the skills transferred through the founders and employees who have worked for their parent corporations and have maintained contacts, whereas the opposite was found regarding academic spinoffs. Science-based firms do not reap the same benefits from their parent universities, which may not be skilled enough to commercialize new technology or develop appropriate capabilities. Indeed, higher education institutions often lack the mindset and commercial skills to actualize their discoveries. Moreover, they may not have a complete awareness of marketing issues related to the success of technology commercialization. Academic spinoffs which reflect the characteristics of their academic founders, instead, are oriented not to the market but to survival.

11.2.2 Institutional-level Determinants of the Establishment of Academic Spinoffs

Another stream of literature addressing enabling factors for academic spinoff creation identifies the overall university system (Clarysse and Moray, 2004; Clarysse et al., 2005; Degroof and Roberts, 2004; Di Gregorio and Shane, 2003; Lockett et al., 2003). The role of the university has been widely debated in the literature. Some scholars argue that university involvement is beneficial for the success of newly established ventures in terms of survival rates, performance, and reputation (Audretsch et al., 2015; Civera and Meoli, 2018; Clarysse et al., 2005; Degroof and Roberts, 2004; Di Gregorio and Shane, 2003; Harrison and Leitch, 2005).

The important role that universities play in generating knowledge spillovers is testified by the locational choice of firms, which prefer to stay close to universities (Audretsch et al., 2004, 2005). Especially academic spinoffs, which represent a larger proportion of companies in parks that are geographically closer to their university and in parks that have a biotechnology focus and in parks that are associated with richer university research environments (Audretsch et al., 2017).

The specific incubation strategies for spinning out companies are explored by Clarysse et al. (2005), who use a two-stage approach to answer two research questions: the first focuses on the modeling of incubation strategies, whereas the second is dedicated to the resources employed and the activities undertaken within the model categorized. In the first step, they identify three distinct incubation models of managing the spinoff process (Low Selective, Supportive, and Incubator) through an in-depth analysis of seven cases in five European countries. In the second phase, they determine differences in resources relating to finance, organization, human resources, technology, network, and infrastructure by validating them through 43 cases. The first model, predominantly service-oriented, supports the creation of self-employment-oriented spinoffs, while the second model stimulates the creation of economic profitability-oriented spinoffs. When the amount of necessary money is too small to be efficient, or the market is simply too small to be attractive to financial investors, the third model is preferable.

At the same time, other scholars underline the negative consequences of strong ties with parent universities, such as dependency, non-beneficial reputation effects, and delayed graduation from incubators (Rothaermel and Thursby, 2005). Rothaermel and Thursby (2005) employ the resource-based view of the firm and the absorptive capacity construct to confirm that knowledge flows enhance incubator firm performance. Drawing on longitudinal and fine-grained firm-level data on 79 technology ventures incubated between 1998 and 2003 at the Advanced Technology Development Center, a technology incubator sponsored by the Georgia Institute of Technology, they provide empirical evidence that incubator firms' absorptive capacity is an important factor in transforming university knowledge into a firm-level competitive advantage. Their analysis focuses on two mechanisms by which incubator firms can access this knowledge: the license to develop and use a university invention and the citations of university research found in the patents associated with incubator firms. The dependent variable investigated is the probability of obtaining venture capital (VC) funding, a mechanism that appears to assess the performance of new technology ventures satisfactorily.

Given that scientist-entrepreneurs are actively involved in knowledge creation and dissemination, and given the importance of the complementary resources they can access, their academic affiliation is expected to influence the performance of their business ventures. Affiliation with a university is beneficial for academic spinoffs because it signals that a firm has access to specific resources and contacts, enhancing the valuation of the firms in the eyes of investors (Cattaneo et al., 2015; Meoli et al., 2013). The signal can be stronger for more eminent universities because the affiliation with a prestigious institution provides a connection to more qualified resources (Civera and Meoli, 2018). Prominent higher education institutions are expected to possess the best faculties, instruments, and equipment and to provide high-quality research that may result in leading-edge innovations (Wang and Shapira, 2012). Stuart and Ding (2006) examine the conditions prompting university-employed life scientists to become entrepreneurs, focusing on biotechnology. The orientation toward entrepreneurship is stronger when academics work in university departments that have employed other scientists who had ventured into the commercial sector and when these scientists are prestigious. Kenney and Patton (2011) contribute to the literature by investigating the role of inventor ownership in generating spinoffs instead of the probability of becoming an entrepreneur. The results of their empirical analysis on 515 academic spinoffs from one Canadian and five US universities with several characteristics in common covering 1957 to 2009 show that eminent institutions have a greater propensity to generate spinoffs at the technological field level. However, low-status universities may be able to increase their performance by stimulating entrepreneurship since inventor ownership seems to be a powerful intervening variable that can mitigate the role of status. More recently, Avnimelech and Feldman (2015) exploit the online professional networking resource LinkedIn to identify researchers at 124 universities (83 public and 41 private) who have established a new firm. The data collection conducted in September 2013 leads to a sample of 18,127 faculty founders of spinoffs. Their evidence confirms that the founder-spawning rate is positively affected by the quality of the institutions and their departments, besides R&D expenditure and the strength of local clusters. From a different perspective, Meoli and Vismara (2016) argue that academics may also choose to launch an independent company by pursuing alternative business engagement activities when their university provides inadequate support. Drawing from hybrid entrepreneurship and necessity entrepreneurship, they model the individual decision to spin off as a reaction to the organizational characteristics of the parent university rather than to its outcome. Through a longitudinal study of 559 spinoffs from 85 Italian universities covering 1999 to 2013, they find that, although stronger administrative support from the parent university leads academics to create more technology spinoffs, a U-shaped relationship exists between the number of administrative staff within a university and the rate of establishment of non-technology spinoffs: when the level of administrative staff is too low or too high relative to the fitted amount estimated using several university-level factors, academic staff react by establishing firms to achieve improved cash and human resource management. Siegel et al. (2004), by relying on an inductive, qualitative approach identify the key organizational issues in promoting successful knowledge transfers. Based on 55 structured interviews of 98 university/industry technology transfer (UITT) stakeholders associated with five US research universities, the authors conclude that there are numerous impediments to effectiveness in UITT: cultural and informational barriers among the three key stakeholder types (university administrators, academics, and firms/entrepreneurs), technology transfer office (TTO) staffing and compensation practices, and inadequate rewards for faculty involvement in UITT.

11.2.3 Contextual-level Determinants of the Establishment of Academic Spinoffs

Several studies highlight the role of holistic conditions in fostering the creation of academic spinoffs, particularly high-tech spinoffs.

Bekkers et al. (2006) analyze and compare data from 23 Dutch universities and 13 other public research organizations (PROs) together with data from 168 US universities and 10 other PROs to better understand the role and relevance of a variety of factors concerning national, sectoral, regional, and managerial institutions and how they interrelate. These authors implement a two-step analysis to, first, determine the key factors affecting the potential to establish a science-based spinoff and, second, to identify the main factors affecting their success. Comparing the Netherlands and the US shows that the combination of national and sectoral institutions is the key influence on the probability of establishing an academic spinoff, whereas managerial institution conditions determine the chances of success.

The characteristics of the universities can also contribute to the social and economic development of regions and countries in their role of fostering entrepreneurial attitudes. Braunerhjelm (2008) detects whether and to what extent universities influence regional productivity, focusing on the age of higher education institutions. The author creates a unique dataset by collecting information from Sweden Statistics, covering 1975 to 1999 and measuring the impact of the university on the region, intended as a correspondence between a university's research specialization and industrial specialization in the region hosting the university, through a Heckman regression analysis. The results suggest the existence of correspondence in the specialization as well as a positive effect in terms of productivity created by the universities. Differences are seen across regions but are not due to the age of the universities.

Patton and Kenney (2010) underline the importance of universities in fostering regional development by arguing that the external environment makes the difference. Analyzing two clusters in which the universities of Wisconsin and Illinois are located, the authors conclude that higher education institutions plant seeds that fructify only if planted in a dynamic entrepreneurial context. The university cluster of Wisconsin (UW-M) performs better by virtue of several characteristics that the cluster of Illinois (UIUC) lacks. The first institution is deeply involved in the management of the cluster, whereas the activity of the second is more recent. Moreover, the University of Wisconsin can count on the support of private organizations, which are almost completely absent in Champaign.

As universities are at the center of knowledge creation and exploitation, they are directly funded to implement a mix of innovation and entrepreneurship policy interventions, or, more often, indirectly supported through partnerships with industry (Civera et al., 2018). In order to conceptualize the structure and function of such "university-industry-entrepreneurship" networks, academic research introduced the metaphor of "ecosystems" (Audretsch, Cunningham et al., 2019; Colombo et al., 2019; Hayter, 2016; Lehmann et al., 2020). In this context, "entrepreneurial ecosystems" gains increasing popularity as vehicles to describe, explain, advertise, and transport thoughts, frameworks, and opinions on how entrepreneurs interact with their environment as economic agents (Acs et al., 2018; Audretsch, Cunningham et al., 2019). An entrepreneurial ecosystem is then characterized by the participation of entrepreneurial firms as an important output of economic systems and an important mechanism to explain the outcome of economic systems, a structure that fosters entrepreneurship, like financiers, sources of knowledge spillovers, suppliers or consumers, and governance, to coordinate and motivate entrepreneurial activities by setting rules and norms (Lehmann et al., 2020).

11.3 REGULATION

The creation of spinoff companies has been allowed in Italian universities since the enforcement of Ministerial Decree 297/1999. This policy change aimed at facilitating knowledge and technology transfer from the universities to the market (1) by abolishing the incompatibility between the academic and entrepreneurial career for scholars working in an Italian university and (2) by introducing financial incentives to support applied research. Several relieves were meted out to these companies and the involved scholars to incentivize the exploitation of academic research results. For instance, professors, PhD students, and/or holders of research grants can gain income subject to tax exemption from the assignment of shares, quotas, or financial instruments granted by the company to its directors, employees, or permanent collaborators as well as option rights for the purchase of financial instruments (Law 221/2012). Law 221/2012 also introduced fiscal deduction that reduces the tax burden for the subjects that invest directly or indirectly in the share capital of spinoffs to foster companies' ability to attract private financial capital that nurtures their growth.

In addition, Law 134/2012 introduced subsidies to facilitate the recruitment of highly qualified personnel as well as permanent employment carried out through an apprenticeship contract. Ministerial Decree 297/1999 devolves also to the individual universities/research institutes the regulation of the relationship between individual scholars and the institution, the issues related to intellectual property, and limitations aimed at preventing conflicts of interest with the spinoff company. Accordingly, every Italian university regulates autonomously some aspects of its own academic spinoff's activity. Nevertheless, common patterns regarding the formal powers/authority of university leaders and governing bodies are identifiable.

The Administrative Board holds the legislative and deliberative powers within the Italian state universities (Boffo and Dubois, 2005; Boffo et al., 2008). It is thus the governing body responsible for the approval of the establishment of every academic spinoff. The Rector – who is elected from among full professors according to a weighted voting system wherein the academic staff hold the absolute majority through administrative staff and students cast a vote as well (Capano et al., 2016; Donina et al., 2015b) – carries out the executive function, that is, implementing change and putting into practice the decisions taken by the legislative and deliberative bodies. S/he as the governor and external legal representative, performs both strategic and managerial tasks with powers regarding strategic direction, financial matters, teaching, and research activities, and is responsible for pursuing university objectives and day-to-day management. In addition, s/he is an *ex officio* member of the Administrative Board and Academic Senate and chairs both bodies, thus holding agenda-setting powers (Donina and Paleari, 2019). Due to these governance arrangements, Boffo and Dubois (2005) stress that the Rector has several means to dominate the decision-making process, while deliberative bodies simply ratify the Rector's decisions. Following the enforcement of Law 240/2010, the Rector is now involved in the appointment of Administrative Board members in most Italian state universities (Donina et al., 2015a). As a result, the Administrative Board of Italian state universities is now more tightly controlled by the Rector. Specifically, regarding the spinoff activity, the Rector is directly involved also in the appointment of at least one member of the ad hoc committee that monitors and reports to her/him at least once per year about the activity of the spinoffs. Accordingly, through a longitudinal study of 1,176 spinoffs from 59 Italian state universities over the period 2001–2015, Civera, Donina et al. (2020) find that the university leaders foster the creation of academic spinoffs, especially the mobility of university leaders positively affects the rate of establishment of academic spinoffs. Rectors that experienced international mobility in their career develop networks and collaborations that facilitate the business involvement of their institutions.

From the legal point of view, academic spinoffs are considered in Italy a fully-fledged startup, according to the report *Restart, Italia!* drawn up in 2012 by the task force established by the Minister of Economic Development and the consequent Legislative Decree 221/2012, commonly called Startup Act. Specifically, Italian startups are defined as enterprises that are not listed on the stock exchange, reside or are subject to taxation in Italy, and fulfill the following criteria:

- Owned directly and for at least a 51 percent share by individuals, also in terms of voting rights.
- Established for no longer than five years.
- No turnover or have a turnover that does not exceed 5 million euros (according to the last approved financial statement).
- No distribution of profits.

- Their social goal is the development of innovative goods or services of high technological value.
- Their bookkeeping is transparent, and they do not use cash – except for the expenses related to reimbursements.

Even more, academic spinoffs are classified as an innovative startup because the R&D dimension required to be an innovative startup had been mainly developed before their establishment. Being a university spinoff is indeed the third out of three criteria to be considered an innovative startup. This criterion is complementary to that concerning the possession of a patent since not all the innovations which were the result of business projects have already been patented (they may never be). In any case, the timeframe for patents is often too long and is not compatible with establishing a startup. The other two are (1) a percentage of R&D expenditures greater than 15 percent; (2) a percentage of PhD students, PhD holders, or staff highly qualified in research activities greater than 30 percent.

Due to their inclusion in the innovative startup category, academic spinoffs can benefit from several fiscal incentives. Especially at the time of foundation, startup companies can freely register on the Chamber of Commerce's Business Register with a consequent reduction of administrative burdens. All startups must pay tax (VAT and corporate tax) depending on their real cash flow rather than fixed taxation rates. They benefit from a very flexible short-term contract, which is exempt from as many taxes as possible (IRAP, payroll income tax). Startups can issue stock options with a tax advantage regime to attract talented people from Italy and abroad into startups. During their growth phase, investments in startups are encouraged by the creation of two ad hoc funds, the Fund of Funds and the Fund for Seed Investments, by tax reductions for private investments in startups, for crowdfunding, and social lending activities. Finally, in the maturity phase, the industrial acquisition of startups as well as the listing of startups on the stock exchange is intencivized by exclusions from corporate taxation, thus lowering the costs of acquiring startups and reducing listing costs. In case of failure, a simplified liquidation process is provided to speed up the liquidation. The startupper himself is appointed as liquidator, and the consequences of the declaration of bankruptcy for the entrepreneur are limited.

This implies that the supportive program policy does not end when startups stop being in their startup phase (i.e. five years since establishment). A new portfolio of fiscal benefits is set up for the so-called innovative small and medium enterprises. Due to the way the incentive is set up, startups in general and academic spinoffs, in particular, are incentivized to survive even in the absence of economic activity, giving birth to the phenomenon of "zombie firms" (Broughman and Fried, 2012; Kane, 1990).

11.4 EMPIRICAL EVIDENCE

The phenomenon of academic entrepreneurship became relevant in Italy from the 1990s, when public researchers were allowed to be involved in technology transfer activities without negative consequences for their jobs or wages, according to Law 297/1999, followed by Ministerial Decree 593/2000. Nonetheless, the phenomenon began to take shape some years before. The first academic spinoff indeed was established in 1981 from the University of Florence.

Table 11.1 presents the evolution in the number of academic spinoffs established annually, from 1981 to 2018, as reported by the Ministry of Education database, *Spin-off Italia.*[1] More than 65 percent of the spinoffs were established after 2008 as, in Italy, like in other countries, academic entrepreneurship has been a response to the financial crisis. The five years between 2010 and 2015 are particularly noteworthy for the high number of companies created. The economic crisis has been enumerated among the push factors associated with necessity entrepreneurship as, after the economic upheaval represented by the economic crisis in 2008, in Italy, entrepreneurship has become a special means to avoid the unemployment of highly skilled human capital, academics included (Civera, Meoli et al., 2020; Horta et al., 2016; Vismara and Meoli, 2016). In particular, the study from Horta et al. (2016) demonstrates that the rate of academic spinoff creation is positively associated with the skilled unemployment rate. That longitudinal study of 559 spinoffs launched between 1999 and 2013 showed that a higher level of unemployment reduced the probability of academic spinoff creation up to a threshold, beyond which the effect was reversed. By contrast, the relatively skilled unemployment level was related positively to the probability of academic spinoff, particularly high-tech spinoff, creation.

Table 11.1 Number of academic spinoffs per year

Year of establishment	Number of spinoffs		Year of establishment	Number of spinoffs
1981	1		2005	69
1984	1		2006	76
1993	2		2007	106
1994	1		2008	97
1995	3		2009	93
1996	3		2010	114
1997	3		2011	94
1998	4		2012	142
1999	13		2013	113
2000	19		2014	129
2001	22		2015	129
2002	18		2016	96
2003	44		2017	110
2004	68		2018	56

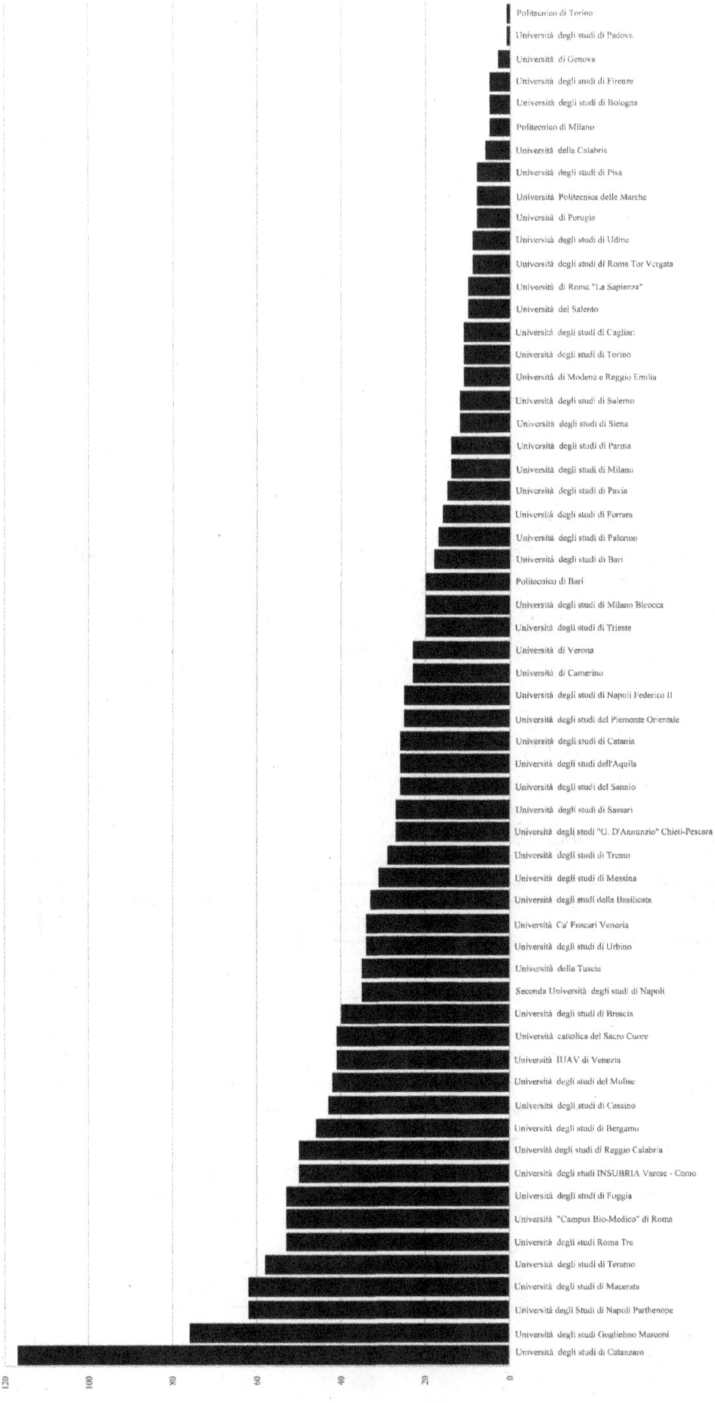

Figure 11.1 Number of academic spinoffs per university

Despite the University of Florence being a pioneer in the establishment of academic spinoffs, the Technical University of Turin is the most prolific, followed by the University of Padova. It is not a coincidence that the first six universities in Figure 11.1 devoted high attention to the technical disciplines in both their teaching and research activity. The presence of control variables at university and regional levels such as technical universities, STEM (Science Technology Engineering and Math) graduates and faculty testifies to the importance of technical knowledge and competencies for the creation as well as the performance of academic spinoffs (Civera et al., 2019; Meoli and Vismara, 2016; Vismara and Meoli, 2016).

The supportive role of universities depends on the nature of academic spinoffs. Through a longitudinal study of 559 spinoffs from 85 Italian universities from 1999 to 2013, Meoli and Vismara (2016) find that although stronger administrative support from the parent university leads academics to create more technology spinoffs, a U-shaped relationship instead exists between the number of administrative staff within a university and the rate of establishment of non-technology spinoffs. When the level of administrative staff is too low or too high relative to the fitted amount estimated using several university-level factors, academic staff react by establishing firms to achieve improved cash and human resource management.

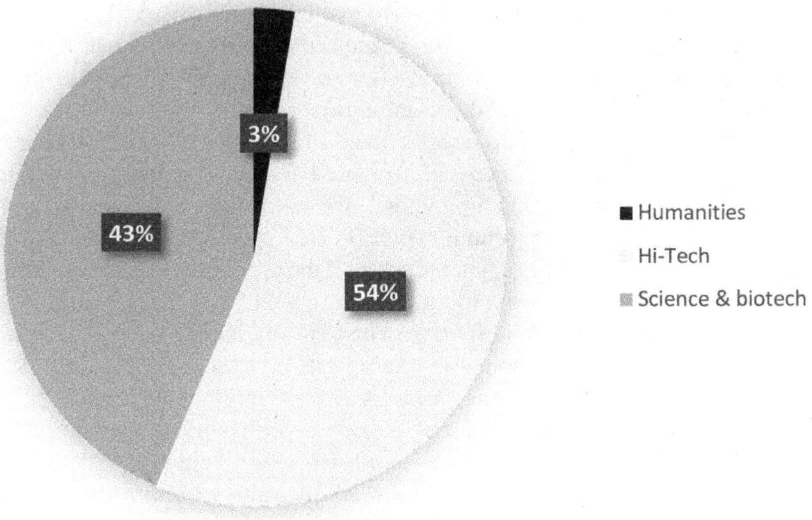

Figure 11.2 *Percentage of academic spinoffs by sector*

With regard to the sectors where the academic spinoffs operate, Italian academic spinoffs can be distinguished in the (a) biomedical and life science sector, which engages 43 percent of the total population; (b) high-tech sector consisting of aerospace, industrial automation, electronics, ICT, nanotech, and automation services, which 54 percent of the total academic spinoffs belong to; (c) humanities, for example, cultural heritage, which involves only 3 percent of academic spinoffs. Figure 11.2 represents graphically the concentration of academic spinoffs in the sectors.

By focusing on the output, the academic spinoffs offer both products and services. Almost half of the academic spinoffs (46 percent) are exclusively service-oriented, 38 percent are product-oriented while the remaining 16 percent are involved in both service and products. An example of the latter is those spinoffs offering consulting activity together with their products (for the vast majority, this refers to products such as software and platforms). The distinction is shown to be interesting when considering the governance of universities in relation to their engagement within the entrepreneurial ecosystem. Thus, when local stakeholders are appointed as lay members of the university's board of directors, they are associated with increased establishments of service-oriented spinoffs (Meoli et al., 2019).

Despite more than half of the Italian academic spinoffs being hi-tech spinoffs, only 2 percent of academic spinoffs are created out of a patent. This may be due to the Italian regulation, according to which being an academic spinoff does not require the possession of a patent since not all the innovations which were the result of business projects have already been patented (they may never be). In any case, the timeframe for patents is often too long and is not compatible with establishing a startup. Moreover, the goals of academic spinoffs are different, and commercializing innovation is only one of them. Academic spinoffs can be established either to exploit an opportunity, based on an economic rationale or out of necessity in order to provide a social service to the community, especially at a time when the state finds it increasingly difficult to respond to all of the citizens' needs (Civera, Meoli et al., 2020). In the same way, necessity-oriented academic spinoffs created to protect academics from eventual unemployment are becoming a diffused reality (Roach and Sauermann, 2010). Opportunity spinoffs are expected to contribute to the added value generated by the regional industrial fabric, whereas necessity spinoffs are expected to be beneficial in avoiding unemployment.

Opportunity academic spinoffs have economic rationales behind so that, according to the neoclassical economic theory, it is expected that growth is their primary post-establishment focus (Gundry and Welsch, 2001; Hessels et al., 2008). Nonetheless, many entrepreneurs, also academic ones, choose not to grow their business (Gundry and Welsch, 2001; Kolvereid, 1992; Wiklund et al., 2003). Individuals pushed into entrepreneurship, looking for an alternative job to make a life out of it (Horta et al., 2016; Rizzo, 2015; Roberts, 1991) are expected to be invested in the necessity academic spinoffs they generate, and they are focused on their survivability.

By relying on a sample of 613 Italian academic spinoffs established over the period 2006–12, Civera, Meoli et al. (2020) find that necessity-oriented academic spinoffs are associated with higher survival profiles, while opportunity-oriented spinoffs exhibit higher post-entry growth rates. Focusing on the performance of academic spinoffs, the survival rate of academic spinoffs is extremely high, as exhibited in Figure 11.3.

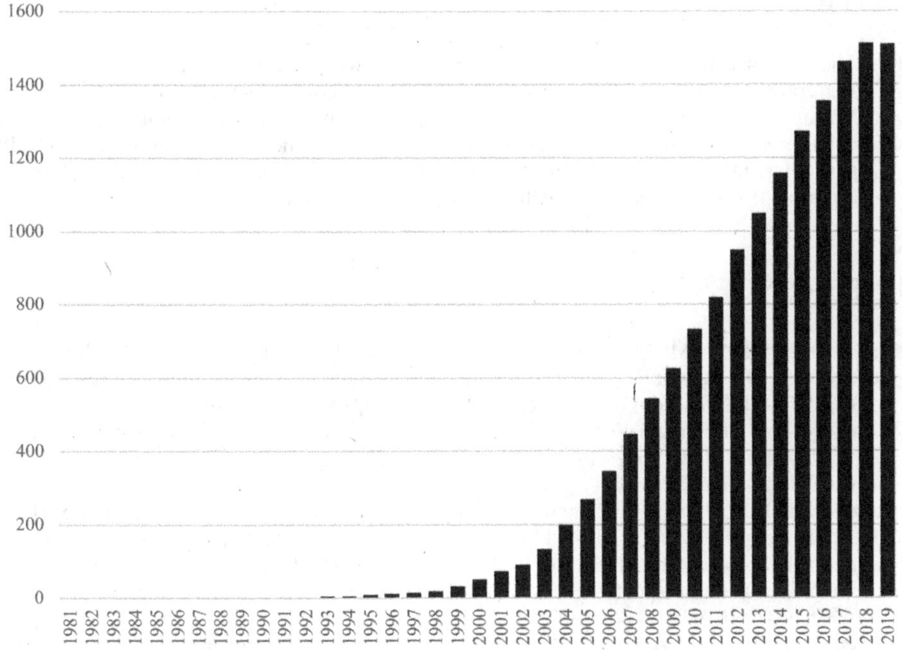

Figure 11.3 *Number of active academic spinoffs per year*

11.5 CONCLUSION

The academic spinoff phenomenon has been extensively studied in several contexts. In Italy, they represent an important share of innovative companies. The technological level and the output they produce as well as their performance in terms of survival rate are extremely correlated to the motivation of their establishment, which in turn is inextricably related to the national regulation. Italian scientists may decide to start a new company out of necessity, overall, to avoid unemployment. In academia, the case of academic spinoffs as a second-best solution is a frequent phenomenon (Horta et al., 2016). Indeed, it is likely that graduates or PhD students create a necessity spinoff by considering the foundation of a new company as a "pushed alternative" considering the lack of stable jobs that promise or entail the possibilities of decent earning and career progress (Hahn et al., 2017). The academic career is a highly selective job path, and those who cannot remain in the system may be interested in spinning out a company to exploit, to some extent, the knowledge previously developed (D'Este and Perkmann, 2011; Perkmann et al., 2013). Necessity-oriented companies are important to the creators because they need to make a life out of them. Necessity-oriented entrepreneurs may be more invested by them in terms of survivability (Hessels et al., 2008). They intend the creation of a spinoff not as a transitory activity but rather as a "for life" option.

NOTE

1. The database is created by the collaboration between the Center for Innovation and Entrepreneurship of the Polytechnic University of Marche, Netval, and the Institute of Management at the Sant'Anna School and contains information on companies, that the collaboration has decided to associate the concept of spinoffs from public research. In particular, in the database, you can select the companies that have been accredited as a spinoff from research parent institutions (http://www.spinoffitalia.it/).

REFERENCES

Acs, Z.J., Estrin, S., Mickiewicz, T., and Szerb, L. (2018). Entrepreneurship, institutional economics, and economic growth: An ecosystem perspective. *Small Business Economics*, 51(2), 501–14. https://doi.org/10.1007/s11187-018-0013-9

Audretsch, D.B., Lehmann, E.E., and Warning, S. (2004). University spillovers: Does the kind of science matter? *Industry and Innovation*, 11(3), 193–206. https://doi.org/10.1080/1366271042000265375

Audretsch, D.B., Lehmann, E.E., and Warning, S. (2005). University spillovers and new firm location. *Research Policy*, 34(7), 1113–22. https://doi.org/10.1016/j.respol.2005.05.009

Audretsch, D.B., Lehmann, E.E., and Paleari, S. (2015). Academic policy and entrepreneurship: A European perspective. *Journal of Technology Transfer*, 40(3), 363–8. https://doi.org/10.1007/s10961-014-9359-6

Audretsch, D.B., Link, A., Link, A.N., and Scott, J.T. (2017). Opening the ivory tower's door: An analysis of the determinants of the formation of U.S. university spin-off companies. In *Universities and the Entrepreneurial Ecosystem* (pp. 37–43). Cheltenham, UK and Northampton, MA, USA: Edward Elgar Publishing. https://doi.org/10.4337/9781786432797.00008

Audretsch, D.B., Link, A.N., and van Hasselt, M. (2019). Knowledge begets knowledge: University knowledge spillovers and the output of scientific papers from U.S. Small Business Innovation Research (SBIR) projects. *Scientometrics*, 121(3), 1367–83. https://doi.org/10.1007/s11192-019-03260-3

Audretsch, D.B., Cunningham, J.A., Kuratko, D.F., Lehmann, E.E., and Menter, M. (2019). Entrepreneurial ecosystems: Economic, technological, and societal impacts. *Journal of Technology Transfer*, 44(2), 313–25. https://doi.org/10.1007/s10961-018-9690-4

Avnimelech, G., and Feldman, M.P. (2015). The stickiness of university spin-offs: A study of formal and informal spin-offs and their location from 124 US academic institutions. *International Journal of Technology Management*, 68(1–2), 122–49. https://doi.org/10.1504/IJTM.2015.068755

Baumol, W.J. (1996). Entrepreneurship: Productive, unproductive, and destructive. *Journal of Business Venturing*, 11(1), 3–22. https://doi.org/10.1016/0883-9026(94)00014-X

Bekkers, R., Gilsing, V., and van der Steen, M. (2006). Determining factors of the effectiveness of IP-based spin-offs: Comparing the Netherlands and the US. *Journal of Technology Transfer*, 31(5), 545–6. https://doi.org/10.1007/s10961-006-9058-z

Boffo, S., and Dubois, P. (2005). The weakness of university legislative bodies: The cases of France and Italy. *International Review of Administrative Sciences*, 71(1), 35–54. https://doi.org/10.1177/0020852305051682

Boffo, S., Dubois, P., and Moscati, R. (2008). Changes in university governance in France and in Italy. *Tertiary Education and Management*, 14(1), 13–26. https://doi.org/10.1080/13583880701814132

Braunerhjelm, P. (2008). Specialization of regions and universities: The new versus the old. *Industry and Innovation*, 15(3), 253–75. https://doi.org/10.1080/13662710802040853

Broughman, B.J., and Fried, J.M. (2012). Do VCs use inside rounds to dilute founders? Some evidence from Silicon Valley. *Journal of Corporate Finance*, 18(5), 1104–20. https://doi.org/10.1016/j.jcorpfin.2012.06.012

Capano, G., Regini, M., and Turri, M. (2016). *Changing Governance in Universities: Italian Higher Education in Comparative Perspective.* London: Palgrave Macmillan.

Cattaneo, M., Meoli, M., and Vismara, S. (2015). Cross-border M&As of biotech firms affiliated with internationalized universities. *Journal of Technology Transfer*, 40(3), 409–33. https://doi.org/10.1007/s10961-014-9349-8

Civera, A., and Meoli, M. (2018). Does university prestige foster the initial growth of academic spin-offs? *Economia e Politica Industriale*, 45(2), 111–42. https://doi.org/10.1007/s40812-017-0072-y

Civera, A., Meoli, M., and Vismara, S. (2018). Do academic spinoffs internationalize? *Journal of Technology Transfer*, 1–23 April. https://doi.org/10.1007/s10961-018-9683-3

Civera, A., Meoli, M., and Vismara, S. (2019). Do academic spinoffs internationalize? *Journal of Technology Transfer*, 44(2), 381–403. https://doi.org/10.1007/s10961-018-9683-3

Civera, A., Meoli, M., and Vismara, S. (2020). Engagement of academics in university technology transfer: Opportunity and necessity academic entrepreneurship. *European Economic Review*, 123, 103376. https://doi.org/10.1016/j.euroecorev.2020.103376

Civera, A., Donina, D., Meoli, M., and Vismara, S. (2020). Fostering the creation of academic spinoffs: Does the international mobility of the academic leader matter? *International Entrepreneurship and Management Journal*, 16(2), 439–65. https://doi.org/10.1007/s11365-019-00559-8

Clarysse, B., and Moray, N. (2004). A process study of entrepreneurial team formation: The case of a research-based spin-off. *Journal of Business Venturing*, 19(1), 55–79. https://doi.org/10.1016/S0883-9026(02)00113-1

Clarysse, B., Wright, M., Lockett, A., Van de Velde, E., and Vohora, A. (2005). Spinning out new ventures: A typology of incubation strategies from European research institutions. *Journal of Business Venturing*, 20(2), 183–216. https://doi.org/10.1016/j.jbusvent.2003.12.004

Clarysse, B., Wright, M., Lockett, A., Mustar, P., and Knockaert, M. (2007). Academic spin-offs, formal technology transfer and capital raising. *Industrial and Corporate Change*, 16(4), 609–40. https://doi.org/10.1093/icc/dtm019

Colombo, M.G., D'Adda, D., and Piva, E. (2010). The contribution of university research to the growth of academic start-ups: An empirical analysis. *Journal of Technology Transfer*, 35, 113–40. https://doi.org/10.1007/s10961-009-9111-9

Colombo, M.G., Dagnino, G B., Lehmann, E.E., and Salmador, M.P. (2019). The governance of entrepreneurial ecosystems. *Small Business Economics*, 52(2), 419–28. https://doi.org/10.1007/s11187-017-9952-9

D'Este, P., and Perkmann, M. (2011). Why do academics engage with industry? The entrepreneurial university and individual motivations. *Journal of Technology Transfer*, 36(3), 316–39. https://doi.org/10.1007/s10961-010-9153-z

Degroof, J.-J., and Roberts, E.B. (2004). Overcoming Weak entrepreneurial infrastructures for academic spin-off ventures. *The Journal of Technology Transfer*, 29(3/4), 327–52. https://doi.org/10.1023/b:jott.0000034126.23592.23

Di Gregorio, D., and Shane, S.A. (2003). Why do some universities generate more startups than others. *Research Policy*, 32(2), 209–27. https://doi.org/10.1016/S0048-7333(02)00097-5

Donina, D., and Paleari, S. (2019). New public management: Global reform script or conceptual stretching? Analysis of university governance structures in the Napoleonic adminis-

trative tradition. *Higher Education*, 78(2), 193–219. https://doi.org/10.1007/s10734-018-0338-y
Donina, D., Meoli, M., and Paleari, S. (2015a). The new institutional governance of Italian state universities: What role for the new governing bodies? *Tertiary Education and Management*, 21(1), 16–28. https://doi.org/10.1080/13583883.2014.994024
Donina, D., Meoli, M., and Paleari, S. (2015b). Higher education reform in Italy: Tightening regulation instead of steering at a distance. *Higher Education Policy*, 28(2), 215–34. https://doi.org/10.1057/hep.2014.6
Fini, R., Grimaldi, R., and Sobrero, M. (2009). Factors fostering academics to start up new ventures: An assessment of Italian founders' incentives. *The Journal of Technology Transfer*, 34(4), 380–402.
Fini, R., Grimaldi, R., Santoni, S., and Sobrero, M. (2011). Complements or substitutes? The role of universities and local context in supporting the creation of academic spin-offs. *Research Policy*, 40(8), 1113–27. https://doi.org/10.1016/j.respol.2011.05.013
Griliches, Z. (1992). *Output Measurement in the Service Sectors*. Chicago, IL: University of Chicago Press.
Gundry, L.K., and Welsch, H.P. (2001). The ambitious entrepreneur: high growth strategies of enterprises. *Journal of Business Venturing*, 16(312), 453–70. https://doi.org/10.1016/S0883-9026(99)00059-2
Hahn, D., Minola, T., Van Gils, A., and Huybrechts, J. (2017). Entrepreneurial education and learning at universities: Exploring multilevel contingencies. *Entrepreneurship and Regional Development*, 29(9–10), 945–74. https://doi.org/10.1080/08985626.2017.1376542
Harrison, R.T., and Leitch, C.M. (2005). Entrepreneurial learning: Researching the interface between learning and the entrepreneurial context. *Entrepreneurship Theory and Practice*, 29(4), 351–71. https://doi.org/10.1111/j.1540-6520.2005.00089.x
Hayter, C.S. (2011). In search of the profit-maximizing actor: Motivations and definitions of success from nascent academic entrepreneurs. *Journal of Technology Transfer*, 36(3), 340–52. https://doi.org/10.1007/s10961-010-9196-1
Hayter, C.S. (2016). A trajectory of early-stage spinoff success: The role of knowledge intermediaries within an entrepreneurial university ecosystem. *Small Business Economics*, 47(3), 633–56.
Hessels, J., Van Gelderen, M., and Thurik, R. (2008). Entrepreneurial aspirations, motivations, and their drivers. *Small Business Economics*, 31(3), 323–39. https://doi.org/10.1007/s11187-008-9134-x
Hmieleski, K.M., and Ensley, M.D. (2005). A comparative study of new venture top management team composition, dynamics and performance between university-based and independent start-ups. *Research Policy*, 34(7), 1091–105. https://doi.org/10.1016/j.respol.2005.05.008
Horta, H., Meoli, M., and Vismara, S. (2016). Skilled unemployment and the creation of academic spin-offs: A recession-push hypothesis. *Journal of Technology Transfer*, 41(4), 798–817. https://doi.org/10.1007/s10961-015-9405-z
Kane, E.J. (1990). Principal-agent problems in S&L Salvage. *The Journal of Finance*, 45(3), 755–64. https://doi.org/10.1111/j.1540-6261.1990.tb05104.x
Kenney, M., and Patton, D. (2011). Does inventor ownership encourage university research-derived entrepreneurship? A six university comparison. *Research Policy*, 40(8), 1100–12. https://doi.org/10.1016/j.respol.2011.05.012
Kolvereid, L. (1992). Growth aspirations among Norwegian entrepreneurs. *Journal of Business Venturing*, 7(3), 209–22. https://doi.org/10.1016/0883-9026(92)90027-O M4 - Citavi
Lam, A. (2011). What motivates academic scientists to engage in research commercialization: "Gold", "ribbon" or "puzzle"? *Research Policy*, 40(10), 1354–68. https://doi.org/10.1016/j.respol.2011.09.002

Landry, R., Amara, N., and Rherrad, I. (2006). Why are some university researchers more likely to create spin-offs than others? Evidence from Canadian universities. *Research Policy*, 35(10), 1599–615. https://doi.org/10.1016/j.respol.2006.09.020

Lehmann, E.E., Meoli, M., Paleari, S., and Stockinger, S.A.E. (2020). The role of higher education for the development of entrepreneurial ecosystems. *European Journal of Higher Education*, 10(1), 1–9. https://doi.org/10.1080/21568235.2020.1718924

Link, AN., Siegel, D.S., and Bozeman, B. (2007). An empirical analysis of the propensity of academics to engage in informal university technology transfer. *Industrial and Corporate Change*, 16(4), 641–55. https://doi.org/10.1093/icc/dtm020

Ljungberg, D., and McKelvey, M. (2012). What characterizes firms' academic patents? Academic involvement in industrial inventions in Sweden. *Industry and Innovation*, 19(7), 585–606. https://doi.org/10.1080/13662716.2012.726808

Lockett, A., Wright, M., and Franklin, S. (2003). Technology Transfer and universities' spin-out strategies. *Small Business Economics*, 20(2), 185–200. https://doi.org/10.1023/A:1022220216972

Lockett, A., Siegel, D.S., Wright, M., and Ensley, M.D. (2005). The creation of spin-off firms at public research institutions: Managerial and policy implications. *Research Policy*, 34(7), 981–93. https://doi.org/10.1016/j.respol.2005.05.010

Meoli, M., and Vismara, S. (2016). University support and the creation of technology and non-technology academic spin-offs. *Small Business Economics*, 47(2), 345–62. https://doi.org/10.1007/s11187-016-9721-1

Meoli, M., Paleari, S., and Vismara, S. (2013). Completing the technology transfer process: M&As of science-based IPOs. *Small Business Economics*, 40(2), 227–48. https://doi.org/10.1007/s11187-012-9416-1

Meoli, M., Paleari, S., and Vismara, S. (2019). The governance of universities and the establishment of academic spin-offs. *Small Business Economics*, 52(2), 485–504. https://doi.org/10.1007/s11187-017-9956-5

Murray, F. (2004). The role of academic inventors in entrepreneurial firms: Sharing the laboratory life. *Research Policy*, 33(4), 643–59. https://doi.org/10.1016/j.respol.2004.01.013

O'Shea, R.P., Allen, T.J., Chevalier, A., and Roche, F. (2005). Entrepreneurial orientation, technology transfer and spin-off performance of U.S. universities. *Research Policy*, 34(7), 994–1009. https://doi.org/10.1016/j.respol.2005.05.011

Patton, D., and Kenney, M. (2010). The role of the university in the genesis and evolution of research-based clusters. In *Emerging Clusters: Theoretical, Empirical and Political Perspectives on the Initial Stage of Cluster Evolution*, 214–38. https://doi.org/10.1080/00343404.2011.608258

Perkmann, M., Tartari, V., McKelvey, M. et al. (2013). Academic engagement and commercialisation: A review of the literature on university–industry relations. *Research Policy*, 42(2), 423–42. https://doi.org/10.1016/j.respol.2012.09.007

Ritsilä, J., and Tervo, H. (2002). Effects of unemployment on new firm formation: Micro-level panel data evidence from Finland. *Small Business Economics*, 19(1), 31–40. https://doi.org/10.1023/A:1015734424259

Rizzo, U. (2015). Why do scientists create academic spin-offs? The influence of the context. *Journal of Technology Transfer*, 40(2), 198–226. https://doi.org/10.1007/s10961-014-9334-2

Roach, M., and Sauermann, H. (2010). A taste for science ? PhD scientists' academic orientation and self-selection into research careers in industry. *Research Policy*, 39(3), 422–34. https://doi.org/10.1016/j.respol.2010.01.004

Roberts, E.B. (1991). *Entrepreneurs in High Technology: Lessons from MIT and Beyond*. New York: Oxford University Press.

Rothaermel, F.T., and Thursby, M. (2005). University-incubator firm knowledge flows: Assessing their impact on incubator firm performance. *Research Policy*, 34(3), 305–20. https://doi.org/10.1016/j.respol.2004.11.006

Siegel, D.S., Waldman, D.A., Atwater, L.E., and Link, A.N. (2004). Toward a model of the effective transfer of scientific knowledge from academicians to practitioners: Qualitative evidence from the commercialization of university technologies. *Journal of Engineering and Technology Management*, 21(1–2), 115–42. https://doi.org/10.1016/j.jengtecman.2003.12.006

Storey, D.J. (1991). The birth of new firms – does unemployment matter? A review of the evidence. *Small Business Economics*, 3(3), 167–78. https://doi.org/10.1007/BF00400022

Stuart, T.E., and Ding, W.W. (2006). When do scientists become entrepreneurs? The social structural antecedents of commercial activity in the academic life sciences. *American Journal of Sociology*, 112(1), 97–144. https://doi.org/10.1086/502691

Vismara, S., and Meoli, M. (2016). Poor opportunities in the academic labor market as a cause of spin-offs. In *Academic Entrepreneurship* (pp. 39–62). Cheltenham, UK and Northampton, MA, USA: Edward Elgar Publishing. https://doi.org/10.4337/9781785363443.00009

Wang, J., and Shapira, P. (2012). Partnering with universities: A good choice for nanotechnology start-up firms? *Small Business Economics*, 38(2), 197–215. https://doi.org/10.1007/s11187-009-9248-9

Wiklund, J., Davidsson, P., and Delmar, F. (2003). What do they think and feel about growth? An expectancy-value approach to small business managers' attitudes toward growth. *Entrepreneurship Theory and Practice*, 27(3), 247–70. https://doi.org/10.1111/1540-8520.t01-1-00003

Zahra, S.A., Van de Velde, E., and Larrañeta, B. (2007). Knowledge conversion capability and the performance of corporate and university spin-offs. *Industrial and Corporate Change*, 16(4), 569–608. https://doi.org/10.1093/icc/dtm018

12. Academic entrepreneurship: the performance and impacts of academic spin-offs in Norway

Marius Tuft Mathisen and Einar Rasmussen

12.1 INTRODUCTION

Scientific knowledge and inventions are important for economic and societal progress (Audretsch and Feldman, 1996) and the application of academic research is seen as a key element in addressing many of society's challenges related to the eradication of disease, national security, job creation, food production, quality of life, protection of the natural environment and sustainable development (Fini et al., 2018). The process of converting academic research to practical application is often referred to as science commercialization (Fini et al., 2019) or technology transfer (Hayter et al., 2020). This process is long, resource-demanding, and with an uncertain outcome, but universities are increasingly promoting academic entrepreneurship, broadly defined as entrepreneurial activity at academic institutions, including patenting, technology licensing, and spinning off new ventures; all with the objective of commercializing research results discovered by academic researchers (Grimaldi et al., 2011; Rothaermel et al., 2007).

This chapter elaborates on a specific pathway of technology transfer, the creation of academic spin-offs (ASOs). ASOs can be defined as new ventures established to commercialize scientific research results, technologies, and inventions from universities and public research institutions (Clarysse et al., 2005). ASOs represent a small fraction of all knowledge and technology transfer from universities, but are still considered as powerful ventures with significant wealth creation potential (Shane, 2004a). In particular, ASOs are expected to enhance regional economic development through new knowledge-based employment, generation of tax revenues, and indirect effects through the dissemination of new knowledge, improving the absorptive capacity of the region (Criaco et al., 2014; Lawton Smith et al., 2008; Vincett, 2010). In addition, scholars argue that ASOs are important firms because they commercialize research results that might have remained undeveloped otherwise (Fontes, 2005; Mathisen and Rasmussen, 2019).

More generally, ASOs often commercialize technologies with radical technical advances typically not suitable for licensing directly to industry (Clausen and Rasmussen, 2013; Shane, 2004a). ASOs also often engage in rapid internationalization to grow their innovative and unique technology offering (Bjørnåli and Aspelund, 2012). In other words, ASOs are new ventures with the potential for engaging in entrepreneurial activity that challenges existing technologies and markets (Mohr and Garnsey, 2011; Schumpeter, 1934; Walter et al., 2006).

The high expectations on the economic and societal impacts of ASOs are evident. Yet, the academic entrepreneurship literature has predominantly been concerned with explaining *why* some universities create more spin-offs than others (Mathisen and Rasmussen, 2019; O'Shea et al., 2008), that is, the process from scientific discovery to venture creation. Far less attention has been paid to the performance and impact of ASOs, that is, the process from venture creation to outcome. Hence, there is a significant gap in our understanding of how ASOs grow to become successful firms and create economic impact (Colombo et al., 2010; Fini et al., 2018; Mathisen and Rasmussen, 2019). ASOs must overcome technological, market, and organizational uncertainties when engaging in the complex process of transforming scientific knowledge into commercial products and services (Mathisen, 2017; Rasmussen and Mathisen, 2017).

The objective of this chapter is to introduce the academic entrepreneurship phenomenon from a firm-level perspective, illustrated by novel empirical evidence on the performance and impact of ASOs in Norway. First, after defining key concepts, a brief historical backdrop of academic entrepreneurship is provided. Second, important conceptual and practical differences between the creation and impact of ASOs are discussed. Further, we empirically analyze the development of the population of ASOs established in Norway in the period of 1999–2011. Finally, implications for research and practice are discussed.

12.2 THE HISTORICAL BACKDROP OF ACADEMIC ENTREPRENEURSHIP

The role of the university has evolved over time and is now increasingly expected to include commercial exploitation of scientific and technological advances (Hayter, 2016b; Pries and Guild, 2007; Rothaermel et al., 2007). Hence, the modern university as a societal institution is best understood within the context of "Academic Capitalism" (Slaughter and Rhoades, 2004) or "Triple Helix" (Etzkowitz and Leydesdorff, 2000) frameworks, which emphasize its role in knowledge-based economic development (Hayter, 2011).

12.2.1 Institutionalization and Growth of Academic Entrepreneurship

Academic entrepreneurship activities have been increasingly institutionalized with more direct and formal involvement by the universities (Geuna and Muscio, 2009). This process accelerated in the 1980s when the US enacted new legislations and policies aiming to formalize and encourage commercial use of scientific discoveries at universities (Grimaldi et al., 2011). Many European countries later introduced similar initiatives. The most important change was to transfer the control and ownership of inventions to the university. In the US, legislation was enacted that transferred the control of discoveries from the federal government to the university. In many European countries, the situation was different because the university scientists owned the intellectual property (IP) rights to their own discoveries. This is often referred to as the "Professor's Privilege" (Fini et al., 2017) and many countries have enacted new legislation which transferred the control of IP from the scientists to the universities. These legislations provide formalized rules regarding IP ownership and a greater incentive for universities to exploit commercial opportunities based on research discoveries.

Although many universities had existing initiatives in place, a large-scale formalization of technology transfer took place both in the US and Europe (Rasmussen et al., 2006). Notably, most universities established technology transfer offices (TTOs) (Siegel et al., 2007) and similar boundary-spanning agents to manage the technology transfer process (Geuna and Muscio, 2009; Villani et al., 2017). Stanford University was one of the first to establish its TTO in the early 1970s. In Europe, the first example was probably at the KU Leuven in the mid-1970s. Over the last 30 years, almost all larger research universities have established some form of TTO, even when no legislation promoted it directly (Link et al., 2007).

These policy initiatives were followed by a rapid increase in academic entrepreneurship. However, some researchers have called the effects of legislation overemphasized, as academic entrepreneurship was on the rise also prior to these reforms (Chiesa and Piccaluga, 2000; Clarysse et al., 2007; Mowery, 2011; Mustar et al., 2008). One proposed explanation is the rise of the biomedical and ICT research fields. These are the major technological areas of formal technology transfer today, and they grew into important fields of research during this time period. There was also the introduction of new patent laws which provided much stronger protection for biomedical discoveries, for instance (Shane, 2004b). Another proposed reason is the increase in industry involvement and sponsoring of university research. This is believed to have generated new interactions and opportunities due to more applied research occurring. Finally, feedback loops are also likely. Increased academic entrepreneurship provided legitimacy and improved the culture and competencies at the universities to engage in more commercial activity (Rasmussen and Wright, 2015).

12.3 ACADEMIC ENTREPRENEURSHIP: THE PROCESS FROM DISCOVERY TO IMPACT

Most research on ASOs has focused on the antecedents of spin-off activity, more specifically how individual (e.g., scientist, entrepreneur), institutional (e.g., university, TTO), and environmental (e.g., legislation, ecosystems) conditions impact spin-off creation (O'Shea et al., 2008). This "black box" approach to ASOs implicitly assumes that firms are homogeneous (Mustar et al., 2008) and that the number of firms corresponds to the economic impact (Harrison and Leitch, 2010). The former is challenged by researchers finding that ASOs are a surprisingly heterogeneous group of firms that differs in terms of resources, business model, and institutional links (Mustar et al., 2008). Scholars have questioned the expected impact of ASOs, arguing that they are predominantly small firms (e.g., Criaco et al., 2014; Salvador, 2011), showing negligible growth on average (e.g., Hayter, 2011; Wright et al., 2006), and are so rare that they have a too limited total economic impact to justify the public support they receive (Borlaug et al., 2009; Harrison and Leitch, 2010). Further, ASOs face exceptional growth challenges (Mustar et al., 2008), being characterized with high levels of innovation and often exploiting novel, early-stage, and general-purpose technologies (Knockaert et al., 2011; Shane, 2001). In addition, ASOs often struggle with attracting critical resources (Patzelt and Shepherd, 2009; Wright et al., 2006) and knowledge (Ensley and Hmieleski, 2005; Hayter, 2016a; Rasmussen et al., 2015). Yet, paradoxically, when successes do occur, they can generate tremendous impact (Shane, 2004a) and many extremely valuable firms, including some of the most important firms globally (e.g., Google), originated by virtue of academic research.

In a recent review of the literature on the performance and outcomes of ASOs, Mathisen and Rasmussen (2019) argue that ASOs typically have very long development paths and that outcomes have to be evaluated over sufficiently long periods of time to capture "real" impacts. Further, ASO outcomes are highly skewed, with most of the impact accounted for by a few highly successful ventures. Also, Mathisen (2017) finds that economic impact is not necessarily confined to what happens within the ASO firm, but can also take place within other organizations as well. For instance, trade sales appear to be a very common successful outcome, and not accounting for this mechanism may significantly underestimate the impact of ASOs (Mathisen et al., 2021). Lastly, Rasmussen (2019) points to the fact that ASOs can also have indirect economic impact when acting as technology transfer agents transferring new technology into application in society. However, these impacts are hard to identify because they depend on the interaction between ASOs, industry, and the economy more generally.

In sum, it seems evident that a multifaceted understanding of the outcomes and economic impacts created by ASOs is needed as governments and universities are investing heavily to encourage and support their creation, including establishing TTOs, incubators, and internal seed funds (Rasmussen et al., 2006). The genuine policy objective is to generate successful firms creating wealth for society, not just to establish many firms. To shed light on the development and potential impacts of academic entrepreneurship, we use a unique dataset tracking ASOs in Norway.

12.4 ACADEMIC ENTREPRENEURSHIP IN NORWAY

A key event in the Norwegian context for academic entrepreneurship is the legislative changes enacted in 2003 providing the universities with a greater responsibility for science commercialization. The "Professor's Privilege" was removed, and the universities were given ownership and control over IP with commercial value developed at the university. Overall, the institutional context for academic entrepreneurship has been significantly professionalized, and the establishment of ASOs has generally become accepted as an important activity at most Norwegian universities.

In a comprehensive analysis of the Norwegian population of ASOs established in the 1999–2011 period, we tracked 373 ASOs longitudinally by following their development and key events until reaching a terminal outcome (or right-censored in 2015 if survived). The database is unique due to its richness and prolonged time horizon and was developed by integrating a range of qualitative and quantitative data sources.[1]

In the following, we provide a brief summary of the development, growth and outcomes of ASOs in Norway (Mathisen, 2017). The objective is to provide a detailed overview of the potential economic impacts created by ASOs.

12.4.1 Firm Origins

The four largest and oldest universities are the most active institutions in establishing ASOs in Norway, followed closely by public research institutions (PRIs). The academic entrepreneurship literature is predominantly concerned with universities, often overlooking PRIs as a source of ASOs (for notable exceptions, see Clarysse et al., 2005; Heirman and Clarysse, 2004; Moray and Clarysse, 2005). Our data show that PRIs represent approximately one-third of total establishments. Hence, not taking PRIs into account would miss a significant part of total commercialization activities. PRIs account for about 25 percent of total public R&D spending in Norway, which is not unique because most European countries have a large number of PRIs (Gulbrandsen, 2011). The university colleges and new universities are significantly less research-intensive and play a quite limited role in commercialization activities. A few institutions are dominant within the traditional university category. NTNU represents 46 percent of ASOs from traditional universities (21 percent of the total portfolio). Three institutions represent half of the ASOs established from PRIs. The distribution of ASO establishments over time and type of institutions is shown in Figure 12.1.

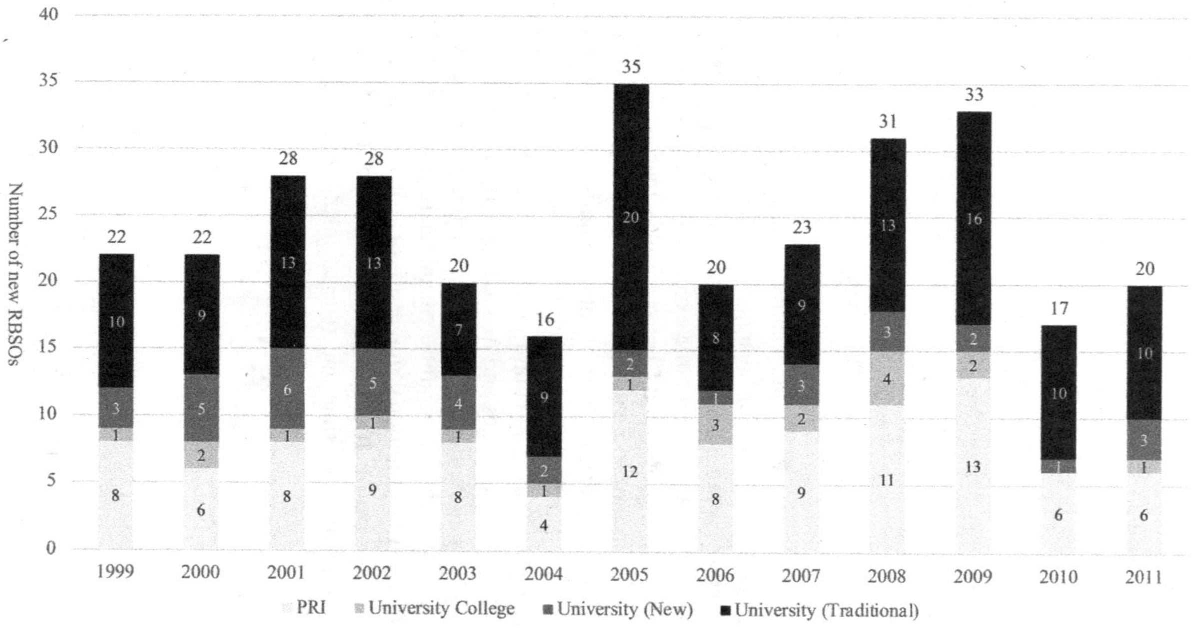

Figure 12.1 Absolute number of new ASOs per year (1999–2011) distributed by type of academic parent institution

12.4.2 Technology

While industry is a common control variable in management research, it is a challenging concept in the context of academic entrepreneurship. ASOs are known to commercialize radical and general-purpose technology, which can be exploited in several industries through multiple application areas (Shane, 2001). Hence, it is more informative to segregate the portfolio based on the type of technologies the ASOs are commercializing. Biomedical and software technologies have generally been found to be the most common for academic entrepreneurship (e.g., Golob, 2006; Lundqvist, 2014; Shane, 2004a). Figure 12.2 illustrates that this is also the case in Norway, where close to two-thirds of the ASOs are primarily based on biomedicine and/or software.

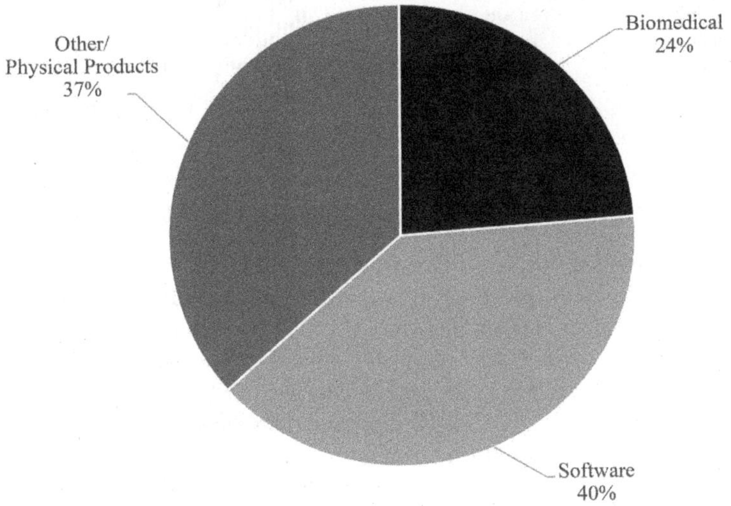

Figure 12.2 Distribution of ASOs in the portfolio by technological field

Figure 12.2 is somewhat misleading because, for example, software technologies can be used for purposes outside what is usually considered as the ICT industry and the *Other/Physical Products* category contains a range of different technologies. Hence, Figure 12.3[2] provides a more detailed classification of technical application areas.[3]

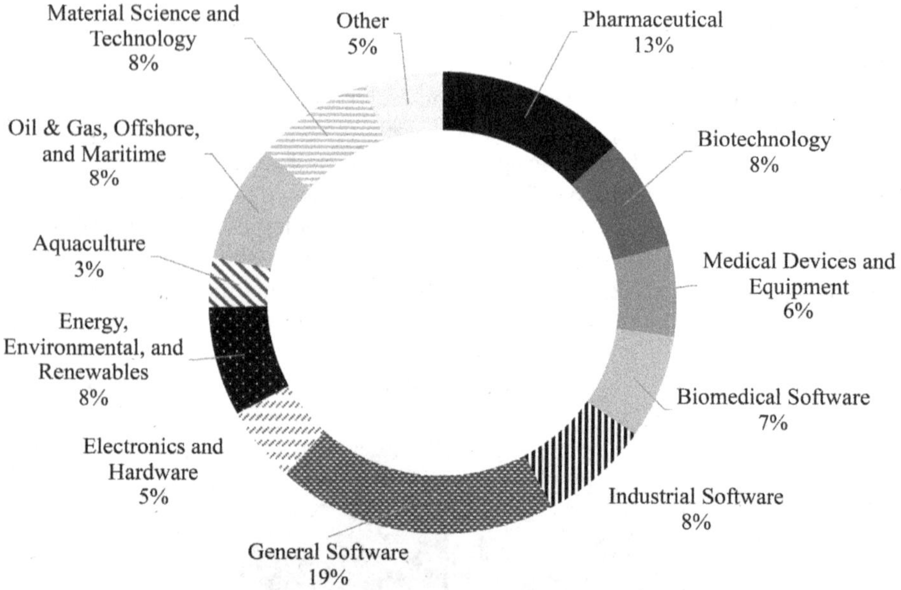

Figure 12.3 Distribution of ASOs in the portfolio by application areas

The 11 application areas in Figure 12.3 can be grouped roughly into three technical domains: Life Sciences, ICT and Electronics, and Physical Products. This provides some interesting new details. For instance, the *Medical Device* and *Biomedical Software* application areas together represent around 13 percent of the portfolio. While these technologies are not based on the chemical or biological sciences, they still represent a significant proportion of the ASOs targeting what could be considered the medical industry. Overall, the distribution into technical application areas is in line with previous academic entrepreneurship research (see, e.g., Lundqvist, 2014; Shane, 2004a). However, the influence of the industry structure in Norway is also evident. The *Oil & Gas*, *Offshore and Maritime,* and *Aquaculture* application areas are related to major industries in Norway, and ASOs commercializing physical products and solutions within these application areas represent 11 percent of all ASOs. This is likely an understatement, as ASOs in other categories also target these industries. For instance, a substantial amount of ASOs commercializing *Industrial Software* are targeting the oil and gas and maritime industries.

Further, the ASOs' technological focus is strongly linked to the research activity of the parent institution. PRIs tend to have a narrow scientific focus and ASOs created by PRIs overwhelmingly inherit and share the technical domain of their parent. Albeit having a much broader scientific research base, a similar situation exists with universities. The University of Oslo represents over 40 percent of the ASOs in the *Pharmaceutical* and *Medical Device* application areas, reflecting their leading scientific position in Norway within the medical area.

The literature has mostly assumed that ASOs are commercializing specific inventions (Karnani, 2013). However, ASOs can also be based on non-formalized scientific expertise and tacit knowledge (Pirnay et al., 2003). In the Norwegian population, 26 percent of the ASOs are established based on scientific expertise, while 74 percent are based on specific inventions.

ASOs commercializing specific inventions can opt to legally protect their IP through patents. In certain areas, such as drug development, patenting is extremely common (Thumm, 2004). In other areas, ASOs may favor secrecy and speed over formal intellectual property rights (IPR) (Hall et al., 2014). For instance, software code can be difficult to patent effectively, and trade secrets are often viewed as a better strategy (Hurmelinna-Laukkanen and Puumalainen, 2007). Overall, 26 percent of the ASOs commercializing specific inventions had applied for a patent(s).[4] However, this figure is pushed downwards by the presence of software firms, where only 6 percent have patented inventions.

12.4.3 VC Financing

ASOs are typically relying on attracting risk capital from venture capitalists (VCs) to finance their development (Mustar et al., 2006). About 45 percent of the Norwegian ASOs are expected to eventually raise VC funding, and most do so quite early in their development. After four years, one-third of the ASOs had raised some form of VC financing and no ASO raised initial VC financing after ten years. Together these findings indicate that ASOs are very frequent beneficiaries of VC funding, but that the possibility to raise VC funding declines dramatically around four to six years after establishment.

The most common VC type is investors specifically focused on seed investments (i.e., the earliest stage of investment). Most of the seed VCs active in the portfolio manage government-sponsored funds (e.g., Brander et al., 2014), where the Norwegian government is an investor and/or provides risk adjustments to the private investors. Traditional VCs (mostly private and early-stage[5] funds) represent about 31 percent while corporate VCs backed fully by a corporate incumbent represent 11 percent. The remaining 19 percent are a group of VCs that invests in a portfolio structure similar to more traditional VCs but, in fact, are fully funded by high-net worth individuals[6] or families.

More broadly, professional investors like VCs are not the only external owners in ASOs. Public research organizations (i.e., both universities and PRIs) are often initial owners in the venture, and other individuals and companies will invest in the firm over time. Figure 12.4 shows the evolution of ownership between four major owner groups.[7]

Founders[8] are the largest owner category maintaining 35–45 percent ownership over time on average. Together with other individuals (i.e., other private owners such as employees and private investors), individual owners control 45–60 percent of the ASOs over time. Individuals other than the original founders become relatively more important over time, although the founders still remain the largest group.

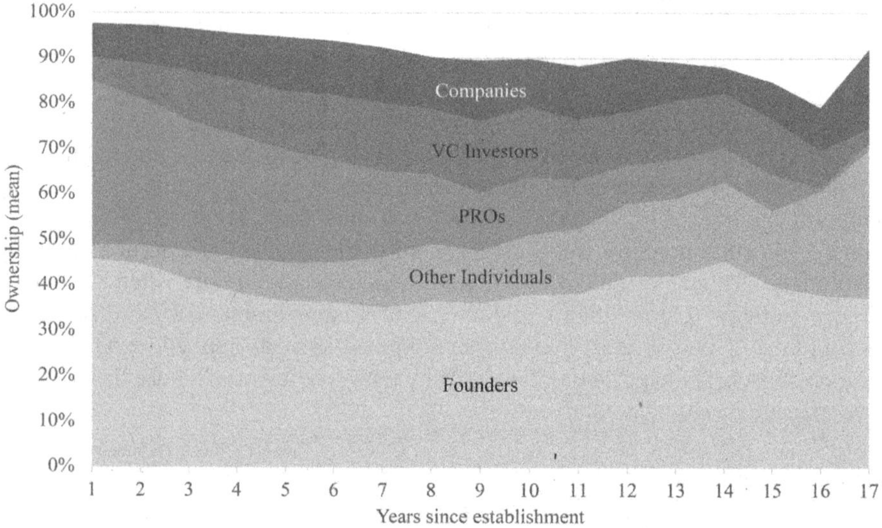

Figure 12.4 *Evolution of ownership structures for ASOs in the portfolio*

The distribution of private ownership is rather bi-modal, where very low (below 5 percent) and very high (above 95 percent) ownership is the most frequent. For instance, the increase in individual ownership in the oldest firms is primarily due to the relatively higher proportion of ASOs fully owned by individuals at that time.

The parent academic institutions (i.e., both universities and PRIs) are typically quite large owners in the early years but decrease their ownership significantly over time. This is expected because these stakeholders receive equity primarily as compensation for transferring IPR into the ASO but typically have limited ability to invest new capital going forward. In addition, there are notable examples in Norway of TTOs selling shares in ASOs to external investors.[9] More detailed analyses reveal two interesting findings. First, the initial ownership by academic institutions is substantially higher in ASOs established more recently, but the same pattern of decreasing ownership is present. This could be expected from the changes in IPR legislation in 2004. Second, academic ownership has a skewed distribution (and increasingly so over time), illustrated by the median academic ownership being zero already in year six. At this point, the majority of ASOs do not have any academic owners and average ownership is affected by a few ASOs with very high academic ownership (dominantly from PRIs).

VCs and company owners (i.e., public and privately held companies with operational business activities) seem to be of about equal importance. These two groups essentially represent financial and industrial investors, respectively, and together these investors increase ASO ownership over time to around 25–30 percent on average. However, VC and company ownership have quite different distributions. Both share a skewed distribution where zero ownership clearly is the modal condition. However, significant differences arise in the cases when either VC or company owners are present. VC investors typically maintain ownership of around 30–50 percent and very low/high ownership is rare. In contrast, moderate company ownership is quite uncommon, and full ownership becomes the modal condition over time. The latter refers to scenarios where a company has acquired an ASO but operates the business independently.

The ownership structure differs significantly based on many factors related to circumstances at establishment, but also how resource-demanding and successful the ASO's development process has been. For instance, ASOs that do not experience significant growth are typically characterized with a high degree of founder/individual ownership.

12.4.4 Survival and Trade Sales

Terminal outcomes of ASOs can be categorized as either failed, transformed,[10] or acquired (i.e., experiencing a trade sale). By definition, surviving ASOs have not achieved either of these outcomes (during the observation period). Figure 12.5 illustrates the evolution of these outcomes, with cumulative incidence curves modeling the three outcomes as competing events[11] (Scrucca et al., 2007).

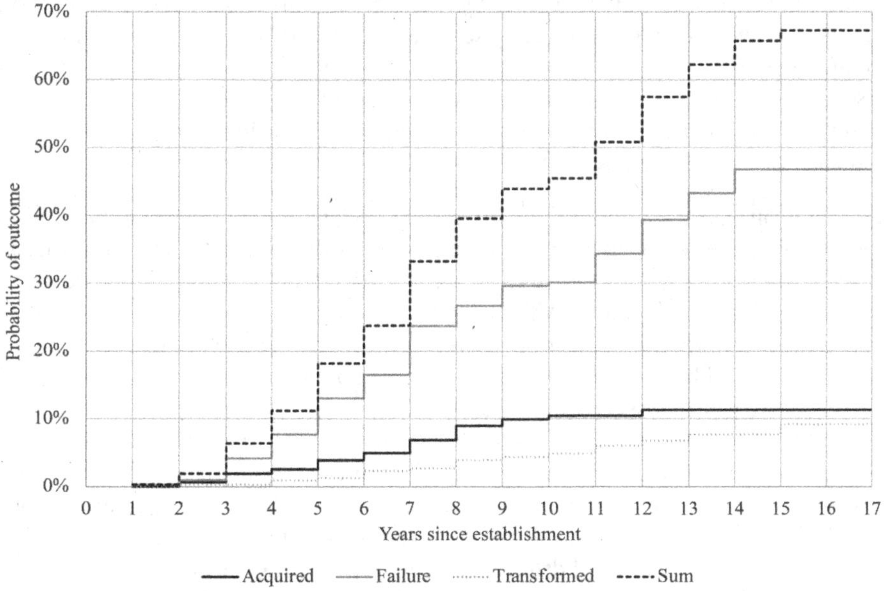

Figure 12.5 *Cumulative incidence plot estimating the probability of terminal outcomes related to acquired, failed, and transformed ASOs*

In sum, one-third of the ASOs are expected to survive (i.e., exist independently) after 15 years. The survival rate appears high, in line with other academic entrepreneurship research (Rothaermel and Thursby, 2005; Zhang, 2009). After ten years, only 30 percent of ASOs are expected to fail. Contrary to what could be expected, failure by bankruptcy is actually quite rare. Only 23 percent of failures are due to bankruptcy (or other forced dissolution), implying that most ASO failures are voluntary dissolutions of the firm. The failure curve does not seem to really flatten out at any point, but interpretation is difficult beyond ten to twelve years due to the limited number of observations. Nevertheless, many ASOs have long development processes before reaching terminal outcomes.

Trade sales seem to be a very important outcome, with about 10 percent of ASOs expected to be acquired within ten years. A trade sale is defined here by two criteria adopted from the technology acquisition literature (Puranam et al., 2003): (1) the buyer is an industry incumbent significantly larger than the new venture, and (2) the buyer assumes 51 percent or more ownership (stock sale) or all productive assets (asset sale). Figure 12.5 illustrates that trade sales occur quite steadily from a very young age, with the following conditions at the time of the trade sale:

- The majority of ASOs are very small: 72 percent have five employees or less; 55 percent have less than 3 million NOK in revenue; 80 percent are unprofitable.
- Only four firms have over ten employees, and three firms have above 20 million NOK in revenue. One firm represents about 43 percent of all revenue and employees.
- Seven firms have an acquisition valuation above 50 million NOK and one firm represents more than half the (known[12]) acquisition value in the portfolio (approximately 500 million NOK).

Only about one-third of the trade sales can be understood as having a substantial value exchange between buyer and seller. The majority of ASOs pursuing a trade sale seem to use it as a mechanism to further develop their technology and business concept in an industrial context. Although acquisitions by large foreign companies do occur in the portfolio, the majority of buyers were domestic firms. Further, many of the ASOs acquired by foreign companies remained domestically and even increased the scope of their business activities locally.

An initial public offering (IPO) is a non-terminal outcome and is often viewed as a financial performance event. IPOs are rare outcomes of academic entrepreneurship in Norway, but still represent a major source of firm value in the portfolio. Four ASOs have gone public on the Oslo Stock Exchange in the observation period, where three are pharmaceutical drug development companies. Two ASOs were later delisted and remain privately owned with modest valuations. The remaining two ASOs have current[13] valuations of approximately 5 billion NOK (approximately €550 million), where one firm (*Nordic Nanovector*) represents 90 percent of the total value. Again, a very skewed distribution is observed. More importantly, the valuation of *Nordic Nanovector* grew around 700 percent in 2016 as the market responded to positive results from the firm's clinical trials. As commonplace in the pharmaceutical context, *Nordic Nanovector* still has not experienced any significant commercial revenue (Gilbert et al., 2006).

12.4.5 Firm Growth

12.4.5.1 Revenue

Firms that fail cannot grow, but survival does not equal growth in the context of academic entrepreneurship. Looking first at revenue, Figure 12.6 shows the development in revenue from founding (ASOs are included in the analysis until reaching a terminal outcome). The mean is always higher than the median revenue, indicating that the distribution of revenue is skewed to the right. The bars show the share of the largest firms over time, illustrating that a small number of ASOs are responsible for a large share of total revenue. Over time, the distribution becomes more skewed and, after ten years, the highest-grossing firm represents over one-quarter of total revenue. More detailed investigation of the distribution finds that very low revenues (i.e., below NOK 500,000) are most common, regardless of firm age. Three ASOs have experienced revenues above 100 million NOK, and four additional firms above 50 million NOK.

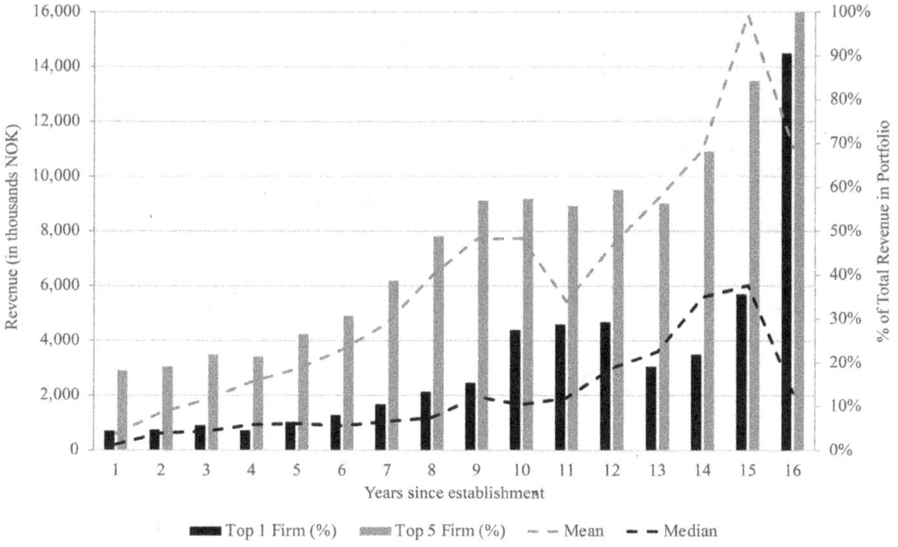

Figure 12.6 *Development in revenues (mean and median; thousands NOK) in years since establishment (lines; measured on left axis). Share of total represented by the largest and five largest firms (bars; measured on right axis)*

12.4.5.2 Employment

Employment often precedes revenue growth and can therefore be a useful indicator of growth (Delmar et al., 2003; Knockaert et al., 2011). For instance, ASOs commercializing new pharmaceutical drugs often spend many years without any sales (Gilbert et al., 2006). Employment follows a similar development as revenue, where a few ASOs represent the majority of employment over time. Revenue and labor cost are also highly correlated ($r = 0.85$, strongly statistically significant), indicating that the biggest employers are, to a large extent, the same ASOs with the highest revenues (similar result as Shepherd and Wiklund, 2009). Most ASOs have very few employees (median increase from one to four employees for the oldest firms), and only seven ASOs have more than 30 employees at any time. The largest number of employees observed for any ASO was 78. Comparable to revenues, most ASOs have no growth in employment. Employment and revenue, however, differ with respect to the dynamics of growth. Labor cost exhibits very strong serial correlation ($r = 0.94$, strongly statistically significant). This indicates that growth in employment is highly path-dependent, with less variability and erratic patterns compared with revenue growth.

12.4.5.3 Profitability

Profitability is a key performance objective; however, losses frequently occur in the period when technologies are developed into marketable products and services. Figure 12.7 shows the development in operating result (earnings before interest and taxes, EBIT) from founding.

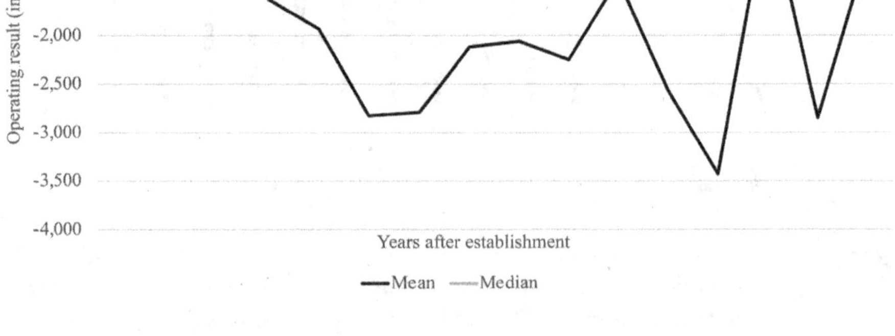

Figure 12.7 Development in operating result (mean and median) in years since establishment (thousands NOK)

Most ASOs sustain consistent negative but very small losses. The major difference between the mean and median indicates a skewed distribution and that some ASOs run major operational losses. While profits above 10 million NOK occur from around year five, losses are far more frequent and substantial. Starting at year four, the activities of certain ASOs generate losses in excess of 50 million NOK. More detailed insight was obtained from analyzing the ASOs at both ends of the profitability extremes. First, the ASOs achieving the highest profits generally do so after running losses (and often quite substantial ones) in earlier years. These ASOs have successfully commercialized their technology and have been able to grow revenues to a point where profitable operations have commenced.

Further analysis reveals that several growth trajectories can lead to profitability. Two of the most profitable ASOs (in accumulative terms) have created consistent, but moderate, positive profits for the last 15 years. Both these ASOs offer consulting services based on unique scientific expertise. In other words, these ASOs are not only commercializing specific inventions. While the highest profits are not generated from these service-based business models, profitability can be achieved much quicker.

Further, the ASOs generating the highest losses generally have major operating costs in addition to employment. Many of these ASOs are developing new pharmaceutical drugs associated with high costs for clinical trials and other development expenses. These ASOs are able to sustain such losses because of success in raising substantial VC funding. Hence, these highly unprofitable firms are receiving high valuations[14] and viewed as the most promising new ventures in the portfolio. The ambition of both entrepreneurs and investors is future high profits or high-value IPOs and/or trade sales. This counterintuitive fact is a distinctive characteristic of this portfolio of ASOs (Rasmussen and Mathisen, 2017).

12.5 CONCLUSIONS AND IMPLICATIONS

This chapter has presented a comprehensive analysis of the origin, development, growth, and ultimate outcomes of ASOs in Norway. In conclusion, three areas seem particularly important for policy and further research. First, commercialization can occur through qualitatively different outcomes; therefore, focusing only on simple or singular performance outcomes has major limitations. While a firm might appear as unsuccessful in one facet of performance (e.g., profits), it might simultaneously be highly successful in another (e.g., firm value). For instance, trade sales with very high valuations do occur, and more generally, trade sales seem to be an important mechanism to access the necessary resources to successfully scale up the commercialization processes. Hence, using survival as an independent firm as the only measure of success is misguided. Moreover, trade sales often have unfavorable connotations in policy circles because of the fear that technologies and firms partly financed by governmental resources and tax dollars will be "sold out of the country," thus surrendering the new knowledge-based employment opportunities which would be created locally. Although acquisitions by large foreign companies do occur, the majority of buyers in the portfolio were domestic firms. Many of the ASOs acquired by foreign companies remained domestically and even increased the scope of their business activities locally. Successful trade sales also provide high returns for entrepreneurs and VC investors, which can be reinvested into new ventures. For these reasons, policies should include the promotion of trade sales. More specifically, policymakers should facilitate for large foreign industry incumbents to interact with domestic universities and PRIs for the mutual benefit of both parties.

Second, skewed distributions and the extreme impact of outliers appear to be the norm across most dimensions of the ASO phenomenon. Under these circumstances, interpreting average values should be done with care as they are mostly misleading. A meaningful evaluation of a portfolio of ASOs must, therefore, pay particular attention to the portfolio's outliers and not only its most common members. This is a general feature of all science commercialization and technology transfer activities, including patenting and licensing.

Finally, the time needed for the commercialization of research is very long, and development and growth can be discontinuous and erratic. Timing of evaluation matters a great deal as the status and prospects of ASOs can be dramatically changed over short periods. For example, in the days preparing this chapter, the portfolio company *Ultimovacs* went public on the Oslo Stock Exchange with a valuation of over 500 million NOK. These findings are fully compatible with phenomena characterized with high uncertainty and skewed outcomes, such as ASOs demonstrably are.

NOTES

1. The population of firms was identified through the FORNY program administrated by the Research Council of Norway (RCN). FORNY was the key governmental policy mechanism for promoting the infrastructure supporting the commercialization of research. The program operated with an incentive model which ensured that the universities were highly incentivized to report all new ASOs. The FORNY portfolio approximates the full population of such firms in Norway, and the annual reporting of ASOs limits survivorship bias typically associated with retrospective entrepreneurship studies. The empirical database was based on integrating several archival data sources primarily from The National Business Registry in Norway. This included complete financial statements for all firms in all observation years. Further, over 12,000 corporate announcements (mandatory notices from firms on a number of key corporate events) and patent data were added. In addition, several less structured and qualitative data sources were structured. Examples include all annual reports for all years, the ASOs' original business plans, and a comprehensive archive of all news articles ever written about the firms. These unstructured sources were coded into structured form and thus synthesizing this qualitative information into a form that allowed for quantitative analysis. See Mathisen (2017) for more details.
2. The apparent minor discrepancy between the two figures in this section is expected. For instance, while ASOs based on software represent 40 percent in Figure 12.2, the combined software categories in Figure 12.3 is only 34 percent. The Life Sciences category is 24 percent in Figure 12.2, and the combined biomedical categories in Figure 12.3 is 21 percent. The reason for this is that some ASOs are partly based on either software or biomedical technology, but *not* regarded as the primary product or service.

3. A technical application area is understood here as the scientific discipline or area that the technology/scientific knowledge stems from.
4. This is most likely an understatement as the available data relate to Norwegian patent applications in the name of the company only. We expect that some firms rely on patent applications in the inventor's or university's name, or have patented directly outside Norway.
5. There is only one example in the portfolio of a private equity group investment focused on the late/expansion stage of a target company.
6. These investors could be labeled as business angels. However, the qualitative difference between a business angel and the investors in this category is that the latter operate with a professional investment organization, including websites listing investment criteria and current portfolios (similar to institutional VCs).
7. Ownership outside the four categories are primarily municipalities/governmental agencies, nonprofit organizations (e.g., foundations), mutual funds (i.e., only relevant for the few public firms in the portfolio), and share repurchases by the firms. Some ownership is unknown due to the requirement of small firms in Norway to only report owners over a certain ownership threshold.
8. Founders are defined as private individuals that hold 5 percent or more in one of the first two operating years. Individuals that do not satisfy these criteria are defined as "Other Individuals." Individual owners are identified regardless of upholding the ownership directly or through financial holding companies.
9. Three of the largest TTOs in Norway have sold their portfolios (i.e., their shares in several ASOs) to an external VC.
10. Sixty-seven percent of the transformed outcomes occur when the ASO merges with another company, but not falling within the criteria of a trade sale (i.e., the buyer is not an industrial incumbent but another small business, or a holding company of an associated stakeholder of the firm). The other transformed outcomes are related to transfer of the core business activities to another, often associated, company, essentially leaving the original entity unrecognizable and typically without any significant activity. There is no evidence that any of the ASOs labeled as transformed outcomes experienced any significant success at a later stage within the observation period.
11. Essentially, the sum of the cumulative incidence functions from all causes is equal to 1 minus the Kaplan-Meier estimate of survival.
12. Acquisition values could not be reliably estimated for about one-third of the trade sales. However, by manual investigation of all available data, all of these trade sales appear to have low or even marginal value exchanges.
13. Downloaded from the websites of the Oslo Stock Exchange, 28 March 2017.
14. For example, *BerGenBio* AS (www.bergenbio.com) is one ASO with major losses in recent years. Yet (this company has raised about 470 million NOK in equity) financing was valued at 1.2 billion NOK at their IPO in 2017.

REFERENCES

Audretsch D.B., Feldman, M.P. R&D spillovers and the geography of innovation and production. *The American Economic Review* 1996;86(3):630–40.

Bjørnåli E.S., Aspelund A. The role of the entrepreneurial team and the board of directors in the internationalization of academic spin-offs. *Journal of International Entrepreneurship* 2012;10:350–77.

Borlaug S.B., Grunfeld L., Gulbrandsen M. et al. Between entrepreneurship and technology transfer: Evaluation of the FORNY programme. NIFU STEP; 2009.

Brander J.A., Du Q., Hellmann T. The effects of government-sponsored venture capital: International evidence. *Review of Finance* 2014:rfu009.

Chiesa V., Piccaluga A. Exploitation and diffusion of public research: The case of academic spin-off companies in Italy. *R&D Management* 2000;30:329–39.

Clarysse B., Wright M., Lockett A., Van de Velde E., Vohora A. Spinning out new ventures: A typology of incubation strategies from European research institutions. *Journal of Business Venturing* 2005;20:183–216.

Clarysse B., Wright M., Lockett A., Mustar P., Knockaert M. Academic spin-offs, formal technology transfer and capital raising. *Industrial and Corporate Change* 2007;16:609–40.

Clausen T.H., Rasmussen E. Parallel business models and the innovativeness of research-based spin-off ventures. *Journal of Technology Transfer* 2013:1–14.

Colombo M., Mustar P., Wright M. Dynamics of science-based entrepreneurship. *Journal of Technology Transfer* 2010;35:1–15.

Criaco G., Minola T., Migliorini P., Serarols-Tarrés C. "To have and have not": founders' human capital and university start-up survival. *Journal of Technology Transfer* 2014:1–27.

Delmar F., Davidsson P., Gartner W.B. Arriving at the high-growth firm. *Journal of Business Venturing* 2003;18:189–216.

Ensley M.D., Hmieleski K.A. A comparative study of new venture top management team composition, dynamics and performance between university-based and independent start-ups. *Research Policy* 2005;34:1091–105.

Etzkowitz H., Leydesdorff L. The dynamics of innovation: from national systems and "mode 2" to a triple helix of university–industry–government relations. *Research Policy* 2000;29:109–23.

Fini R., Fu K., Mathisen M.T., Rasmussen E., Wright M. Institutional determinants of university spin-off quantity and quality: A longitudinal, multilevel, cross-country study. *Small Business Economics* 2017;48:361–91.

Fini R., Rasmussen E., Siegel D., Wiklund J. Rethinking the commercialization of public science: From entrepreneurial outcomes to societal impacts. *Academy of Management Perspectives* 2018;32:4–20.

Fini R., Rasmussen, E., Wiklund, J., Wright, M. Theories from the lab: How research on science commercialization can contribute to management studies. *Journal of Management Studies* 2019;56:865–94.

Fontes M. The process of transformation of scientific and technological knowledge into economic value conducted by biotechnology spin-offs. *Technovation* 2005;25:339–47.

Geuna A., Muscio A. The governance of university knowledge transfer: A critical review of the literature. Minerva 2009;47:93–114.

Gilbert B.A., McDougall P.P., Audretsch D.B. New venture growth: A review and extension. *Journal of Management* 2006;32:926–50.

Golob E. Capturing the regional economic benefits of university technology transfer: A case study. *Journal of Technology Transfer* 2006;31:685–95.

Grimaldi R., Kenney M., Siegel D.S., Wright M. 30 years after Bayh-Dole: Reassessing academic entrepreneurship. *Research Policy* 2011;40:1045–57.

Gulbrandsen M. Research institutes as hybrid organizations: Central challenges to their legitimacy. *Policy Sciences* 2011;44:215–30.
Hall B., Helmers C., Rogers M., Sena V. The choice between formal and informal intellectual property: A review. *Journal of Economic Literature* 2014;52:375–423.
Harrison R.T., Leitch C. Voodoo institution or entrepreneurial university? Spin-off companies, the entrepreneurial system and regional development in the UK. *Regional Studies* 2010;44:1241–62.
Hayter C.S. In search of the profit-maximizing actor: Motivations and definitions of success from nascent academic entrepreneurs. *Journal of Technology Transfer* 2011;36:340–52.
Hayter C.S. Constraining entrepreneurial development: A knowledge-based view of social networks among academic entrepreneurs. *Research Policy* 2016a;45:475–90.
Hayter C.S. A trajectory of early-stage spinoff success: The role of knowledge intermediaries within an entrepreneurial university ecosystem. *Small Business Economics* 2016b;47:633–56.
Hayter C.S., Rasmussen, E., and Rooksby, J.H. Beyond formal university technology transfer: Innovative pathways for knowledge exchange. *Journal of Technology Transfer* 2020;45(1):1–8.
Heirman A., Clarysse B. How and why do research-based start-ups differ at founding? A resource-based configurational perspective. *Journal of Technology Transfer* 2004;29:247–68.
Hurmelinna-Laukkanen P., Puumalainen K. Nature and dynamics of appropriability: Strategies for appropriating returns on innovation. *R&D Management* 2007;37:95–112.
Karnani F. The university's unknown knowledge: Tacit knowledge, technology transfer and university spin-offs: Findings from an empirical study based on the theory of knowledge. *Journal of Technology Transfer* 2013;38:235–50.
Knockaert M., Ucbasaran D., Wright M., Clarysse B. The relationship between knowledge transfer, top management team composition, and performance: The case of science-based entrepreneurial firms. *Entrepreneurship Theory and Practice* 2011;35:777–803.
Lawton Smith H., Romeoa S., Bagchi-Senb S. Oxfordshire biomedical university spin-offs: An evolving system. *Cambridge Journal of Regions, Economy and Society* 2008;1:303–19.
Link A.N., Siegel D.S., Bozeman B. An empirical analysis of the propensity of academics to engage in informal university technology transfer. *Industrial and Corporate Change* 2007;16:641–55.
Lundqvist M.A. The importance of surrogate entrepreneurship for incubated Swedish technology ventures. *Technovation* 2014;34:93–100.
Mathisen M.T. The growth of research-based spin-offs: Unleashing the value of academic entrepreneurship. Trondheim, Norway: Department of Industrial Economics and Technology Management, Norwegian University of Science and Technology; 2017.
Mathisen M.T., Rasmussen E. The development, growth, and performance of university spin-offs: A critical review. *Journal of Technology Transfer* 2019; 44(6):1891-1938.
Mathisen, M. T., Shankar, R. K., Widding, Ø., Rasmussen, E., and McKelvie, A. Enablers of exit through trade sale: the case of early-stage research-based spin-offs. *Small Business Economics* 2021:1-15.
Mohr V., Garnsey E. How do high-growth firms grow? Evidence from Cambridge, UK. *Economics, Management and Financial Markets* 2011;6:29.
Moray N., Clarysse B. Institutional change and resource endowments to science-based entrepreneurial firms. *Research Policy* 2005;34:1010–27.
Mowery D.C. Learning from one another? International policy "emulation" and university-industry technology transfer. *Industrial and Corporate Change* 2011;20:1827–53.
Mustar P., Renault M., Colombo M.G. et al. Conceptualising the heterogeneity of research-based spin-offs: A multi-dimensional taxonomy. *Research Policy* 2006;35:289–308.

Mustar P., Wright M., Clarysse B. University spin-off firms: Lessons from ten years of experience in Europe. *Science and Public Policy* 2008;35:67–80.
O'Shea R.P., Chugh H., Allen T.J. Determinants and consequences of university spinoff activity: A conceptual framework. *Journal of Technology Transfer* 2008;33:653–66.
Patzelt H., Shepherd D.A. Strategic entrepreneurship at universities: Academic entrepreneurs' assessment of policy programs. *Entrepreneurship Theory and Practice* 2009;33:319–40.
Pirnay F., Surlemont B., Nlemvo F. Toward a typology of university spin-offs. *Small Business Economics* 2003;21:355–69.
Pries F., Guild P. Commercial exploitation of new technologies arising from university research: Start-ups and markets for technology. *R&D Management* 2007;37:319–28.
Puranam P., Singh H., Zollo M. A bird in the hand or two in the bush?: Integration trade-offs in technology-grafting acquisitions. *European Management Journal* 2003;21:179–84.
Rasmussen E., Mathisen M.T. Science-based entrepreneurial firms as real options: Assessing the outcomes of the Norwegian firm population from 1995 to 2012. In: Fini R., Grimaldi R. (eds), *Process Approach to Academic Entrepreneurship: Evidence from the Globe*. World Scientific, Singapore; 2017:215–40.
Rasmussen E., Wright, M. How can universities facilitate academic spin-offs? An entrepreneurial competency perspective. *Journal of Technology Transfer* 2015;40(5):782–99.
Rasmussen E., Moen O., Gulbrandsen M. Initiatives to promote commercialization of university knowledge. *Technovation* 2006;26:518–33.
Rasmussen E., Mosey S., Wright M. The transformation of network ties to develop entrepreneurial competencies for university spin-offs. *Entrepreneurship and Regional Development* 2015;27:430–57.
Rasmussen, E. The regional impacts of university spin-offs: in what ways do spin-offs contribute to the region?. In *Handbook of Universities and Regional Development*. Edward Elgar Publishing, Cheltenham; 2019:151–65.
Rothaermel F.T., Thursby M. Incubator firm failure or graduation? The role of university linkages. *Research Policy* 2005;34:1076–90.
Rothaermel F.T., Agung S.D., Jiang L. University entrepreneurship: A taxonomy of the literature. *Industrial and Corporate Change* 2007;16:691–791.
Salvador E. Are science parks and incubators good "brand names" for spin-offs? The case study of Turin. *Journal of Technology Transfer* 2011;36:203–32.
Schumpeter J.A. *The Theory of Economic Development: An Inquiry into Profits, Capital, Credit, Interest, and the Business Cycle*. New Brunswick, NJ: Transaction Publishers; 1934.
Scrucca L., Santucci A., Aversa F. Competing risk analysis using R: An easy guide for clinicians. *Bone Marrow Transplantation* 2007;40:381–7.
Shane S. Technological opportunities and new firm creation. *Management Science* 2001;47:205–20.
Shane S. *Academic Entrepreneurship University Spinoffs and Wealth Creation*. Cheltenham, UK and Northampton, MA, USA: Edward Elgar Publishing; 2004a.
Shane S. Encouraging university entrepreneurship? The effect of the Bayh-Dole Act on university patenting in the United States. *Journal of Business Venturing* 2004b;19:127–51.
Shepherd D., Wiklund J. Are we comparing apples with apples or apples with oranges? Appropriateness of knowledge accumulation across growth studies. *Entrepreneurship Theory and Practice* 2009;33:105–23.
Siegel D.S., Veugelers R., Wright M. Technology transfer offices and commercialization of university intellectual property: Performance and policy implications. *Oxford Review of Economic Policy* 2007;23:640–60.
Slaughter S., Rhoades G. *Academic Capitalism and the New Economy: Markets, State, and Higher Education*. Baltimore, MD: Johns Hopkins University Press; 2004.
Thumm N. Strategic patenting in biotechnology. *Technology Analysis & Strategic Management* 2004;16:529–38.

Villani, E., Rasmussen, E., and Grimaldi, R. How intermediary organizations facilitate university–industry technology transfer: A proximity approach. *Technological Forecasting and Social Change* 2017;114:86–102.

Vincett P.S. The economic impacts of academic spin-off companies, and their implications for public policy. *Research Policy* 2010;39:736–47.

Walter A., Auer M., Ritter T. The impact of network capabilities and entrepreneurial orientation on university spin-off performance. *Journal of Business Venturing* 2006;21:541–67.

Wright M., Lockett A., Clarysse B., Binks M. University spin-out companies and venture capital. *Research Policy* 2006;35:481–501.

Zhang J.F. The performance of university spin-offs: An exploratory analysis using venture capital data. *Journal of Technology Transfer* 2009;34:255–85.

13. Universities' ownership of intellectual property: focus on Canada

Shiri M. Breznitz, Samaa Kazerouni and Qiantao Zhang

13.1 INTRODUCTION

Universities are important contributors to national and regional economies. This contribution is both direct – via employment, purchasing, and taxes – and indirect – via the future impact of students and university research (Breznitz 2014; Uyarra 2010). Since the enactment in the US in 1980 of the Bayh-Dole Act, which transferred the ownership of federally funded research to universities, a worldwide debate has been ongoing about the benefit of universities' ownership of intellectual property (IP). In recent decades, governments have implemented policies related to the ownership of intellectual property rights (IPR) to foster technology transfer by universities and public research organizations (Kochenkova et al. 2016). It is widely accepted that legislative changes have acted as a key catalyst for the commercialization of university research and university-industry collaboration (Geuna and Rossi 2011). However, existing studies obtain conflicting results about university ownership of IP and suggest that ownership is not enough on its own. Other government programs that support university-industry relationships, services such as incubators and accelerators, and patent services, including funding and strategies, have a role to play when it comes to the commercialization of basic research from universities. In this chapter, we review different policies regarding universities' ownership of IP and their impact. We then focus on the case of Canada. Canada does not have an IP ownership policy at either the federal or provincial levels. However, some universities have created their own ownership policies. The outcome of specific universities' IP policies has a positive impact on income from licenses and patents. However, the IP ownership policy does not seem to make much difference when it comes to the regional economic impact through the creation of universities' spinouts. This is particularly evident in the case of the University of Waterloo.

13.2 THE TRAJECTORY OF IP LEGISLATION FOR UNIVERSITY RESEARCH IN WESTERN EUROPE AND THE US

Despite the shift toward institutional ownership commonly associated with the US Bayh-Dole Act in 1980, legislation giving IP ownership to universities in the UK predates Bayh-Dole. In the UK, since 1948 academic IP rights have been managed by the National Research Development Corporation (NRDC), a government organization (Breznitz 2011; Geuna and Rossi 2011). The UK's Patent Act of 1977 states that an employee's invention is owned by the employer, and it has no special provision for academics (Geuna and Rossi 2011). In 1981, these IP rights were given to the British Technology Group (BTG), allowing it to commercialize publicly funded research. Then, in 1985, UK universities were given the option to own and manage their patents independently or rely on the services of the BTG (Della Malva et al. 2011; Geuna and Rossi 2011). However, this legislation on institutional ownership was not widely enforced until 1993 (Geuna and Rossi 2011). At some UK universities, such as the University of Cambridge, university ownership of IP did not commence until 2006 (Breznitz 2011).

The widespread commercialization of academic research before 1980 was facilitated by the decentralized structure of the US higher education system, which consists of a heterogeneous collection of institutions with high levels of competition, incentivizing the pursuit of commercialization by university researchers (Mowery and Sampat 2005). In 1980, the Bayh-Dole Act was implemented in the US to foster knowledge and technology transfer from publicly funded university research to industry (Della Malva et al. 2011; Giuri et al. 2013; Kochenkova et al. 2016; Mowery and Sampat 2005). This was implemented as part of an effort to bring property rights nearer to the inventor (Geuna and Rossi 2011). This served as a landmark policy reform for academic IPR. Spain was also an early implementer of institutionally owned IP. In 1986, the Law for the Promotion and General Coordination of Scientific and Technological Research (the "Science Law") stated that universities and public research organizations held ownership of their research (García-Aracil et al. 2015; Geuna and Rossi 2011). The US and Spanish policies were very similar, though the latter was not directly inspired by the former (Geuna and Rossi 2011).

After the implementation of Bayh-Dole, a wave of IP legislation change occurred across Europe, shifting IPR ownership from the inventor or government funder to the research institution. In 2000 the Lisbon European Council Summit set the agenda for the next decade, and one of its objectives was to improve key interfaces in the innovation landscape by improving technology transfer from universities to industry (Baldini 2009). This agenda attempted to address the "European paradox": that, compared to the US, Europe performs better in academic research than in technological application (Baldini 2009). However, whereas in the US, Bayh-Dole shifted ownership from the government to universities, in Europe, IPR ownership was transferred from researchers to their employers (Cunningham et al. 2019; Damsgaard and Thursby 2013). Since 2000, most European countries have replaced the traditional professor's privilege, in which inventions from publicly funded research are owned by academic researchers, with institutional ownership (Geuna and Rossi 2011). This new framework gives the university ownership of all research outcomes developed by their academic employees.

In France, universities have historically lacked autonomy and been heavily controlled by the central government (Della Malva et al. 2011). University employees were considered civil servants, meaning that the IPR of their inventions belonged to their employers. However, these rights were not usually retained by universities, as this was considered counterproductive for knowledge diffusion (Geuna and Rossi 2011). In 1999, the French government introduced the Innovation Act, with the aim of increasing university awareness of IPR and increasing the rate of commercialization of academic inventions via technology transfer infrastructure (Della Malva et al. 2011; Geuna and Rossi 2011). The Innovation Act implemented provisions to encourage universities and public research organizations to retain IPR or at least share it with partners in the industry (Della Malva et al. 2011). This included the addition of commercial exploitation of patents and licenses to the universities' mission, the potential to create internal technology transfer offices (TTOs), and a favorable taxation rule encouraging IPR ownership by public research organizations or universities (Della Malva et al. 2011).

In the early 2000s, Germany, Denmark, Finland, and Norway modified their IP regulations to introduce institutional ownership (Geuna and Rossi 2011; Grimaldi et al. 2011). Until then, these countries all had the professor's privilege in place, giving university researchers full rights to their inventions (Cunningham et al. 2019). In 2002, the German Employees' Inventions Act forced the transfer of ownership of most inventions to universities, giving them the freedom to use research outcomes for technology transfer and commercialization efforts, with 30 percent of the revenues given to the inventor (Cunningham et al. 2019).

In contrast to the general trend toward institutional ownership since the 1980s, Italy has done the opposite, moving from institutional to inventor ownership. The original IP legislation in Italy was implemented in 1939, giving the employer ownership of any invention developed by an employee while at work (Baldini et al. 2014). However, at that time, universities in Italy were fully public and highly centralized, giving them little autonomy (Baldini et al. 2006). This meant that the government had the authority to determine how much money should be invested in each activity at universities, including IP management (Baldini et al. 2014). Hence, universities could not receive any income from the exploitation of their IPR, so they had little incentive to promote commercialization. This began to change in 1989, thanks to a series of new policies giving universities greater administrative and financial autonomy (Baldini et al. 2006). Specifically, these universities could plan and control their own budget for the first time, allowing them to keep the surplus generated by their activities. Most of these institutions responded by creating mechanisms to commercialize their research, such as TTOs and university incubators (Baldini et al. 2014; Lissoni et al. 2013).

In 2001, Law 383 in Italy granted IPR ownership to public researchers and entitled universities to between 30 and 50 percent of the revenue from the patents (Baldini et al. 2006). This was due to an assumption that individual inventors would be better placed to profit from their discoveries because universities did not have the competence to commercialize inventions (Geuna and Rossi 2011). This policy gave the university a free, non-exclusive license to use the invention if the inventor did not use the patent within five years (Geuna and Rossi 2011). In 2005, Italy implemented another legislative change in partial institutional ownership, giving ownership to the researcher only for research fully financed by the institution that employs the researcher. If the research is partially funded by the private sector or the government, the IPR belongs to the university (Geuna and Rossi 2011). This policy has reduced the researchers' ability to own the IPR of their inventions, because of the huge reliance on external funding (Geuna and Rossi 2011).

Like Italy, Sweden has maintained its professor's privilege, which was first implemented in 1949 (Geuna and Rossi 2011). Under this policy, researchers bear the costs and receive the benefits of their IPR. Between 2004 and 2009, Sweden considered the possibility of university ownership to address its relatively slow growth in comparison with the average across the member countries in the Organisation for Economic Co-operation and Development (OECD). The Swedish government ultimately decided against adopting this policy because of the importance of the professor's privilege in incentivizing researchers to commercialize their inventions (Damsgaard and Thursby 2013).

Canada, another anomaly, failed in its efforts to design a national policy on IPR commercialization. IPR in Canada remains highly decentralized at the individual university level (Hepburn and Wolfe 2015). Universities have no legislative obligation to create policies for mandatory disclosure or for the creation of TTOs. Instead, ownership is decided a priori, often enabling universities to retain some rights for research and teaching purposes, while giving a research partner significant ownership (Hepburn and Wolfe 2015).

Countries in Eastern Europe began to update their general patent systems in the early 1990s, assigning ownership of academic IPR to universities, in a shift away from government ownership, which was typically of the communist era (Geuna and Rossi 2011). Japan shifted from inventor or government ownership to university ownership in 1999 (Damsgaard and Thursby 2013; Mowery and Sampat 2005).

13.3 IMPACTS OF LEGISLATIVE CHANGES

Empirical evidence on the consequences of university IPR ownership legislation on commercialization is limited (Giuri et al. 2013). The impact of Bayh-Dole-type regulations is evaluated in terms of two schools of thought. Its proponents argue that these policy changes have been successful, with a positive effect on university patenting outputs, whereas its opponents are skeptical about the necessity of these policies (Weekowska et al. 2018).

Lissoni et al. (2013) conducted one of the first longitudinal analyses of academic patenting in Italy, finding that between 1996 and 2006, the share of academic patenting has declined since the implementation of the professor's privilege in 2001. However, they attribute this to a lack of funding rather than legislative changes. Overall, the research concludes that the positive effects of legislation on university ownership outweigh any potential negative impacts, and many studies hold that the legislation helps increase the commercialization of university research by creating greater incentives for commercialization of the research (Grimaldi et al. 2011; Kochenkova et al. 2016; Weekowska et al. 2018).

The results in Germany are mixed. Some studies conclude that the law and associated financing had no impact on the number of university startups. The new policy actually reduced the number of patents approved and the amount of technology transfer, with the greatest effect on faculty-firm relationships (Czarnitzki et al. 2016). At the same time, Grimm and Jaenicke (2012) find that the new law facilitated an increase in patents, but that, in order for the policy to be effective, expertise needs to be bolstered and financial and organizational support improved, especially because since the new law was put in place, universities have continued to report low levels of commercialization through entrepreneurship. Moreover, Cunningham et al. (2019) find that the change had no definitive effect on innovation. At first, the number of patent applications from universities greatly increased, but subsequently, it plateaued and then declined, decreasing innovation. However, they question whether the number of patents is an appropriate measure of innovation. In addition, after the new legislation went into effect, business registrations initially decreased, but in the long run, they increased, implying a rise in entrepreneurship. These outcomes are similar to those measured in the US as a result of Bayh-Dole. Ultimately, it is difficult to achieve positive outcomes in both entrepreneurship and innovation through university-based technology transfers, which can be particularly influenced by environmental and regional factors.

Cross-country evidence shows that the number of university-owned patents has increased overall but not solely due to changes in IPR legislation (Geuna and Rossi 2011). Mowery and Sampat (2005) argue that the growth in licensing and university-based spinoffs that has occurred since Bayh-Dole almost certainly would have happened even in the absence of the law. Della Malva et al. (2011) find that France's Innovation Act increased the likelihood that universities would retain their IPR over their scientists' inventions;, however, the data is too limited to see whether greater technology transfer has resulted. Weekowska et al. (2018) argue that national legislative frameworks may stimulate the development of local practices for the management and exploitation of IP by signaling legitimacy of academic patenting and emphasizing the importance of local IP management and commercialization.

Comparisons of the outcomes of legislative changes are often based on the US and Europe, but have little consideration of their legal and institutional differences (Della Malva et al. 2011). In Europe, many inventions from university professors are patented by other institutions, even though universities are legally allowed to retain IPR to these inventions (Grimaldi et al. 2011). For instance, in France, Italy, and Sweden,

11 percent of the academic patents are university-owned, whereas 60–80 percent are owned by businesses (Della Malva et al. 2011).

Thus, it is important to highlight the numerous other cultural and political changes that have occurred and could have increased university patenting. Numerous policies have been implemented to support the creation of a knowledge transfer infrastructure (Geuna and Rossi 2011). For instance, in the US, the University and Small Business Patent Procedures Act was passed at the same time as Bayh-Dole, enabling the federal government to arrange for the licensing of patents not exploited by academic administrations (Kochenkova et al. 2016). In Italy, Lissoni et al. (2013) argue that increased autonomy by Italian universities enabled them to introduce explicit IP regulations on their employees' inventions, leading to a rise in university-owned patents. These parallel policy changes should not be overlooked in assessments of the success of university IPR ownership policies.

Several studies indicate that the key determinant of increasing enforcement of university-owned IPR is the creation of TTOs (Grimaldi et al. 2011). This is echoed by Weekowska et al. (2018), who conducted a within-country analysis, considering the impact of local practices on academic patenting under the same legislative framework. They find that universities with a wider range of activities to support research commercialization have higher rates of patenting (Breznitz 2014; Weekowska et al. 2018). For instance, problems related to the enforcement of Bayh-Dole suggest that local organizational practices (e.g. business plan competition, entrepreneurship education, incubators) are equally important in the commercialization of academic research (Weekowska et al. 2018).

The importance of the university's TTO is highlighted in many studies (e.g. Breznitz 2011). The level of its resources, reflected by the quality and number of personnel, allows more flexibility and trust at the university, and encourages inventors and investors to work with it (Lockett and Wright 2005; O'Shea et al. 2005; Shane 2002). Because universities and industries have different business perspectives, the availability of TTO employees, who are highly educated and have business experience as well as knowledge of the correct technical and business jargon, reassures inventors and investors that their product is getting the best available treatment. Clarysse et al. (2005) and Lockett and Wright (2005) find that the TTO's business development capabilities have a positive influence on the formation of startups. These capabilities include the marketing, technological, and negotiating skills of the TTO staff, a clear process for spinning out companies, clear due diligence procedures, and sufficient staff to manage these processes (Clarysse et al. 2005; Lockett and Wright 2005). However, very few studies measure these capabilities; instead, most examine best practices. Another factor related to the availability of resources is the use of outside lawyers. Siegel et al. (2003) find that spending more on outside lawyers reduces the number of licensing agreements but increases revenue. They confirm that hiring external lawyers allows TTO staff to spend more time connecting the invention to the right firm, which results in a successful license and higher revenue. O'Shea et al. (2005) conclude that the historical background and prior technology transfer success by each university are related to its future capabilities and options with regard to spinout capability. When a TTO has successfully commercialized an invention and obtains returns in the form of royalties, it is strengthened and motivated to continue its involvement with commercialization. Hence, commercialization of university technology will not be increased by moving IPR ownership to universities in the absence of adequate support services, expertise in patent registration, a strategy for applying this expertise, and business experience.

13.4 THE CASE OF CANADA: NO IP OWNERSHIP AND LIMITED SUPPORT

As mentioned above, by itself, ownership of IP is insufficient for inducing technology transfer from higher education. Other mechanisms of support such as the quality of the university technology transfer are important catalysts in this process. Canada has no IP policy at the federal or provincial level or at most universities – therefore, by default, at most institutions IP ownership is held by the inventor, regardless of the source of funding. Moreover, less than 50 percent of universities in Canada have a TTO, 41 percent provide IP information, and only 18 percent have IP training workshops.[1]

According to Hepburn and Wolfe (2015), not only does Canada have no national IP policy, but it does not require its universities to set policies regarding mandatory disclosure or a requirement that researchers work with the university TTO. Canadian university practice is a mixture of various IP regimes, and some universities follow more than one. The result is that Canada lacks a uniform IP regime, and firms that are negotiating licenses for IP that originated at more than one institution may have to deal with each institution's IP regime. In particular, universities have their own policies regarding IP ownership and are often included as part of faculty employment and collective agreements. The ownership of IP that results from a collaborative research project with an industry or a public research lab is determined before the project starts as part of the research agreement. Generally, in a collaborative project with an industry in which that industry also is the main funder, the IP is generally owned by the partner or determined through an exclusive license. Much of the research conducted on university campuses, however, is not covered by pre-existing research agreements and is funded by public research grants.

In their review, Hepburn and Wolfe (2015) address the fact that the IP regimes at some of Canada's top universities are highly diverse: the University of British Columbia (UBC) owns the IP, whereas the University of Alberta, Waterloo University, and Queen's University have inventor's rights. The University of Toronto (UofT) has a hybrid model. A review of these universities by output indicates that, after research expenditures are controlled for, ownership of IP does make a difference. Both UBC and UofT lead in licensing income and in patents granted. However, this impact is not visible in a comparison of the number of university startups, in which the University of Waterloo leads, with 24 (versus 23 for UofT and 8 for UBC) (Figures 13.1 to 13.3).

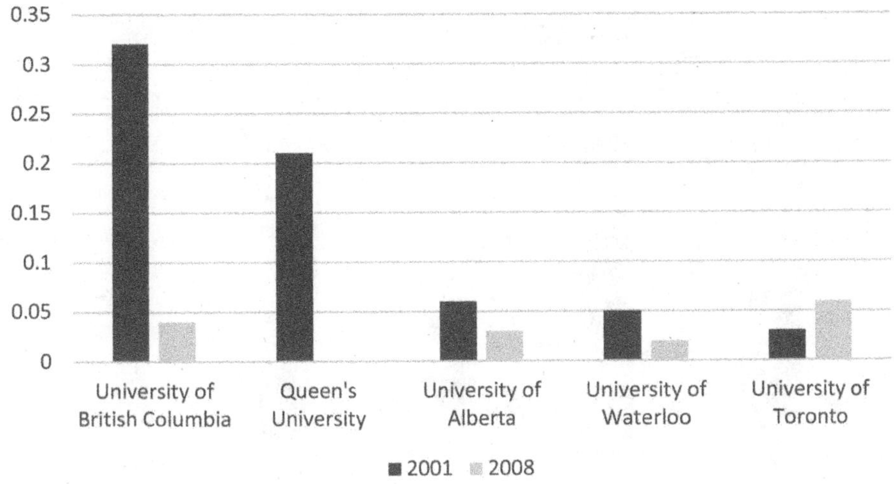

Source: AUTM (2022).

Figure 13.1 *Number of patents per million US dollars in research expenditure*

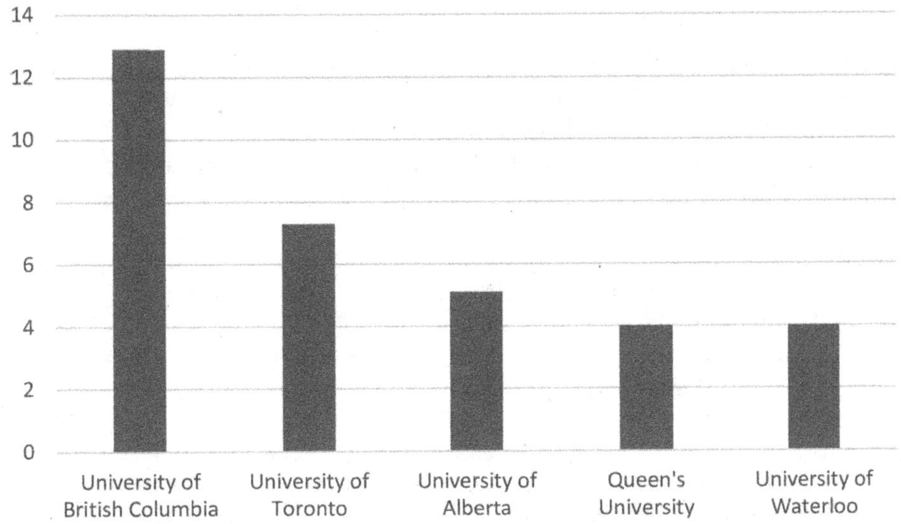

Source: AUTM (2022).

Figure 13.2 *Number of licenses per million dollars in research expenditure*

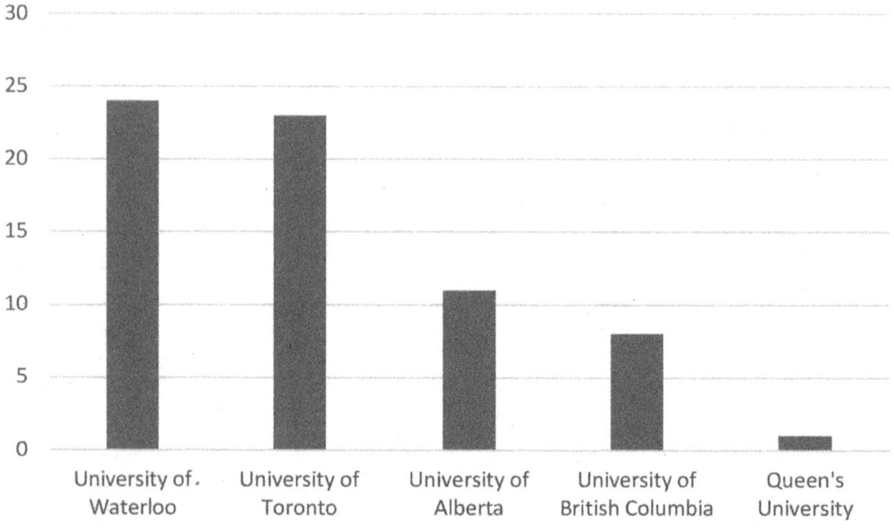

Source: AUTM (2022).

Figure 13.3 *Number of startups formed in 2018*

13.5 THE IMPACT OF THE UNIVERSITY OF WATERLOO ON INNOVATION IN CANADA'S TECHNOLOGY TRIANGLE

The University of Waterloo (UW) is an entrepreneurial research university that is actively engaged in local economic development (Bramwell and Wolfe 2008). It was founded in 1959 as an engineering-focused university (Bathelt et al. 2010). At the time, three factors distinguished UW from other Canadian universities. First, its emphasis on basic and applied research made UW Canada's biggest research university in the late 1960s (Bathelt et al. 2010). Second, its IP policy gave full ownership of all IP to the inventor without requiring payment of revenue to the university if the IP was successfully commercialized (Bathelt et al. 2010; Bramwell and Wolfe 2008). Finally, the university's Cooperative Education Program (co-op) was highly unique at the time and has grown to be the largest and most successful of its kind in the world, allowing students to work in industry for a period as part of their curriculum (Bramwell and Wolfe 2008). As a result, UW has generated and attracted talent that underpins academic and applied excellence in science, math, and engineering while providing support for local firm-based R&D (Bramwell and Wolfe 2008). UW has fostered close ties with industry, creating a highly entrepreneurial community.

UW is at the center of what is known as Canada's Technology Triangle (CTT), comprising the municipalities of Waterloo, Cambridge, Kitchener, and Guelph. This region is commonly touted for its innovative capacity generated by university spinoffs, leading to economic change and growth (Bathelt et al. 2010; Wolfe 2009). Since the 1970s, CTT has increased its spinoffs from university research, which resulted in exponential growth in its high-tech sector (Bathelt et al. 2010; Bramwell and Wolfe 2008). University spinoffs transfer knowledge from the university to the market, enabling the formation of a firm (Bathelt et al. 2010). UW has spun off 59 high-tech firms since 1973, and its Technology Transfer and Licensing Office identified 106 spinoff companies, employing over 2,000 people by the mid-1990s (Wolfe 2009). Many of these firms are in information and communications technology, representing a significant shift from traditional manufacturing in this region historically (Bathelt et al. 2010). The other main subsectors in this cluster are scientific and engineering services, advanced manufacturing, and life sciences biotech and the environment (Wolfe 2009). The region is characterized by its large number of university spinoff firms, with frequent licensing and patenting activity (Bramwell and Wolfe 2008; Wolfe 2009). Consequently, both the university and the region attract a significant amount of public and private research funding (Bramwell and Wolfe 2008). Because such rapid innovation occurs at the university, a similar culture has developed across the CTT that supports entrepreneurial risk taking (Spigel and Vinodrai 2021).

Although UW is known for its success in fostering spinoffs, this rapid pace of commercialization has declined (Bathelt et al. 2010; Bramwell and Wolfe 2008; Wolfe 2009). Over time, existing university spinoffs have pivoted to incremental innovation, and few of the R&D relationships with UW have persisted (Bathelt et al. 2010). Some studies on UW's spinoffs find that fewer firms have emerged as a direct result of university research (Bathelt et al. 2010). Instead, most spinoffs are "decentralized unsponsored," meaning the university did not sponsor the firm's core technologies or contribute specific knowledge inputs through university research (Bathelt et al. 2010). In most cases, their ties to UW are in the form of human capital. For instance, they hire students or have informal research relationships with faculty (Bramwell and Wolfe 2008; Bramwell et al. 2008). Most commonly, the founders of the firm were either trained at UW or met while students at UW or while working in the region after graduation (Bathelt et al. 2010). Hence, the role of UW in the formation of spinoffs has become less direct. Although UW has many formal relationships with local high-tech firms, the majority of knowledge exchange is informal (Wolfe 2009). The main contributions of the university are the provision of basic technical education, the development of an aptitude for product development and business thinking by its students, and the opportunity to develop a social network among its students and graduates (Bathelt et al. 2010).

Additionally, UW plays an integral role in knowledge generation and transfer in the region. It still attracts, retains, and trains top caliber graduates and researchers, connecting them with employers (Bramwell and Wolfe 2008). The majority of firms continue to cite the importance of UW as a provider of highly skilled talent (Bramwell et al. 2008; Wolfe 2009). It provides R&D support to local firms and exchanges tacit knowledge both locally and globally (Bramwell and Wolfe 2008). UW has developed an international cachet that attracts students, faculty, and firms

(Bramwell et al. 2008; Wolfe 2009). Its co-op program serves as a critical intermediary for human capital and knowledge transfer, with students attuned to the needs of industry and curriculum reflecting the rapidly changing nuances in the high-tech sector (Bramwell and Wolfe 2008; Bramwell et al. 2008; Spigel and Vinodrai 2021; Wolfe 2009). Moreover, the local talent pool in the CTT is known to be competitively priced, offering high quality and productivity relative to its cost (Bramwell et al. 2008; Wolfe 2009).

Despite the decrease in informal relationships between UW and spinoff firms, the majority of firms still see the university as critical to the dynamics of the technology cluster in the CTT (Wolfe 2009). UW plays an active role in supporting entrepreneurial education and activities, in line with its emphasis on entrepreneurship. Its Centre for Business, Entrepreneurship, and Technology (CBET) was created to research issues surrounding entrepreneurship and university-industry relationships (Wolfe 2009). In addition, UW offers a Master of Business, Entrepreneurship and Technology (MBET) degree, which teaches business skills that are critical for innovation and entrepreneurship (Wolfe 2009). Finally, UW's Institute for Innovation Research generates and disseminates applied interdisciplinary research to advance understanding of and promote entrepreneurship in technology-based enterprises (Wolfe 2009). These endeavors to promote entrepreneurship through education might affect the rate of university spinoffs in the future.

13.6 CONCLUSIONS

In this chapter, we try to clarify the connections between country- or regional-level IP policies and university policies, but the picture remains murky. Creating a general policy on IP ownership has proven important. However, what studies also show is that merely having an IP ownership policy is not enough – having other policies and services to increase university-industry links in general and to help startups in particular are equally important. Moreover, we find that a distinction is needed between IP ownership and the regional contribution of universities through university spinouts and social networks. Universities cannot create licensing income if they do not own the technology, nor can they apply for patents. However, through their services, they can contribute to the creation of firms, which is exemplified by the UW. Hence, the question that needs to be posed is: What do we want to achieve? If the goal is to make a return on investment in public research funding or protect the IPR of a country, then IP ownership is important. At the same time, IP ownership should not obscure the important role played by universities in building an innovation ecosystem.

NOTE

1. Based on data collected by the authors.

REFERENCES

AUTM. (2022). *Technology Transfer Databases*. Available at: https://autm.net/surveys-and-tools/databases (accessed 22 May 2022).

Baldini, N. (2009). "Implementing Bayh-Dole-like laws: Faculty problems and their impact on university patenting activity." *Research Policy* 38(8): 1217–24.

Baldini, N., Grimaldi, R., and Sobrero, M. (2006). "Institutional changes and the commercialization of academic knowledge: A study of Italian universities' patenting activities between 1965 and 2002." *Research Policy* 35(4): 518–32.

Baldini, N., Fini, R., and Grimaldi, R. (2014). "Organisational change and the institutionalisation of university patenting activity in Italy." *Minerva* 52(1): 27–53.

Bathelt, H., Kogler, D.F., and Munro, A.K. (2010). "A knowledge-based typology of university spin-offs in the context of regional economic development." *Technovation* 30(9–10): 519–32.

Bramwell, A., and Wolfe, D.A. (2008). "Universities and regional economic development: The entrepreneurial University of Waterloo." *Research Policy* 37(8): 1175–87.

Breznitz, S.M. (2011). "Improving or impairing? Following technology transfer changes at the University of Cambridge." *Regional Studies* 45(4): 463–78.

Breznitz, S.M. (2014). *Fountain of Knowledge? The Role of Universities in Economic Development*. Paolo Alto, CA: Stanford University Press.

Clarysse, B., Lockett, A., Velde, E., and Vohora, A. (2005). "Spinning out new ventures: A typology of incubation strategies from European research institutions." *Journal of Business Venturing* 20: 183–216.

Cunningham, J., Lehmann, E.E., Menter, M., and Seitz, M. (2019). "The impact of university focused technology transfer policies on regional innovation and entrepreneurship." *Journal of Technology Transfer* 44(5): 1451–75.

Czarnitzki, D., Doherr, T., Hussinger, K., Schliessler, P., and Toole, A.A. (2016). Knowledge creates markets: The influence of entrepreneurial support and patent rights on academic entrepreneurship. *European Economic Review* 86(C): 131–46.

Damsgaard, E.F., and Thursby, M.C. (2013). "University entrepreneurship and professor privilege." *Industrial and Corporate Change* 22(1): 183–218.

Della Malva, A., Lissoni, F., and Llerena., P. (2011). "Institutional change and academic patenting: French universities and the Innovation Act of 1999." *Journal of Evolutionary Economics* 23(1): 211–39.

García-Aracil, A., Castro-Martínez, E., Azagra-Caro, J. M., and D'Este, P. (2015). "University technology transfer: The Case of Spain." In Breznitz, S.M. and Etzkowitz, H. (eds), *University Technology Transfer: The Globalization of Academic Innovation*. New York: Routledge.

Geuna, A., and Rossi, F. (2011). "Changes to university IPR regulations in Europe and the impact on academic patenting." *Research Policy* 40(8): 1068–76.

Giuri, P., Munari, F., and Pasquini. M. (2013). "What determines university patent commercialization? Empirical evidence on the role of IPR ownership." *Industry and Innovation* 20(5): 488–502.

Grimaldi, R., Kenney, M., Siegel, D.S., and Wright, M. (2011) "30 years after Bayh-Dole: Reassessing academic entrepreneurship." *Research Policy* 40(8): 1045–57.

Grimm, H.M., and Jaenicke, J. (2012). "What drives patenting and commercialisation activity at East German universities? The role of new public policy, institutional environment and individual prior knowledge." *Journal of Technology Transfer* 37(4): 454–77.

Hepburn, N., and Wolfe, D. (2015). "Commercialization and tech transfer policies and intellectual property regimes in Canada." In Breznitz, S.M. and Etzkowitz, H. (eds), *University Technology Transfer: The Globalization of Academic Innovation*. New York: Routledge. 226–50.

Kochenkova, A., Grimaldi, R., and Munari, F. (2016). "Public policy measures in support of knowledge transfer activities: A review of academic literature." *Journal of Technology Transfer* 41(3): 407–29.

Lissoni, F., Pezzoni, M., Poti, P., and Romagnosi, S. (2013). "University autonomy, the professor privilege and academic patenting." *Industry and Innovation* 20(5): 399–421.

Lockett, A., and Wright, M. (2005). "Resources, capabilities, risk capital and the creation of university spin-out companies." *Research Policy* 34(7): 1043–57.

Mowery, D.C., and Sampat, B.N. (2005) "The Bayh-Dole Act of 1980 and university-industry technology transfer: A model for other OECD governments?" *Journal of Technology Transfer* 30(1–2): 115–27.

O'Shea, R.P., Allen, T.J., Chevalier, A., and Roche, F. (2005). "Entrepreneurial orientation, technology transfer and spinoff performance of U.S. universities." *Research Policy* 34(7): 994–1009.

Shane, S. (2002). "Selling university technology: Patterns from MIT." *Management Science* 48(1): 122–37.

Siegel, D.S., and Waldman, D. (2003). "Commercial knowledge transfers from universities to firms: Improving the effectiveness of university–industry collaboration." *Journal of High Technology Management Research* 14(1): 111–33.

Siegel, D.S., Waldman, D., and Link, A. (2003). "Assessing the impact of organizational practices on the relative productivity of university technology transfer offices: An exploratory study." *Research Policy* 32(1): 27–48.

Spigel, B., and Vinodrai, T. (2021). "Meeting its Waterloo? Recycling in entrepreneurial ecosystems after anchor firm collapse." *Entrepreneurship and Regional Development* 33(5–6): 475.

Uyarra, E. (2010). "Conceptualizing the regional role of universities, implications and contradictions." *European Planning Studies* 18(8): 1227–46.

Weekowska, D.M., Molas-Gallart, J., Tang, P. et al. (2018). "University patenting and technology commercialization: Legal frameworks and the importance of local practice." *R&D Management* 48(1): 88–108.

Wolfe, D. 2009. "The ICT cluster of Waterloo, Canada." In *Clusters, Innovation and Entrepreneurship*. Paris: OECD Publishing. https://doi.org/10.1787/9789264044326-10-en.

14. Technology transfer and frugal social innovations: looking inside an emerging economy

Claudia Yáñez-Valdés and Maribel Guerrero

14.1 INTRODUCTION

Innovation has become the central focus when creating new products and services and reformulating the business model (Baden-Fuller and Haefliger, 2013; Chesbrough, 2010). However, innovation can be very difficult, expensive, and uncertain, making it complex for organizations that do not have the resources to finance long research and costly development (Kline and Rosenberg, 2010). In emerging economies where resources are limited, building an initial resource base is a unique challenge (Krohn and Herstatt, 2018). Every choice of resources has significant implications for survival and growth. These choices can even have negative consequences if the wrong resources are acquired, do not fit into the opportunity, or waste other productive resources (Brush et al., 2001).

Frugal innovation appears in context, which involves emerging solutions with scarce resources but quality and cheaper than existing products (Zeschky et al., 2014). Frugal innovations consist of essential features such as low cost, emphasis on local use, local and discarded materials, ease of use, and minimum resources (Rao, 2013). Therefore, the challenge of frugal innovation is to introduce something new or different (innovate) while saving resources (frugal). The concept of frugal innovation is a recent one and addresses problems in several sectors, such as health, water, energy, transport, and communications, involving the private sector (Levänen et al., 2016). Although the origin of frugal innovation is not known, the frugal innovation literature became well known in 2010 through *The Economist* (Zeschky et al., 2011). There is no consensus about the definitions and metrics of frugal social innovations (Hossain, 2018). Little is known about the factors associated with resource scarcity that determine the existence of frugal social innovations and the relationship with knowledge transfer policy frameworks.

Inspired by this gap, this chapter investigates the scarcity of resources and the technology transfer framework in frugal innovation ecosystems. Our proposed conceptual model was tested with a probit analysis in the Chilean context. Chile is an emerging country characterized by income inequality that generates scarcity of public/private resources associated with innovation. In the technology transfer policy framework, the resources are distributed in large conglomerates' hands, and it is not easy to generate/transfer knowledge among organizations. Our results show that, despite favorable regulations, there are no innovations that can be transferred. This chapter contributes to the innovation and entrepreneurship literature by explaining the relationship between frugal social innovation and innovation ecosystem constraints. One of the main lines of research is the incorporation of technology transfer processes.

The chapter is organized as follows. First, we describe the concept and relevant aspects of frugal ecosystems. Then, the research setting and methodological design are described. Finally, the conclusions and implications are summarized.

14.2 THEORETICAL FRAMEWORK

14.2.1 Frugal Innovation

According to Hossain et al. (2016), frugal innovation is understood as the solution to resource scarcity (i.e., product, service, process, or business model) to meet the basic customers' needs who would otherwise be underserved. The importance of frugal innovation lies in its ability to be economically efficient under severe scarcity (Basu et al., 2013). Emerging economies are characterized by either lacking in resources or, more specifically, not abundant in nature (Hoskisson et al., 2000). Frugal innovation is of particular interest since it is a means of meeting a large population's needs with access to disproportionately low resources (Prahalad, 2005). Its practice involves doing the best with the resources available. For countries with limited resources and low incomes, the development of innovative ventures is a challenge that represents a possibility for growth and a chance to solve population problems (Meagher, 2018). In this sense, frugal innovation focuses on engaging with the informal economy to create low-cost, decent-quality goods and services for the most vulnerable population groups (Knorringa et al., 2016).

Table 14.1 shows the outstanding examples of frugal social innovation (i.e., medical devices, health services, solar energy, water purifiers, and others). The key to frugal innovation is to start from the expected result, for example, a product with certain characteristics and a target price (or cost) (Hossain, 2017). The entire process must then be reformulated to meet all existing constraints. It requires a willingness to change many things to meet customer expectations (Belkadi et al., 2016; Mourtzis et al., 2016). Therefore in the practical sense, frugal innovation is not improvisation. It requires a lot of creativity and ingenuity to rethink what is being done by proposing innovative solutions with several constraints.

The approach to frugal innovation is to drive change and move out of the comfort zone by a leap of faith (Lehner et al., 2018). The objectives are to create products and services accessible in price, easy to obtain, and manage to solve a real need valued by people (Khan, 2016). These products or services should be low-cost, sustainable, simple, and offer basic quality. Product frugality can be measured as a formative

Table 14.1 Examples of frugal innovation

Business	Country	Description
M-PESA	Kenya	Mobile phone product of Safaricom (a subsidiary of Vodafone) that offers various services to users making payments with the mobile phone, sending and receiving money between users and non-users, booking hotels, withdrawing cash from ATMs, among others. It is a branchless banking service, which means that it is designed to allow users to perform basic banking transactions without the need to visit a bank branch.
M-KOPA	Kenya	It is a solar panel on the roof, three LED bulbs, a solar radio, and a cell phone charger. The complete kit costs $200, which is expensive for most Kenyans, and this is where mobile phones make it the most affordable. After making an initial deposit of $30, you pay the balance by making a 50-cent daily micropayment with your mobile phone. Once you have made 365 micropayments, you will own the product and can set it up and start receiving a clean payment, free electricity.
Nickel	France	For an annual maintenance fee of 20 euros, it allows people to enter a corner store and activate a service that provides two products: an international bank account number and an international debit card. You can perform all banking transactions, such as sending and receiving money and paying with a debit card without incurring additional charges. It is a low-cost banking service, but without the bank.
Be-Bound	United States	It has developed a unique and patented technology that optimizes existing networks to provide greater connectivity to mobile phones and IoT worldwide without any additional infrastructure investment. Be-Bound technology can increase coverage to reach 95% of the population immediately. Opening this technology to businesses of all sizes enables secure end-to-end encryption and real-time transactions everywhere and ensures consistent connectivity wherever there is a mobile network. Therefore, free markets are increased as well. In emerging markets, where problematic connectivity inhibits development, it contributes to creating inclusive growth and sustainable development.
Hapinoy	Philippines	A social enterprise launched by MicroVentures, provides business training and microfinance to sari-sari store owners to upgrade their businesses and offer essential services such as mobile finance and basic health care.
Tata Nano Car	India	It is a low-cost segment car developed by Indian manufacturer Tata Motors. It offers with Hatchback (original and SUV) four doors, two doors sedan, four doors convertible, and two doors pickup. In India, the sales price without taxes of the Nano with the basic equipment level initially was approximately 100,000 Indian rupees (about US$ 1,500–2,000). As "one hundred thousand" is translated into the Hindi language as "lakh," this car has been called by the media "the lakh car."
TAHMO	Netherlands/ US/Ghana/ Kenya	The Trans-African Hydro-Meteorological Observatory (TAHMO) aims to develop an extensive network of weather stations across Africa. Current and historical weather data are important for agriculture, climate monitoring, and many hydro-meteorological applications. This project aims to develop a dense network of hydro-meteorological monitoring stations in sub-Saharan Africa, one every 30 km. It involves the installation of 20,000 stations across the continent. The stations are placed in schools and integrated into educational programs, adding richness to the curriculum and fostering a new generation of scientists.

Source: Authors.

construction of four dimensions: sustainability, cost of consumption, simplicity, and basic quality (Von Janda et al., 2020).

The frugality associated with the product is appreciated by the market, which increases consumers' willingness generate innovation diffusion through word of mouth. These innovations could have a huge impact on addressing unmet and ignored consumer needs and long-term effects through increased productivity, sustainability, poverty reduction, and inclusion (Pansera and Sarkar, 2016). Frugal innovation, therefore, improves the welfare of the poor in four dimensions: (1) income generation and security, (2) education, (3) infrastructure, and (4) distribution (Kahle et al., 2013). Based on the literature review and the examples of frugal innovation, it is possible to distinguish the organization's fundamental purpose associated with the value it incorporates to the product or service it provides to the community. Therefore, we can define frugal social innovation as innovative activities and services that are motivated by the objective of satisfying a social need (Mulgan, 2006) created from the optimization of scarce resources in a frugal environment (Hossain, 2017).

Frugal social innovation is based on the frugality of the means employed, flexibility in thinking and acting, and the pillar of collaboration between people inside and outside the organization (Soni and Krishnan, 2014). The latter is the pillar that determines an ecosystem's formation and how frugal innovation emerges from adverse contexts. Significant organizational gaps and resource constraints generally characterize the organizations associated with frugal social innovation, but where the collective relational connections of participants are channeled into the development of affordable and sustainable solutions to fill organizational gaps and overcome constraints (Pisoni et al., 2018). These include the major organization, its suppliers, its network of customers and suppliers, government agencies, business support services, logistics, delivery providers, alliance partners, and often rival organizations (Ahuja and Chan, 2016). While constraints represent a shortage of resources and infrastructure, organizational gaps can be an even greater problem. These are characterized by the absence of intermediary organizations, weak regulatory systems, and poor contract enforcement mechanisms (El Ghoul et al., 2017; Khanna and Palepu, 1997). Thus, there is still a gap between government policies to encourage innovation and desirable results (Soni and Krishnan, 2014).

14.2.2 Resources

All the factors mentioned above determine the limitations that an organization faces in a frugal environment (Figure 14.1), which can be summarized in restrictions associated with the cost of incorporating innovation, market restrictions due to other participating actors' actions, and knowledge restrictions. These are associated with innovation development policies as instruments of financing and regulations in favor of innovation.

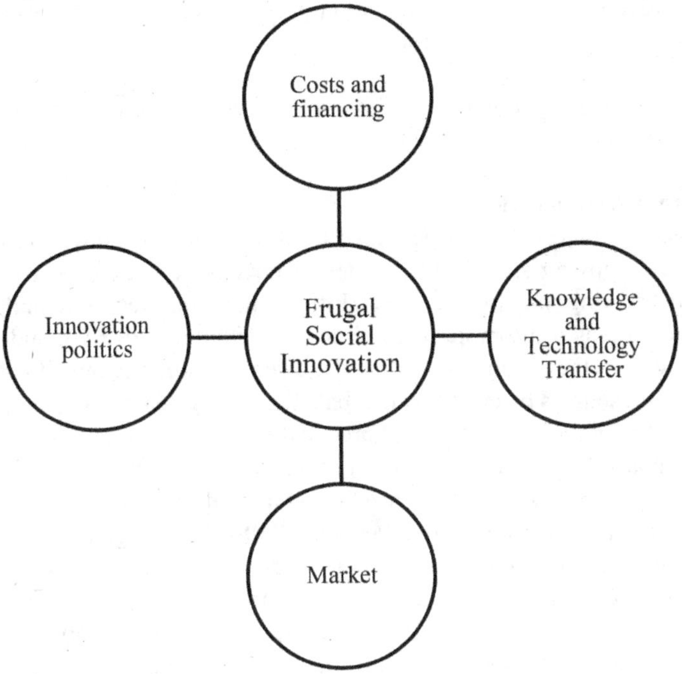

Source: Authors

Figure 14.1 Constraints determining the frugal ecosystems

14.2.2.1 Costs and financing

The costs associated with implementing innovation are for organizations one of the main restrictions to implement innovation policies. An innovative organization facing financial constraints has two options: it can invest effort in finding additional resources to alleviate the resource constraints (Bouty, 2000) directly. The organization can operate under this budget constraint by doing what it has available (Baker and Nelson, 2005). Because of the lack of resources, frugal environments are difficult for normal business operations and innovation operations. Resource-constrained contexts deal with the scarcity of physical resources and a policy and infrastructure gap that hinders innovation diffusion (Pansera and Owen, 2015). However, a positive relationship has also been identified between the scarcity of financial resources and the probability of innovation. Financial constraints appear to act as a stimulus to successful innovation, whether in product development or business strategy (Hewitt-Dundas, 2006). In an environment with financial constraints, teams with a creative approach that are more inclined towards limited methods will find it not only more natural but also easier to innovate. Thus, financial constraints cause individuals to look for ways to avoid the same constraint: the bottleneck. They can then identify creative solutions to an innovation problem (Gibbert and Scranton, 2009). We propose the following hypothesis:

Hypothesis 1: A high cost and lack of financing stimulate the development of frugal social innovations.

14.2.2.2 Innovation policies

In recent years, emerging countries have tried to establish innovation policies by observing and copying successful policies in developed countries. These policies focused on creating organizational facilities, including research and education organizations, funds to promote research and development (R&D), and innovation clusters. However, these policies neglected the demand side and only emphasized the supply side of resources by creating an unbalanced ecosystem (Kaplinsky, 2011). For this reason, at both the conceptual and practical levels, there is a lack of congruence between innovation ecosystems and desired innovation outcomes (Krishnan, 2010). Therefore, there is a lack of congruence between the desired direction of innovation and government policies (Foxon and Pearson, 2008). Also, regulatory difficulties in emerging countries are hampering innovation development, so organizations must create innovative solutions with additional restrictions that limit the possibilities (Hossain, 2018). It is in these adverse situations that the organization expresses its full innovation potential. We propose the following hypothesis:

Hypothesis 2: Poor innovation policies and a weak regulatory system stimulate the development of frugal social innovations.

14.2.2.3 Knowledge and technology transfer

Most technological, organizational, and social innovations arise from a combination of knowledge, learning by doing, use and interaction, and interactions between individuals and organizations (Foster and Heeks, 2013). In this regard, educational organizations play a key role in the process of knowledge and technology transfer. They create "cultural capital" by helping young people develop their competencies and support learning, knowledge transfer, and innovation processes to foster inclusive development (Arocena et al., 2015). However, qualified professionals who can be agents of change are seen as scarce and expensive resources. On the other hand, the lack of information about technology and markets puts the organization in a scenario where other resources such as creativity must be tapped. The transfer of know-how and the local market are also important factors. To succeed with frugal innovations, organizations need to think frugally and bring about a change in employees' mindset by modifying the organizational culture (Agnihotri, 2015; Donate and Sánchez de Pablo, 2015). So in addition, the lack of knowledge can be replaced by strong ties with key partners, but there is also the difficulty in finding the cooperation of partners for innovation, with the organization having to search among its knowledge base for the solutions to remedy this gap. We propose the following hypothesis:

Hypothesis 3: Lack of knowledge and technology transfer stimulate the development of frugal social innovations.

14.2.2.4 Market

The organization facing a competitive environment with few resources must strive to excel, for which they must acquire innovative features if they expect to compete with large organizations (Rosenbusch et al., 2011). Although established organizations can discourage competition, frugal innovation is related to dealing with this type of market constraint since their strategic focus is the customer-facing solution (Cai et al., 2019). Its efficiency and effectiveness are achieved by restructuring the traditional systems of business models, designs that are frugally innovative to meet the demand that is already present, ensuring a certain degree of viability of the initiation of a project (Zeschky et al., 2014), and, therefore, the uncertainty regarding the demand for innovative goods or services. We propose the following hypothesis:

Hypothesis 4: Lack of market experience and resources to compete with other organizations stimulate the development of frugal social innovations.

14.3 METHODOLOGY

14.3.1 Research Setting

Technology transfer is one of the fundamental challenges that emerging economies have to face to advance to be a developed country. The advance of technology in the last decades has become an opportunity to develop complex artifacts or experimentation and give solutions to society's needs (Bercovitz and Feldman, 2006). Technology transfer is how countries use R&D activities in a cross-cutting manner, giving practical use to technological knowledge generated in the productive system. Chilean higher education organizations are the main generators of the knowledge that forms part of this process. The transfer of technological knowledge from universities is supported by the transfer and licensing offices and the Technology Transfer Centers (HUBs), organizations that, in recent years, have received state support to strengthen the intermediation of the results of R&D activities (Yáñez and Guerrero, 2021). Therefore, various incentive mechanisms and instruments have been used to promote transfer and foster links between knowledge-generating centers and enterprises. The latter incorporates knowledge and technology into their work to provide new or improved products and services. Therefore, organizations are in charge of putting into value the research activity results by transforming the knowledge into goods or services that are useful for society and allow the development of relevant innovations and solutions activities (Yáñez and Guerrero, 2021). However, Chile is still in debt with the growth and innovation models that have not achieved the expected results, and the necessary policies to promote research development are not yet strengthened. According to the Organisation for Economic Co-operation and Development (OECD), it is the member that invests the least in R&D.[1] Even though in the last few years, state funds have been promoted for the development of startup incubators and investment in R&D, this is still insufficient and is reflected in the low figures for patenting. It seems that current policies and regulations have not stimulated the development of this area in organizations or have not reached an appropriate level to allow the transmission of knowledge (Ministerio de Ciencia, 2020). This organizational weakness also reflects the lack of coordination between the public and private agents involved in the process. Three main approaches have been adopted in the OECD countries to encourage the transfer of advances in science and technological knowledge in the production sector: promoting the management of intellectual property, encouraging the formation of strategic alliances for the development of applied research, and fostering the creation of new technology-based enterprises (OECD, 2003). These three complementary axes can only be developed under a robust organizational framework that guarantees the processes' efficiency. The organizational framework change may have to do more to raise awareness of the importance of protection, taking a new, more modern, effective, and efficient look at industrial property rights.

The study focuses on Chile because of the particular levels of inequality in its inhabitants' income, which shows that despite the existence of resources, they are not always available due to the configuration of the ecosystem. According to data obtained from the OECD, Chile's average per capita available income was US$ 9,500 in 2016 PPP, which represents the second-lowest income level in the OECD and is equivalent to nearly 40 percent of the OECD average. However, Chile's average disposable income is (considerably) higher than that of the other Latin American countries included in the OECD data source (Brazil, Colombia, Costa Rica, and Mexico). Chile's income inequality is the second-highest among OECD countries, after Mexico. The average income of the richest 10 percent of the Chilean population is 19 times higher than the poorest 10 percent income, compared to the average of 9.3 in all OECD countries. High levels of inequality tend to hinder income and social mobility: in Chile, it may take six generations for the descendants of a family in the lowest 10 percent of the income distribution to reach the average income, compared to four to five generations in all OECD countries. Chile's employment rate has increased slowly but steadily over the last few decades. In 2017, about 63 percent of working-age people were employed, up from 58 percent in 2007 but still below the OECD average of 69 percent. The gender gap in employment persists: only 53 percent of Chilean women aged 15 to 64 were employed, compared to 73 percent of men. About one in two workers in Chile has skills that do not match their employment, the second-highest percentage among OECD countries after Mexico. Thirty percent of workers are over-qualified, and 17 percent are under-qualified. In comparative terms, the average skill mismatch in OECD is 36 percent.

14.3.2 Data Collection

Frugal approaches to innovation cannot be understood without considering the country's particular characteristics in which they arise. Among the characteristics are the socio-economic, organizational, and environmental ones. Emerging countries face considerable challenges in combining skills, knowledge, and innovation to put knowledge into practice and deliver products and services that work for people (Zhang and Mahadevia, 2014). This is relevant, as studies focusing on emerging countries have linked innovation with social inclusion concerns to the logic of solving poverty problems (Pansera and Owen, 2015).

The Innovation Survey captures data bi-annually and aims to provide information on the structure of the innovation process of organizations in Chile (inputs and outputs). It also intends to establish the relations between this process and the organizations' innovation strategy, the innovative effort, the factors that influence their capacity to innovate, and the economic performance of the organizations. The survey measures variables such as innovation (product, process, organizational management and/or marketing), degree of novelty, intellectual property rights, innovative activities (including R&D), which Chilean organizations carry out in the different productive sectors and regions of the country. The survey form and methodology design follow the general guidelines suggested by the OECD and Eurostat's Community Innovation Survey (CIS) for this type of survey set out in the Oslo Manual. These guidelines are applied in most OECD member countries to make the results and statistics on innovation comparable. Within the modules that are compiled to determine the factors that characterize the innovation activities are: product innovation, process innovation, effects of innovation, organizational innovation, marketing innovation, social innovation, innovative activities and expenditure, sources of information, and cooperation in innovative activities, human resources, public financing for innovation, obstacles to innovation, intellectual property rights, and perspectives of future innovations. In addition, the factors that hinder innovation associated with the lack of resources are cost factors, associated with the lack of own funds, lack of external financing to the organization, and very high cost of innovation; factors linked to knowledge, lack of qualified personnel, lack of information on technology, lack of information on markets and difficulty in finding cooperation of partners for innovation; market factors, a market dominated by established organizations, uncertainty regarding the demand for innovative goods or services; other factors, not necessary due to previous innovations, not necessarily due to lack of demand for innovations and regulatory difficulty. For the study, the factors associated with frugal innovation's characterization are associated with the obstacles or disincentives to innovation in the organization and the technology transfer. Table 14.2 summarizes the variables.

Table 14.2 Description of variables

Dependent Variable	Description
Frugal Social Innovation	During the years 2017–18, did your organization carry out any activity related to social innovation?
Independent Variable	
Lack of Resources Technology Transfer	a. Costs and financing: – Lack of equity – Lack of external financing to the organization – The very high cost of innovation b. Knowledge – Lack of qualified personnel – Lack of information about technology – Lack of market information – Difficulty in finding innovation partners c. Market factors – A market dominated by established organizations – Uncertainty regarding the demand for innovative goods or services d. Innovation politics – Regulatory difficulty – Public Financial – Acquisition of external knowledge for innovation

14.3.3 Statistical Method

A probit is carried out to evaluate the factors that determine future organizations' possibility to carry out Frugal Social Innovation. A probit model is a specification for a binary response model (Rencher, 2003). It addresses the same set of problems as logistic regression using similar techniques. The probit model, which employs a probit-link function, is usually estimated using the standard maximum likelihood procedure called a probit regression (Doornik and Hansen, 2008).

$$\Pr(Y = 1|X) = \Phi\left(X^{T}\beta\right) \tag{14.1}$$

To calculate the marginal effects of variables (dy/dx), the marginal effect of how y varies (dy) when x increases by 1 unit (dx). The probit coefficients have no direct interpretation. However, they can quantify the marginal effects of the explanatory variables on the probability of high school graduation.

14.4 RESULTS AND DISCUSSION

14.4.1 Results

Table 14.3 shows the probit frugal social innovations analysis.

Table 14.3 Probit frugal social innovation

	Probit		Marginal Effects	
Frugal Innovation	Coef.	P >|z|	dy/dx	P >|z|
Technology transfer	0.8903607	0.000	0.0296617	0.000
Lack of equity	−0.2050649	0.177	-0.0068316	0.179
Lack of external financing to the organization	0.0288937	0.855	0.0009626	0.855
Very high cost of innovation	0.3125869	0.043	0.0104136	0.045
Lack of qualified personnel	0.131494	0.353	0.0043806	0.358
Lack of information about technology	0.138733	0.375	0.0046218	0.376
Lack of market information	−0.3562834	0.030	−0.0118693	0.031
Difficulty in finding innovation partners	−0.1576491	0.307	−0.005252	0.309
Market dominated by established organizations	−0.1027692	0.478	−0.0034237	0.479
Uncertainty regarding demand for innovative goods or services	−0.0983568	0.510	−0.0032767	0.510
Regulatory difficulty	0.3062969	0.009	0.0102041	0.010
Public Financial	1.043.189	0.000	0.0347531	0.000
Constant	−234.561	0.000		

By increasing public funding by one unit, the average probability of emerging frugal social innovation increases by 3.4 percent. Increasing the cost of innovation by one unit increases the probability of emerging frugal innovation by 1.04 percent. An increase in the lack of market information decreases frugal social innovation by 1.1 percent. Concerning technological transformation, there is a 2.9 percent increase in frugal social innovation. Indeed, the difficulties associated with the technology transfer regulation increase the probability of investing in frugal innovation by 1.02 percent. *Concerning H1*, our results show that resource constraints trigger two major developments in the enterprise: first, increased entrepreneurial activity and second, greater attention to a new recombination of existing knowledge assets (Keupp and Gassmann, 2013). Indeed, in organizations that develop this innovation, resource constraints stimulate innovation (Hoegl et al., 2008). The cost associated with the development of innovation increases the probability of emerging frugal social innovation. *Concerning H2*, our results show that a weak regulatory system increases frugal innovation. Previous studies have validated this relationship since there is a lack of coherence between the organizations involved in innovation ecosystems and the desired outcomes in this area (Krishnan, 2010). It explains why the lack of coherence between the desired direction of innovation and the government policies applied by the authority drives organizations to develop alternative solutions (Foxon and Pearson, 2008). *Concerning H3*, our results show that the lack of critical resources for innovation in small and large enterprises is a latent problem. It is because organizations in frugal environments have difficulty in obtaining specialized human resources and technology for innovation. These relationships reflect that in order to succeed with frugal innovations, organizations need to think frugally and bring about a change in employees' mindset by modifying the organizational culture to adopt resources from outside the organization (Agnihotri, 2015; Donate and Sánchez de Pablo, 2015). Therefore, the lack of market information variable is significant and negative, which indicates that in the face of a lack of market knowledge, innovation acts with a compensatory effect by emerging creative processes within the organization. *Concerning H4*, our results do not provide enough elements about the lack of knowledge and market experience to compete with other organizations. A plausible explanation is the dynamic and evolutionary embryonic stage in the university-industry relationship and interactions with the entrepreneurial innovation ecosystem (Fischer et al., 2020). Figure 14.2 summarizes the results in the Chilean context.

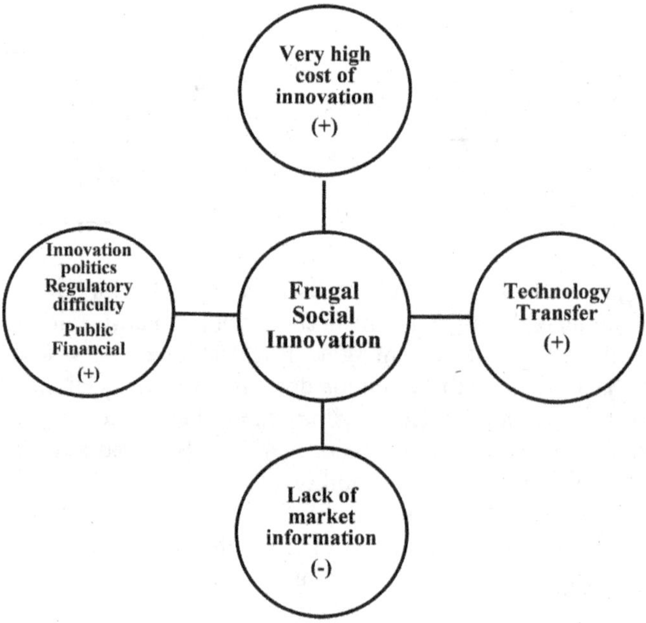

Source: Authors.

Figure 14.2 *Constraints determining the frugal environments results*

14.4.2 Discussion

The literature on frugal innovation has allowed an evaluating of the phenomenon's confirmation in descriptive terms. In this sense, our study contributes with a proposal of measure, evaluating the factors associated with the scarcity of resources. Adding and highlighting the social factor that underlies these initiatives, interesting relations are appreciated that test some hypotheses and establish frugal social innovations. The challenge for organizations is to overcome the limitations. This dynamic can induce managers to discover new possibilities by recombining existing knowledge and resources in new ways. Constraints can encourage managers to find various innovative ways to achieve their goals (Cai et al., 2019). Previous literature supports the relationships to the extent that frugal social innovation fostered by scarce resources drives organizations to use their capabilities and creativity to optimize using these resources to generate community-friendly solutions (Numminen and Lund, 2017; Pansera and Sarkar, 2016). The limitation of available resources is the driving factor, so the contribution discussed in these articles is very useful for measuring the constraints that lead to a frugal ecosystem that decreases in frugal innovation.

Concerning inequality in access to resources, the results suggest that there could be a relationship between frugal social innovation and lack of financing, which indicates that progress should be made in this type of policy to guide the development of innovation towards sustainability. Future empirical research needs to incorporate other variables associated with territoriality and to be able to compare by geographical areas in order to generalize the results with other OECD countries. Also, it is necessary to investigate other policies that may be relevant to this phenomenon, especially those related to the cost of implementing an innovation process, as well as to investigate the impact of other organizations such as universities and incubators on this type of business whose results directly benefit the community.

Despite the existence of a favorable policy towards R&D, there is a shortage of innovation in Chile. The technology transfer system involves various actors with roles and functions that allow the chain to be completed from the generation of knowledge to its incorporation into successful market innovations. The linkage of these actors is key to fulfilling the system's function; however, weaknesses are still perceived among the actors, especially in the link between organizations and the transfer and licensing offices. Likewise, entrepreneurs facing a frugal environment have difficulties linking their activities with the knowledge being developed in universities and making their technological needs known. It has a double negative effect: on the one hand, entrepreneurs are neither beneficiaries nor creators of knowledge, and, on the other hand, they lose the funding opportunities associated with technology transfer programs (Yáñez and Guerrero, 2021).

14.5 CONCLUSIONS

Frugal social innovation is a topic of interest in emerging countries since regulatory and organizational factors can determine the conditions of environmental frugality and also, as in this case, the prevailing inequality. Innovative solutions created for the community and designed for the poorest or marginalized groups provide opportunities to create devices, services, and platforms that surprise their ingenuity and influence their reach. Frugal social innovation is not specific to a territory but can be feasible anywhere. Frugal social innovation and its determinants is a phenomenon that needs to be further researched, especially using empirical analysis and beyond the specific cases, where a considerable amount of literature already exists. Understanding the determinants of frugal innovation and its success can be very useful for developing public policy in countries seeking sustainability in an environment of increasingly scarce resources.

Although the lack of resources may seem adverse, the study of frugal social innovation gives a positive vision in the face of crises and difficulties of any kind. The organization uses the available resources to satisfy the demands of its community or interest group. In this sense, organizations must continuously adapt to different scenarios and shocks that can only be managed with the available resources' creative administration. The development of a frugal social innovation does not guarantee by itself the success of an innovation. However, it can be an indicator of efficient management, so the organization's good performance stands out as a desirable result.

The development of a knowledge-based economy requires a focus on the entities that generate innovation and technology. To this end, organizations that develop R&D must continuously evaluate their processes to ensure that they align with technological development. Implementing new organizational and management forms of these organizations is fundamental for an effective transfer of technology and the subsequent development of innovation from the organization. In general, the regulations implemented in Chile and the organizational framework are part of the strengths of the R&D system, but they are not sufficient to realize technology transfer processes that increase the competitiveness of enterprises. The enterprises' contribution most valued is the initial financing, but in the development of the enterprise, the link with an efficient technology transfer system suffers.

ACKNOWLEDGMENTS

Claudia Yáñez-Valdés acknowledges the financial support provided by the Universidad del Desarrollo during her PhD studies in Business Economics.

NOTE

1. For further details, visit https://data.oecd.org/chile.htm.

REFERENCES

Agnihotri, A. (2015). Low-cost innovation in emerging markets. *Journal of Strategic Marketing*, 23(5), 399–411.

Ahuja, S., and Chan, Y.E. (2016). *Digital Innovation: A Frugal Ecosystem Perspective*. Dublin: Thirty-Seventh International Conference on Information Systems.

Arocena, R., Göransson, B., and Sutz, J. (2015). Knowledge policies and universities in developing countries: Inclusive development and the "developmental university." *Technology in Society*, 41, 10–20.

Baden-Fuller, C., and Haefliger, S. (2013). Business models and technological innovation. *Long Range Planning*, 46(6), 419–26.

Baker, T., and Nelson, R.E. (2005). Creating something from nothing: Resource construction through entrepreneurial bricolage. *Administrative Science Quarterly*, 50(3), 329–66. https://doi.org/10.2189/asqu.2005.50.3.329

Basu, R.R., Banerjee, P.M., and Sweeny, E.G. (2013). Frugal innovation. *Journal of Management for Global Sustainability*, 1(2), 63–82.

Belkadi, F., Buergin, J., Gupta, R.K. et al. (2016). Codefinition of product structure and production network for frugal innovation perspectives: Towards a modular-based approach. 26th CIRP Design Conference, 589–94. Elsevier.

Bercovitz, J., and Feldman, M. (2006). Entrepreneurial universities and technology transfer: A conceptual framework for understanding knowledge-based economic development. *The Journal of Technology Transfer*, 31(1), 175–88.

Bouty, I. (2000). Interpersonal and interaction influences on informal resource exchanges between R&D researchers across organizational boundaries. *Academy of Management Journal*, 43(1), 50–65.

Brush, C.G., Greene, P.G., and Hart, M.M. (2001). From the initial idea to unique advantage: The entrepreneurial challenge of constructing a resource base. *Academy of Management Perspectives*, 15(1), 64–78.

Cai, Q., Ying, Y., Liu, Y., and Wu, W. (2019). Innovating with limited resources: The antecedents and consequences of frugal innovation. *Sustainability*, 11(20), 5789.

Chesbrough, H. (2010). Business model innovation: Opportunities and barriers. *Long Range Planning*, 43(2–3), 354–63.

Donate, M.J., and Sánchez de Pablo, J.D. (2015). The role of knowledge-oriented leadership in knowledge management practices and innovation. *Journal of Business Research*, 68(2), 360–70. https://doi.org/10.1016/j.jbusres.2014.06.022

Doornik, J.A., and Hansen, H. (2008). An omnibus test for univariate and multivariate normality. *Oxford Bulletin of Economics and Statistics*, 70, 927–39.

Economist, The. (2010). First break all the rules: The charms of frugal innovation. Special report on innovation in emerging markets. *The Economist*, 17 April, 3–5. http://www.economist.com/ node/15879359 (accessed 7 April 2021).

El Ghoul, S., Guedhami, O., and Kim, Y. (2017). Country-level institutions, firm value, and the role of corporate social responsibility initiatives. *Journal of International Business Studies*, 48(3), 360–85.

Fischer, B., Guerrero, M., Guimón, J., and Schaeffer, P.R. (2020). Knowledge transfer for frugal innovation: Where do entrepreneurial universities stand? *Journal of Knowledge Management*, 25(2), 360–79.

Foster, C., and Heeks, R. (2013). Conceptualizing inclusive innovation: Modifying systems of innovation frameworks to understand the diffusion of new technology to low-income consumers. *The European Journal of Development Research*, 25(3), 333–55.

Foxon, T., and Pearson, P. (2008). Overcoming barriers to innovation and diffusion of cleaner technologies: Some features of a sustainable innovation policy regime. *Journal of Cleaner Production*, 16(1), S148–S161.

Gibbert, M., and Scranton, P. (2009). Constraints as sources of radical innovation? Insights from jet propulsion development. *Management and Organizational History*, 4(4), 385–99.

Hewitt-Dundas, N. (2006). Resource and capability constraints to innovation in small and large plants. *Small Business Economics*, 26(3), 257–77.

Hoegl, M., Gibbert, M., and Mazursky, D. (2008). Financial constraints in innovation projects: When is less more? *Research Policy*, 37(8), 1382–91.

Hoskisson, R.E., Eden, L., Lau, C.M., and Wright, M. (2000). Strategy in emerging economies. *Academy of Management Journal*, 43(3), 249–67.
Hossain, M. (2017). Mapping the frugal innovation phenomenon. *Technology in Society*, 51, 199–208.
Hossain, M. (2018). Frugal innovation: A review and research agenda. *Journal of Cleaner Production*, 182, 926–36.
Hossain, M., Simula, H., and Halme, M. (2016). Can frugal go global? Diffusion patterns of frugal innovations. *Technology in Society*, 46, 132–9. https://doi.org/https://doi.org/10.1016/j.techsoc.2016.04.005
Kahle, H.N., Dubiel, A., Ernst, H., and Prabhu, J. (2013). The democratizing effects of frugal innovation. *Journal of Indian Business Research*, 5(4), 220–34.
Kaplinsky, R. (2011). Schumacher meets Schumpeter: Appropriate technology below the radar. *Research Policy*, 40(2), 193–203.
Keupp, M.M., and Gassmann, O. (2013). Resource constraints as triggers of radical innovation: Longitudinal evidence from the manufacturing sector. *Research Policy*, 42(8), 1457–68.
Khan, R. (2016). How frugal innovation promotes social sustainability. *Sustainability*, 8(10), 1034.
Khanna, T., and Palepu, K. (1997). Why focused strategies may be wrong for emerging markets. *Harvard Business Review*, 75, 41–54.
Kline, S.J., and Rosenberg, N. (2010). An overview of innovation. In *Studies on Science and the Innovation Process: Selected Works of Nathan Rosenberg* (pp. 173–203). Singapore: World Scientific Publishing.
Knorringa, P., Peša, I., Leliveld, A., and Van Beers, C. (2016). Frugal innovation and development: aides or adversaries? *The European Journal of Development Research*, 28(2), 143–53.
Krishnan, R.T. (2010). From jugaad to systematic innovation: The challenge for India. Utpreraka Foundation, Bangalore.
Krohn, M., and Herstatt, C. (2018). The question of a frugal mindset in Western MNCs: Exploring an emerging phenomenon with a systematic literature review. Hamburg University of Technology (TUHH), Institute for Technology and Innovation Management Working Papers 103.
Lehner, A.-C., Koldewey, C., and Gausemeier, J. (2018). Approach for a pattern-based development of frugal innovations. *Technology Innovation Management Review*, 8(4), 14–27.
Levänen, J., Hossain, M., Lyytinen, T., Hyvärinen, A., Numminen, S., and Halme, M. (2016). Implications of frugal innovations on sustainable development: Evaluating water and energy innovations. *Sustainability*, 8(1), 4.
Meagher, K. (2018). Cannibalizing the informal economy: Frugal innovation and economic inclusion in Africa. *The European Journal of Development Research*, 30(1), 17–33.
Ministerio de Cienca. (2020). Caracterización de los participantes de la Transferencia Tecnológica en Chile. Retrieved from http://ctie.economia.cl/wp-content/uploads/2020/01/20200110-Boletintransferencia-tecnológica-1.pdf (accessed 11 April 2021).
Mourtzis, D., Vlachou, E., Giannoulis, C., Siganakis, E., and Zogopoulos, V. (2016). Applications for frugal product customization and design of manufacturing networks. *Procedia CIRP*, 52(1), 228–33.
Mulgan, G. (2006). The process of social innovation. *Innovations: Technology, Governance, Globalization*, 1(2), 145–62.
Numminen, S., and Lund, P.D. (2017). Frugal energy innovations for developing countries – a framework. *Global Challenges*, 1(1), 9–19.
OECD. (2003). Organisations, turning science into business: Patenting and licensing at public research. Retrieved from https://www.oecd-ilibrary.org/ (accessed 11 April 2021).

Pansera, M., and Owen, R. (2015). Framing resource-constrained innovation at the "bottom of the pyramid": Insights from an ethnographic case study in rural Bangladesh. *Technological Forecasting and Social Change*, 92, 300–11.
Pansera, M., and Sarkar, S. (2016). Crafting sustainable development solutions: Frugal innovations of grassroots entrepreneurs. *Sustainability*, 8(1), 51.
Pisoni, A., Michelini, L., and Martignoni, G. (2018). Frugal approach to innovation: State of the art and future perspectives. *Journal of Cleaner Production*, 171, 107–26.
Prahalad, C.K. (2005). *The Fortune at the Bottom of the Pyramid: Eradicating Poverty through Profits*, Upper Saddle River, NJ: Wharton School Publishing.
Rao, B.C. (2013). How disruptive is frugal? *Technology in Society*, 35(1), 65–73. https://doi.org/10.1016/j.techsoc.2013.03.003
Rencher, A.C. (2003). *Methods of Multivariate Analysis* (Vol. 492). New York: John Wiley and Sons.
Rosenbusch, N., Brinckmann, J., and Bausch, A. (2011). Is innovation always beneficial? A meta-analysis of the relationship between innovation and performance in SMEs. *Journal of Business Venturing*, 26(4), 441–47.
Soni, P., and Krishnan, R.T. (2014). Frugal innovation: Aligning theory, practice, and public policy. *Journal of Indian Business Research*, 6(1), 29–47.
Von Janda, S., Kuester, S., Schuhmacher, M.C., and Shainesh, G. (2020). What frugal products are and why they matter: A cross-national multi-method study. *Journal of Cleaner Production*, 246, 118977. https://doi.org/https://doi.org/10.1016/j.jclepro.2019.118977
Yáñez, C. and Guerrero, M. (2021). Technology transfer policy framework in Chile. In Guerrero, M. and Urbano, D. (eds), *Technology Transfer and Entrepreneurial Innovations: International Studies in Entrepreneurship* (Vol. 51, pp. 103–13). Cham: Springer.
Zeschky, M., Widenmayer, B., and Gassmann, O. (2011). Frugal innovation in emerging markets. *Research-Technology Management*, 54(4), 38–45.
Zeschky, M.B., Winterhalter, S., and Gassmann, O. (2014). From cost to frugal and reverse innovation: Mapping the field and implications for global competitiveness. *Research Technology Management*, 57(4), 20–7.
Zhang, L., and Mahadevia, D. (2014). Translating science and technology policies and programs into grassroots innovations in China. *Journal of Science and Technology Policy Management*, 5(1), 4–23.

Index